Powder Metallurgy
Science, Technology and Materials

T0358902

Anish Upadhyaya
Associate Professor
Department of Materials Science and Engineering
Indian Institute of Technology Kanpur, India

G S Upadhyaya
Former Professor
Department of Materials and Metallurgical Engineering
Indian Institute of Technology Kanpur, India

Foreword by
Prof. Dr Ken-ichi Takagi,
Tokyo City University, Tokyo

Editor-in-Chief: Baldev Raj

Series in Metallurgy and Materials Science

CRC Press
Taylor & Francis Group
Boca Raton London New York

CRC Press is an imprint of the
Taylor & Francis Group, an **informa** business

POWDER METALLURGY: SCIENCE, TECHNOLOGY AND MATERIALS

UNIVERSITIES PRESS (INDIA) PRIVATE LIMITED

Registered Office
3-6-747/1/A & 3-6-754/1, Himayatnagar, Hyderabad 500 029, Telangana, India
info@universitiespress.com; www.universitiespress.com

Distributed in India, China, Pakistan, Bangladesh, Sri Lanka, Nepal, Bhutan,
Indonesia, Malaysia, Singapore and Hong Kong by
Orient Blackswan Private Limited
Registered Office
3-6-752 Himayatnagar, Hyderabad 500 029, Telangana, India

Other Offices
Bengaluru, Bhopal, Chennai, Guwahati, Hyderabad, Jaipur, Kolkata, Lucknow, Mumbai,
New Delhi, Noida, Patna, Vijayawada

Copublished and distributed in the rest of the world by
CRC Press LLC, Taylor and Francis Group
6000 Broken Sound Parkway, NW, Suite 300, Boca Raton, FL 33487, USA

CRC Press is an imprint of Taylor & Francis Group, an Informa business

© Universities Press (India) Private Limited 2011
First published in India by
Universities Press (India) Private Limited 2011

First issued in paperback 2018

ISBN 13: 978-1-138-07501-6 (pbk)
ISBN 13: 978-1-4398-5746-5 (hbk)

© Universities Press (India) Private Limited 2011
Cover and book design

Typeset in Times New Roman 11/13 *by*
MacroTex Solutions, Chennai 600 088, India

Published by
Universities Press (India) Private Limited
3-6-747/1/A & 3-6-754/1, Himayatnagar, Hyderabad 500 029, Telangana, India

501417

Disclaimer
Care has been taken to confirm the accuracy of information presented in this book. The authors and the publisher,
however, cannot accept any responsibility for errors or omissions or for consequences from application of the
information in this book and make no warranty, express or implied, with respect to its contents.

Dedicated to

Professor Dr. rer. nat. Dr. h.c. mult. Guenter Petzow
Founder Director, Powder Metallurgy Laboratory,
Max Planck Institute for Metals Research, Stuttgart, Germany,
who successfully attempted to bridge the gap between
the science and technology of powder metallurgy

Foreword

Powder metallurgy has a very long history. It is considered that the powder metallurgical technique was initiated to fabricate early porcelain goods in the beginning of human civilisation. The origin of metal products was probably when ferrous wares were produced by powder metallurgy in about 3000 BC in the Egyptian civilisation. In the suburbs of New Delhi, India, the iron pillar of Chandragupta at Qutab Minar with 99.7% purity, constructed in the early 4th century, also has its origins in powder metallurgy. The pillar recognised as one of the world's wonders has been free from rust for more than 15 centuries!

It is not wrong to say that Dr William D Coolidge opened the door to the modern era of powder metallurgy by the production of ductile tungsten wire in the early 20th century. Since then powder metallurgy has been applied to produce a wide variety of practical materials such as automotive parts, cutting tools, electronic parts, etc. After the Second World War, intensive research on the sintering mechanism was conducted by many famous researchers such as Kuczynski, Lenel, Coble, Kingery, etc.; but there still remains a lot that is empirical and unexplained in powder metallurgy.

The authors with their wide range of experience in powder metallurgy, both education and research, are well qualified to write this book. The book covers both fundamental aspects of powder metallurgy, such as classic sintering theory, and various industrial aspects of metallic materials, ceramics and their based composites. Chapters 1 to 10 include general powder metallurgical concepts from powder production to full density consolidation and secondary treatment. Distinct from other powder metallurgy textbooks, pyrophoricity and toxicity in Chapter 3, and nanostructured materials and the electronic theory of sintering in Chapter 7, are new approaches. Furthermore, various kinds of actual testing and quality control methods in Chapter 11, advanced ceramic materials in Chapter 12, and a wide range of practically applied parts in Chapter 13 are very informative and instructive to practical powder metallurgical engineers. As mentioned in Chapter 14, techno-economics of PM processing powder metallurgy inherently provides not only various economical and cost saving processes, but also provides environment conscious aspects such as material saving and recycling.

Nowadays a wide range of industrial products relate to powder and powder metallurgy. Powder metallurgy fulfils a technological role appropriate for this era with various hard social demands. This book, from traditional technology to the most recent and developed materials and technologies in powder metallurgy, contains a lot of worked examples and exercises. These illustrate the structure–property–processing–performance relationships in varied classes of powder metallurgy materials and products in a holistic style. The book will be one of the desk-books for not only undergraduate and postgraduate students, but also researchers and engineers who are engaged in powder metallurgy.

Prof. Dr Ken-ichi Takagi
Head of Mechanical Engineering Department
Faculty of Engineering, Tokyo City University, Tokyo

Preface

Any forming process, including powder metallurgy (PM), has the following goals:

- Achieve the desired geometry (size, shape and tolerance) of a part or component with adequate defect control;
- Develop a controlled microstructure to yield the desired properties and in-service performance;
- Optimise economic aspects of production, including the conservation of materials and energy.

The aim of the present book is to discuss in detail the above concurrent issues. The book serves as a textbook, both for undergraduate and postgraduate courses in engineering, and also as a handy reference book for serving engineers in the PM industry.

PM processing in the global sense is inclusive for both metal/ceramic powders and composites based on them. From this point of view, the book is suitable for metallurgical and ceramic engineering students, or, to all materials and manufacturing engineers. The prerequisite for using this book is an introductory knowledge of chemical and physical metallurgy. This requirement is obviously met, as the PM course is invariably not taught at the beginning of the engineering curriculum.

Chapter 1 introduces the subject and provides a holistic view of PM science and technology. It is mandatory for newcomers in the area of PM to read this chapter first. It provides sufficient background information in PM to be able to read any later chapter on its own. Chapter 2 describes the theory and practice of powder production. Various methods like chemical, physical and mechanical are covered at length. Powder characterisation is discussed in Chapter 3, where a section on pyrophoricity and toxicity, generally not covered in other textbooks, has been added. Chapter 4, Powder Treatment, covers various means of powder treatment, including diffusion alloying, powder mixing/milling, granulation, coating and degassing. Particle size reduction methods are not included in this chapter, but are found in Chapter 2. This is the logical sequence, since these methods form one part of powder production techniques. Chapter 5, Powder Compaction, highlights the basic and technological aspects, where powder mass is shaped under pressure. Chapter 6 includes a description of pressureless powder shaping methods. Chapter 7, Sintering Theory, elaborates on the 'why' and 'how' of sintering powders (loose or compact). It covers both pressureless and pressure-assisted sintering. The microstructural features at each stage of sintering are adequately covered, since a composite visualisation of the processing–structure relation is of utmost importance. The electronic theory of sintering is a novel addition in this chapter. Sintering of nanostructured materials has been especially included as they are a unique and challenging group of materials. We considered it proper to deal with sintering theory in rather more detail than other textbooks, particularly for the benefit of postgraduate students. As a natural follow up of the previous chapter, Chapter 8 gives a detailed description of the technological aspects of conventional sintering (pressureless sintering), including sintering furnaces and atmospheres, and process and materials variables. Chapter 9, Full Density Consolidation, highlights dynamic compaction, hot pressing, hot isostatic pressing, spark sintering, and methods where deformation processing (extrusion, forging) is superimposed on powder preforms. For serving engineers, a reasonable amount of description of equipment and process variables has been provided. Readers without a background in metallurgy and ceramics may skip Chapter 7 in the first reading.

Secondary Treatment (sizing/coining, machining, impregnation, surface engineering, heat treatment and joining) is covered in Chapter 10. Chapter 11, Testing and Quality Control of PM Materials and Products, is not found in many PM books. In the two subsequent chapters—Metallic and Ceramic PM Materials and PM Applications—there is no discrimination between metallic and ceramic systems. In fact, a basic knowledge of both metals and ceramics is essential in developing new classes of cermets. The last chapter (Chapter 14) describes the techno-economics of PM processing, an area generally neglected in many books. We were particular that this topic be placed as a separate chapter and not tagged on at the end of each chapter.

Powder injection moulding has emerged as a major offshoot of PM processing. We strongly felt that pedagogically it is better to integrate it in various conventional PM unit processes concurrently, and have therefore discussed it alongside in various chapters.

The book is based on years of teaching undergraduate and postgraduate engineering students who studied PM as a core subject or as an elective. The constant interaction with them has been a great asset. At the end of each chapter, a variety of exercises are included. Some of them are simple while others require in-depth thinking on the part of both instructor and student. In order to aid and broaden the problem-solving capability of students, some worked-out examples are also included in each chapter. These examples are both objective and numerical, so that students can appreciate the fine intricacies of the subject matter. Depending on the class and level of students, the instructor can take the decision to emphasise/de-emphasise certain portions of the book. Further, at the end of each chapter, citations for further reading in the form of books, proceedings and monographs are provided. Engineers serving in PM plants may find the technological information in the book more useful than the basic aspects. In brief, this book attempts to impart details related to both basic and applied aspects of PM on a composite plank. We have felt that a scientific/technological interactive presentation is far better than to separate the two.

We thank Dr Baldev Raj (Director, Indira Gandhi Centre for Atomic Research, Kalpakkam, India), Editor-in-Chief of the Series in Metallurgy and Materials Science, Indian Institute of Metals, for inviting us to write this book. The authors also thank the Indian Institute of Metals (IIM) and the Centre for Development of Technical Education (CDTE), Indian Institute of Technology Kanpur, for providing partial financial support to write this book.

The authors sincerely thank Prof. Dr Ken-ichi Takagi, an eminent Japanese powder metallurgist, for accepting our request to write the Foreword. We also thank Ms Javanthi Singaram of Universities Press, Chennai for meticulously editing the manuscript. The publisher deserves all praise for timely completion of the publication.

We have attempted to make the text devoid of any errors. In spite of our best efforts, it is possible that some may have crept in. We shall be obliged if the readers bring them to our attention.

Authors

About the Series

The study of metallurgy and materials science is vital for developing advanced materials for diverse applications. In the last decade, the progress in this field has been rapid and extensive, giving us a new array of materials, with a wide range of applications, and a variety of possibilities for processing and characterising the materials. In order to make this growing volume of knowledge available, an initiative to publish a series of books in Metallurgy and Materials Science was taken during the Diamond Jubilee year of the Indian Institute of Metals (IIM) in the year 2006. As part of the series we have already brought out four books, and all of them have been copublished by CRC Press, USA, for distribution overseas. This is the fifth book in the series and the first textbook to be published.

The IIM is a premier professional body representing an eminent and dynamic group of metallurgists and materials scientists from R&D institutions, academia and industry in India. It is a registered professional institute with the primary objective of promoting and advancing the study and practice of the science and technology of metals, alloys and novel materials. The institute is actively engaged in promoting academia–research and institute–industry interactions.

Universities Press, an associate of Orient Blackswan, with its long tradition of publication of quality books in engineering and sciences, has come forward to undertake the publication of this series, thus synergising the professional expertise of IIM with the publishing experience of Universities Press towards effective knowledge dissemination. This book series shall include different categories of publications: textbooks to satisfy the requirements of undergraduates and beginners in the field, monographs on select topics by experts in the field, and proceedings of select international conferences organised by IIM after mandatory peer review. To increase the readership and to ensure wide dissemination, some of the books in the series will be copublished with international publishers.

The international character of the authors and editors has helped the books command a global readership. An eminent panel of international and national experts acts as the advisory body in overseeing the selection of topics, important areas to be covered, and the selection of contributing authors. These publications are expected to serve as a source of knowledge to a wide spectrum of students, engineers, researchers, and industrialists in the field of metallurgy and materials science. I look forward to receiving your valuable response to the present book as well as the other books in the series.

Baldev Raj
Editor-in-Chief

Contents

1

Introduction

LEARNING OBJECTIVES

- Historical perspective of powder metallurgy
- Metal and ceramic powder production methods
- Powder compaction
- Purpose of sintering
- Full density processing of powder metallurgy preforms
- Post-sintering treatments

The technology of pressing metal powders into specific shape is not new; ancient civilisations practiced the art. The iron pillar in Delhi, certain Egyptian implements and articles of precious metals made by the Incas bear testimony to this. The largest historical piece constructed using powder metallurgy is undoubtedly the Delhi Iron Pillar (400 AD) (Fig. 1.1), which was fabricated out of solid-state reduced ferrous granules by sinter-forging technique. Modern powder metallurgy (PM) technology commenced in the 1920s with the production of cemented carbides and mass production of porous bronze bushes for bearings. During the Second World War, further development took place in the manufacture of a great variety of ferrous and non-ferrous materials, including many composites. There was a period of steady growth in development during the post-war years until the early 1960s. Since

Fig. 1.1 Delhi iron pillar

then, the growth of PM has expanded rapidly, mainly because of three potential reasons—economical processing, unique properties and captive processes. The PM process is a rapid, economical and high-volume production method for making precision components from powders. However, there are a number of related consolidation techniques whereby powders can be rolled into sheets, extruded into bars, etc., or compacted isostatically into parts of more involved geometry. Over the last few decades, the technology of powder hot forging has established itself for fabricating powders into precise engineering parts which have properties comparable with those of conventional forgings. Figure 1.2 shows the general flow sheet of powder metallurgy processing.

During the past decade, there have been significant advances in powder manufacturing techniques. The production of larger and stronger materials has become possible by the use of new powders with superior properties. Careful control of the structure of the particles of the original powder has made it possible to have more intelligent manipulation of the structure of final sintered materials. The PM route can be used to produce difficult-to-process materials, where fully dense high-performance alloys can be processed with uniform microstructure. Multiphase composites with a wide combination of properties can economically be produced by the PM technique. Non-equilibrium materials such as amorphous, microcrystalline or metastable alloys are also processed by PM methods. Today, the part size has increased many times and large parts, including billets, are produced in large quantities. Materials with mechanical properties far exceeding those of

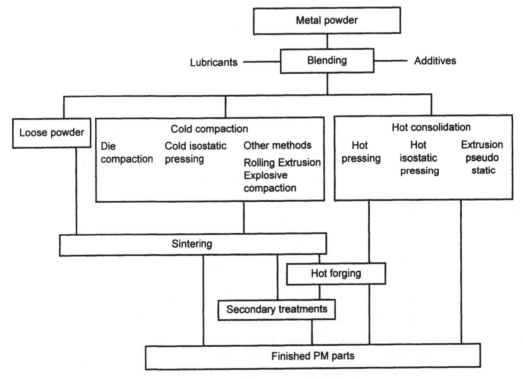

Fig. 1.2 Basic steps of the powder metallurgy process

more conventional materials have been developed by using new alloying elements, by improved heat treatment and by achieving higher densities.

At present, the major consumer of PM parts is the automotive industry. Hardware, tools, cameras, farm and garden equipment, industrial and business machines, sporting goods and defense equipment are a few more areas where their use is on the upswing. The PM process is many times more competitive than other fabrication methods such as casting, stamping or machining. It is the choice when requirements for strength, wear resistance or high operating temperatures exceed the capabilities of die casting alloys. It offers greater precision, eliminating most or all of the finish machining operations required for castings. It avoids casting defects such as blow holes, shrinkage and inclusions. Powder injection moulding (PIM) has come as a big challenge for investment casting. PM is highly competitive with fine blanking, which runs at a slower cycle than conventional stamping, and has higher equipment cost. Screw machines use bar stock as raw metal and the process is characterised by very poor material utilisation, sometimes les than 50%. However, the PM process is economical only when production rates are higher, since the tooling cost is quite appreciable.

As processing of metal and ceramic powders is very similar, it would not be inappropriate to describe a few distinctive features of the latter. Ceramics can be broadly divided into *traditional ceramics* (whiteware structural clay products and refractories) and *technical ceramics* or *advanced ceramics*. The latter offers precision making parts, similar to conventional metal powder parts. Sometimes, modern refractories are also grouped into technical ceramics. Ceramic processing commonly begins with one or more ceramic materials, one or more liquids, and one or more special additives known as processing aids. Some of the processes involved are crushing, milling, washing, dissolution, settling, flotation, magnetic separation, dispersion mixing, classification, de-airing, filtration and spray drying. It is not necessary that all the steps be involved. The next operation is forming material-based slurry, paste, plastic body or a granular material. Additional operations may include green machining, surface grinding, surface cleaning and application of surface coating as per the need. The final and major, but critical, operation is sintering in ambient or controlled atmosphere as per the chemistry of the material. For example, in the case of non-oxide ceramics, the complete avoidance of oxygen is essential. The sintered parts may be of single phase (e.g., WC, TiC, Si_3N_4) or a multiphase microstructure (e.g., industrial refractories). Lastly, the sintered parts may have to be given some surface finishing/ coating. Figure 1.3 shows a general processing flow diagram for processing ceramics from raw materials to the sintered product. In the past many ceramists used the term 'firing' instead of 'sintering'. Currently, with the firm establishment of sintering science, the term 'sintering' is preferred.

The *powder injection moulding* (PIM) process, used for ceramic processing, has recently been embraced by the powder metallurgy community with great enthusiasm. The process is similar to plastic injection moulding, with the difference that the polymer is filled with dispersed metallic or ceramic powders. This technology permits the production of stronger, more uniform and more complex PM parts. This process is more expensive than conventional PM techniques. A flow chart illustrating the steps of the PIM process is shown in Fig. 1.4.

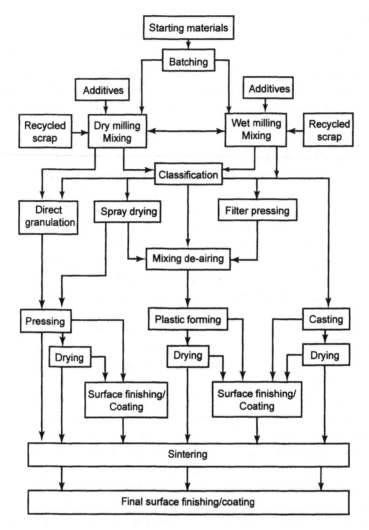

Fig. 1.3 A general processing flow diagram illustrating processing paths from ceramic raw materials to the final product

The process steps involved in injection moulding are:

- Selection and production of metal/ceramic powders
- Mixing
- Moulding
- Debinding
- Sintering

As PIM products are made from powders of engineering materials (metal/ceramic/polymers/composites), it is but natural that the internal structure and properties of materials will have a substantial effect on the performance of the end products. During the 1980s, much attention

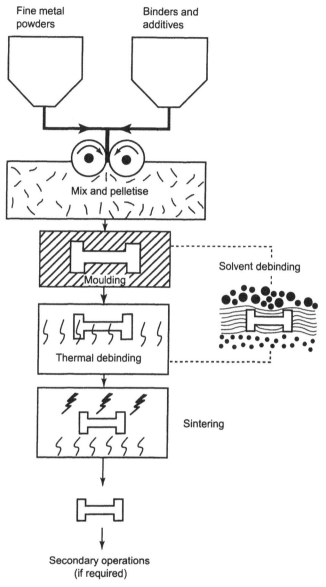

Fig. 1.4 Flow diagram of the PIM process

was focussed on 'new materials' (otherwise called 'advanced materials' or 'engineered materials'). These materials exhibit high strength, great hardness, and superior thermal, electrical, optical and chemical properties. They have dramatically altered communication technologies, reshaped data analysis, advanced space travel, restructured medical devices and transformed industrial production processes. In addition, there is considerable interest in making functionally graded materials. The gradient may be chemical composition, density or coefficient of thermal expansion of the material, or it may involve microstructural features, such as, a partial arrangement of second phase particles or fibres in a matrix. Instead of having a step function, we strive to achieve a gradual change. PM processing is one of the major routes to produce such materials.

Another group of materials of recent origin is 'nanomaterials', which contain a controlled morphology with at least one nano-scale dimension. As we reduce the size of the bulk material, the percentage of atoms at the surface increases. Apart from application in cutting tools, nanotechnology has made possible small-scale devices that perform vital functions in an automobile's operations. Powder metallurgy is one of the methods for producing nanoparticulates and nanostructured bulk materials. However, caution towards toxicity against the use of nanoparticulates is called for, although this has not yet been established conclusively.

Amorphous (non-crystalline) metallic alloys have been developed by rapid cooling of a melt or by very-high-energy mechanical milling. The details of the production of powders of such materials is described in the next chapter. Another material obtained through rapid solidification is 'quasicrystals', which are neither crystalline nor amorphous, but form an ordered structure somewhere between the two recognised structures. Such materials are expected to exhibit far-reaching electrical and other engineering properties.

1.1 PRODUCTION AND CHARACTERISATION OF POWDERS

The major methods of production of metal powders can be classified as:

- Chemical reaction and decomposition
- Electrolytic deposition
- Atomisation of molten metals
- Mechanical processing of solid materials.

Reduction of iron oxide is a classic example of the first method, producing a sponge-like iron powder. Fine powders of nickel and iron are produced by decomposing their carbonyls. A number of metals can be precipitated on the cathode of an electrolytic cell as a sponge, powder, or in a form which can be mechanically disintegrated easily. Copper, beryllium and iron powders are made in considerable quantities by this technique.

Most metal and alloy powders are fabricated by atomisation techniques. The use of molten alloy and high-velocity water or gas jet provides a major amount of powder. Gas and centrifugal techniques produce spherical powders, while water atomisation gives an irregular shape. An area of considerable interest is the generation of rapidly solidified powders (RSP)—an American industrial development stemming directly from the techniques of splat quenching of alloy ribbons from the melt. In these materials, cooling rates in excess of $10^{6}{}^{\circ}C$ s^{-1} are possible, thus giving rise to amorphous, microcrystalline and metastable materials. The advantage of such powders is that macro-segregation is entirely eliminated and large amounts of solute can be held in metastable equilibrium, which makes possible new levels of precipitation hardening; the consolidated products can have very fine grain sizes.

The mechanical method of powder making produces coarse, irregular or angular particles via machining, milling, crushing or impacting routes. With increased concerns for energy efficiency, the mechanical fabrication approaches are not preferred. A major exception is the mechanical alloying approach to the formation of oxide dispersion-strengthened alloys. This relies on repeated milling–fracture welding events on a macroscopic scale.

Pure ceramic powders are mainly prepared by chemical methods like (i) solid-state reactions, (ii) liquid solutions and (iii) vapour phase reactions. The descriptions of these methods are given in the next chapter.

Apart from particle appearance, other important powder characteristics are:

- Particle size and its distribution
- Particle shape and its variation with particle size
- Surface area
- Interparticle friction
- Flow and packing
- Internal particle structure
- Chemical gradients, surface films and admixed materials.

Unfortunately, the particle size is not a simple parameter—it is dependent on the powder shape and the technique used in the analysis of size.

Since the bulk of powder metallurgy processing relies on uniform powder flow into die cavities of automatic compaction press, other characteristics such as density and flow are also important. Another measure of interparticle friction is given by *tap density*—the maximum density that can be achieved under repeated vibration.

Beyond the bulk chemical information, knowledge of the surface condition of the powder is also necessary. It is important to differentiate between the absorbed oxygen, moisture and oxides to determine appropriate cleaning treatments.

1.2 TREATMENT OF POWDER

After the powder has been fabricated, several modifications in structure, chemistry and size might be necessary for best compaction. This includes removal of oxide and inclusions, mixing, addition of lubricants and other sintering aids. Care must be exercised at this point to avoid segregation of components or particle sizes. Control in each handling step is crucial to successful powder processing.

As shown in Fig. 1.4, mixing is an important operation in powder injection moulding.

1.3 COMPACTION OF POWDER

The compaction of metal powders has the following major functions:

- to consolidate the powder into the desired shape
- to impart as high a degree as possible of the desired final dimensions
- to impart the desired level and type of porosity
- to impart adequate strength for subsequent handling.

Die compaction is the most widely used method and is the 'conventional' technique. It involves rigid dies and special mechanical or hydraulic presses. Densities of up to 90% of full solid density can be achieved following the compaction cycle, the duration of which may be of the order of just a few seconds for very small parts.

1.4 SINTERING

Often, a pre-sintering heating operation is necessary for removing the pressing lubricant or binder. This may be done in an ambient or controlled atmosphere, as per the requirement. Debinding in powder injection moulding processing of PM parts is an essential stage.

In case of metallic systems, sintering is carried out in a protective atmosphere, within a furnace at a temperature below the melting point of the base metal. This temperature is generally 75% of the absolute melting temperature. The process leads to a decrease in the surface area, an increase in compact strength, and in many cases, a shrinkage in the compact. With prolonged high-temperature sintering there will be a decrease in the number of pores; the pore shape will become smooth and grain growth can be expected. It is possible that some melting or liquid phase sintering of other powder additives may take place during sintering, an example being sintering ferrous powders mixed with copper.

The process of sintering is generally the result of atomic motion stimulated by high temperatures. The initial strains, surface area and curvatures existing in a pressed powder compact drive the atomic motions responsible for sintering. Several different patterns of atomic motion can contribute to the effect, including evaporation and condensation, volume diffusion, grain boundary and surface diffusion and plastic flow. In most cases the kinetics of sintering are determined by several parameters, including pressed density, material, particle size, sintering atmosphere and temperature.

The presence of a liquid phase in minor quantity during all or part of the sintering cycle of the material also represents an enhanced sintering method. This method is very common, as the sintering temperature is rather low. During liquid phase sintering, different stages—rearrangement or 'liquid flow', 'accommodation' or 'dissolution and precipitation' and coalescence or 'solid-state bonding'—take place. These stages follow in the approximate order of their occurrence, but there may be significant overlapping for any specific system. A special variant of liquid phase sintering is super-solidus sintering, where the starting powder is invariably pre-alloyed.

In case of sintering of injection moulded PM parts, the operation is similar. However, in general, PIM sintering occurs at higher temperature, where densification is more likely to occur. The main concerns with PIM materials are the removal of the residual porosity which can be accomplished by methods adopted for full densification, i.e. hot isostatic pressing. This is described in the next section. Another method is infiltration, which may be carried out after sintering.

1.5 FULL DENSITY PROCESSING

To enhance the overall densification process, it is common to impose an external force through pressure or to provide a highly active kinetic path through the addition of a second phase.

Example 1.1: Explain why, historically, press and sintered technology of PM products came into practice much later than the technology for traditional ceramic products.

Solution: Clay, the raw material for ceramine products, which is commonly known as kaolin, is abundant in nature. The mineral in kaolin is kaolinite, which is a hydrated aluminium silicate $Al_2Si_2O_5(OH)_4$. The grinding of kaolin into powder was not a major hurdle. Addition of water in kaoline makes it plastic. In other words, the resulting thick paste can be deformed and moulded and it will hold its new shape. This unique characteristic of kaolin cannot be duplicated with metal powders. The wet clay could be hand formed for the production of ceramic products. The degree to which a particular clay shows plastic behaviour, is no doubt a complex function of particle size, particle shape and impurity content. Our ancestors were skilled in handling different clays. After drying and firing (sintering), the finished ceramic products were used for different purposes.

Modern powder metallurgy techniques, i.e., pressing and sintering, came into vogue only after the industrial revolution. The ease of mass production and dimensional control in the finished parts were the main driving force for the wide acceptance of PM technology.

Example 1.2: Figure Ex. 1.2 illustrates a needle bar joint that has been mass produced from grey cast iron after machining. For PM processing the same product, cast iron was substituted by a ferrous Fe-1.5Cu-2Ni alloy. Justify that PM processing is far more profitable.

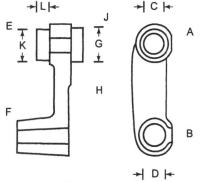

Fig. Ex. 1.2

Solution: In machining operations, there are 7 operations to be performed:

1. Milling surfaces A and B
2. Rough-boring C and D, milling surfaces E and F, reaming bore C
3. Turning hub G, surfaces H and J, chamfering bore C
4. Rough-turning collar K, finish-turning height L, chamfering bore C
5. Finish-turning collar K
6. Finish-reaming bore C, counter-sinking bore D on both sides
7. Fine-boring D

Apart from the above, melting, moulding and casting costs also need to be included.

However, if the same part is produced by PM method, the number of steps involved are only four:

1. Pressing
2. Sintering
3. Coining
4. Tumbling

In conclusion, the cost saving and productivity will be considerable in PM part. The saving will be even more attractive in case the batch size is increased.

Hot pressing techniques use superimposed temperature and pressure to effect rapid controlled densification. The form of hot pressing can be manifested as either uniaxial or hydrostatic forces. In recent years, hot isostatic pressing (HIP) has been the most rapidly expanding commercial form of pressure enhanced sintering.

Among other full density processing methods, mechanical working at room or high temperature with different strain rates may be applied: for example, hot extrusion and hot forging. In *hot extrusion*, large hydrostatic compression forces occur and a unidirectional force component makes the compact flow through the die. *Powder forging* can be broadly divided into two classes: conventional and preform. Conventional forging as applied to original compacting of loose powders consists

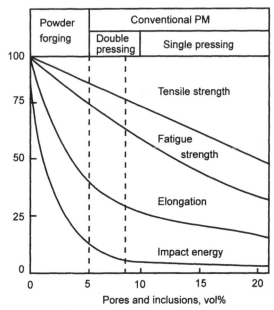

Fig. 1.5 Variation of mechanical properties for conventional PM and powder forging

of canning the powder in some type of metal containers, which after forging is removed by chemical or mechanical means. Powder preform forging, on the other hand, is a combination of powder metallurgy and forging. The process has developed into different variants.

Figure 1.5 illustrates how mechanical properties vary with densification obtained by pressing/sintering, double pressing/sintering and hot forging. It is evident that pores are completely eliminated in powder forging.

1.6 SECONDARY TREATMENT

Sintered compacts in most cases can be subjected to treatments analogous to wrought metals. The compact can be adjusted for surface finish or final dimensions using coining or sizing operations. Another operation may be coating. The pore structure must be sealed before coating in order to prevent intrusions of the fluid. Heat treatment is often desirable to improve the mechanical properties.

Other post-sintering treatments for powder metallurgy compacts include machining, welding, brazing and surface finishing. These operations are generally dependent on sintered density.

1.7 APPLICATIONS

Metal and ceramic powders or particulates are mostly used in the as-produced condition. Examples are aluminium granules for thermit welding, iron powder for flame cutting, fine

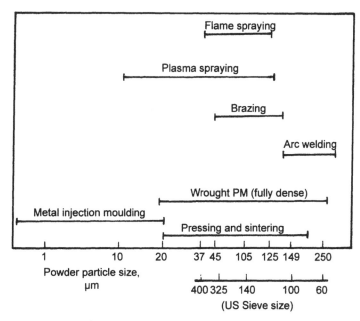

Fig. 1.6 Application of metal powders as a function of particle size and size range

powders as catalysts and cheap filler material. These applications require different size ranges of powders (Fig. 1.6).

Sintered parts are used widely in various engineering industries. Table 1.1 shows their respective percentage breakdowns. It is evident that the maximum contribution is in the automotive industry (Fig. 1.7).

One of the greatest advantages of PM processed parts is that a wide range of porosity levels can be effectively utilised in various engineering applications. However, in determining the cost effectiveness of a PM product against a conventionally processed alternative, the cost benefits of eliminating machining operations need to be considered carefully. The production volume of a particular PM part also dictates its application. However, there are situations in which the PM product is the only alternative and economies can only be obtained through materials selection.

The potential of powder injection moulding for cost effectively forming small, complex, precision parts is finding application in the production of fire arms, business machines and printers, hand tools, aircraft, automotive, ordnance, medical and dental parts, and cameras and controls. The process is

Table 1.1 PM parts distribution in the engineering industry

Application	%
Automotive	73
Recreation, hand tools and hobbies	10.5
Household appliances	4.3
Hardware	3.1
Industrial motors, controls and hydraulics	1.9
Business machines	1.2
All others	6.0

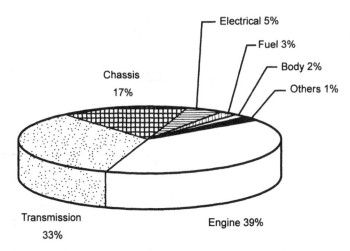

Fig. 1.7 PM automotive applications by weight

finding acceptance on a part-by-part basis, and is expected to penetrate almost all market segments in competition with investment castings.

1.8 HOLISTIC VIEW OF PM SCIENCE AND TECHNOLOGY

Rapid advances in modern powder metallurgy, which started purely as a technological base, can be attributed to the enthusiasm with which powder metallurgists adopted the scientific aspects of structure, plastic deformation behaviour, diffusion and thermodynamics. Thus, physical metallurgy historically came into existence much earlier than physical ceramics. Materials properties and phase equilibria are dependent on the core material structure, whether on macro-, micro- or nano-scale. All these structures are ultimately attributed to the electronic structure and chemical bonding in the material. Powder morphology is in turn based on the crystal structure of the metal/ceramic concerned. The flaky shape of graphite or BN (hex) ceramic powder is a direct manifestation of the crystal structure. Oriented textures in powder shaping are again attributed to the powder particles that are anisotropic in shape. This feature is directly reflected in processes such as tape casting.

In Section 1.2, the versatility of the atomisation method for metal/alloy powder production has been mentioned. The phase equilibria study of various types of metallic alloys is significant in the microstructure evaluation of the resultant atomised alloy powders. Under non-equilibrium cooling conditions the appearance or disappearance of a new phases affects the properties of the end powders. The composition of the powder compact contributes significantly in the proper selection and manipulation of the sintering atmosphere. Even the phase changes in a system, e.g., steel, during heating and cooling can be combined with the sintering process, which offers economic advantage to the overall processing. The stabilisation of ferrite phase in iron by addition of phosphorus is an important milestone in the incipient liquid phase sintering of Fe-P premixes.

The properties of the materials influence the different processing stages. For example, powder compaction of magnetic material is performed in a magnetic field to achieve better end properties.

Momentum transfer is again an invaluable aspect in powder processing, i.e., communition of powders. The mechanical behaviour of the material has direct relation to the efficiency of various communition processes.

The surface chemistry of various binder/solvent additives during powder treatment also has a definitive role in powder shaping processes.

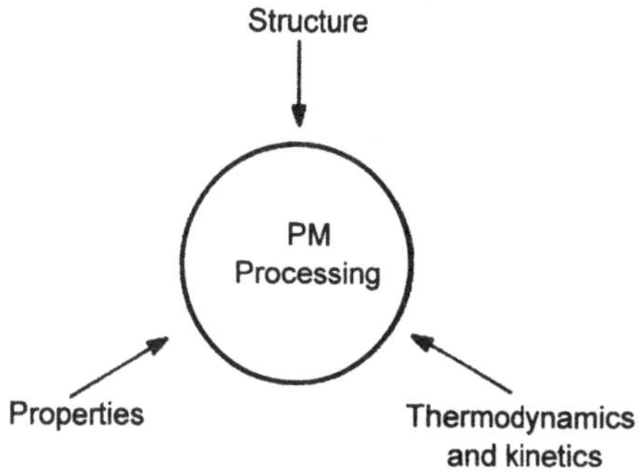

Figure 1.8 summarises the schematics of the role of basic materials science in PM processing. It is hoped that the readers will appreciate these facts, while studying the subject in later chapters of this book.

Fig. 1.8 Schematics of basic aspects of materials influencing PM processing

SUMMARY

- Metal powders are produced by chemical, electrolysis, atomisation or mechanical processing methods.

- Ceramics are broadly divided into traditional and advanced ceramics. The latter offer precision parts similar to conventional metal powder parts.

- Powder characteristics play an important role in powder processing.

- Powder compaction methods include die, isostatic and hot isostatic compactions.

- Sintering is the result of atomic motion stimulated by high temperatures. This can be done either in solid state or in the presence of a minor amount of low melting point metals and alloys.

- Powder injection moulding process is similar to plastic injection moulding, with the difference that the polymer is filled with dispersed metallic or ceramic powders.

Further Reading

Chin GY, (ed), *Advances in Powder Technology*, American Society for Metals, Metals Park, OH, 1992.

German RM, *Powder Metallurgy and Particulate Materials Processing*, Metal Powder Industries Federation, Princeton, 2005.

Hirschhorn JH, *Introduction to Powder Metallurgy*, American Powder Metallurgy Institute, Princeton, 1969.

Jenkins I and Wood JV, (eds), *Powder Metallurgy: An Overview*, The Institute of Metals, London, 1991.

Jones WD, *Fundamental Principles of Powder Metallurgy*, Edward Arnold, London, 1960.

Lenel FV, *Powder Metallurgy: Principles and Applications*, Metal Powder Industries Federation, Princeton, 1980.

Schatt W and Wieters KP, *Powder Metallurgy Processing and Materials*, European Powder Metallurgy Association, Shrewsbury, UK, 1997.

Thümmler F and Oberacker R, *Introduction to Powder Metallurgy*, The Institute of Materials, London, 1994.

Upadhyaya GS, *Sintered Metallic and Ceramic Materials*, John Wiley and Sons, Chichester, UK, 2000.

Upadhyaya GS and Upadhyaya A, *Materials Science and Engineering*, Viva Books Pvt. Ltd., New Delhi, 2006.

Upadhyaya GS and Vajpei AC, *History of Metals*, Uttar Pradesh, Hindi Sansthan, Lucknow, India, 1997 (in Hindi).

EXERCISES

1.1   Powder metallurgy is one of the material forming processes. Enumerate other processes.

1.2   Did powder processes technology begin historically with metals and alloys? Comment.

1.3   What are the advantages of PM processing?

1.4   What are the limitations of PM processing?

1.5   Making engineering parts of varying compositions based on metallic systems is possible by PM processing, which is not always possible in solidification processing. Comment on the statement.

1.6   PM processing originated as a technology, but has now become a science. Comment.

2

Powder Production

LEARNING OBJECTIVES

- Chemical methods of production of Fe, Cu, Ni, Co, Ti and W powders
- Electrolytic method of powder production
- Atomisation methods of powder production
- Evaporation method
- Mechanical methods of powder production
- Various size reduction equipments

In powder metallurgical processing, the methods adopted for powder production are vital, since they affect powder characteristics and other related processing steps. In this chapter, the most important principles and production methods are described. The method of choice is related to the type of application and desired properties of the final product.

2.1 CHEMICAL METHODS

Metal powders: At present metals are not often found as large deposits in their free or uncombined state. Gold often occurs as free metal, but vast quantities are also dissolved as compounds in sea water. Oxide ores are common for iron and tungsten powder production, while copper powder is produced from the mill scale (mostly oxide) from copper fabrication industries. Production of metal powder invariably attempts to achieve chemical reaction in the solid state, unlike in the casting practice where molten metal is used.

Chemical reduction involves a chemical compound which is most frequently an oxide;sometimes a halide or any other salt of the metal. This may be carried out from:

1. *Solid state:* For example, reduction of iron or tungsten oxide (WO_3) with a reducing gas.

2. *Gaseous state:* For example, reduction of titanium chloride ($TiCl_4$) vapour with molten magnesium.

3. *Aqueous solution:* For example, the precipitation of cement copper from copper sulphate solution with iron, or the reduction of an ammoniacal nickel salt solution with hydrogen under pressure (hydro-metallurgical method).

Chemical decomposition of compounds is the other method for powder production. The common examples are: decomposition of metal hydrides or metal carbonyls. High temperatures are used to decompose or change the composition and structure of metal compounds. The relative stability of compounds can change markedly with temperature.

The use of high temperatures also enables the reaction to be carried out in a short time. The main examples of metal powder production by chemical decomposition are:

- Decomposition of metal hydrides: e.g., Ti, Zr, Hf, V, Th or U.
- Decomposition of metal carbonyls MCo_5: e.g., Fe, Ni

Ceramic powders

Chemical methods of ceramic powder preparation are classified as: solid-state reactions, liquid solutions and vapour phase reactions.

Examples of the first category are production of powders of simple oxides from carbonates, hydroxides, nitrates, sulphates, acetates, oxalates, alkoxides and other metal salts. Chemical reactions between solid starting materials—usually in the form of mixed powders—are common for the production of powders of complex oxides such as titanates, ferrites and silicates.

The preparation of liquid solutions involves the dissolution of soluble salts, usually in water, or the dissolution of metals in acid solutions. The purity of the final product is limited by the purity of the starting materials. The sol–gel process is another important method in which sols, or suspensions of collidal particles, are formed by controlled precipitation from aqueous solutions. In some cases a stable sol is made by peptising a slurry using pH control. Gelling to a semi-rigid body is carried out by evaporating the water or changing the pH. In the case of pH control, an ammonia donor such as hexamethylene tetramine is added to the sol and droplets are passed through a hot organic fluid which causes the evolution of ammonia. The gelled sol is further dried to give low-density spheres of high surface area and high chemical homogeneity. The gels can also be made by polymerisation of metal alkoxides. Another category of solution preparation is precipitation or co-precipitation in which the desired cations are precipitated from the solutions as an insoluble salt using anions such as hydroxide, carbonate or oxalate.

The vapour phase reaction process is described later in Section 2.4 of this chapter.

2.1.1 Solid-state reduction

A common feature of the solid-state reduction process is the need for gas flow over the particles to remove product species, provide reactant gases or to provide efficient heat transfer between solids and gas. The gas/solid contact is achieved in practice by mechanical stirring of the particulate bed by raking, by tumbling within the reactor, allowing the particles to fall through the gas phase or suspending the particles in the gas (fluidised bed reactor).

The most common iron ore for producing iron powder is magnetite (Fe_3O_4). In the presence of the CO/CO_2 gas mixture ($p_{CO} + p_{CO_2}$ = 1 atm) the reduction of higher oxide to lower oxide and then to iron may be expressed in the following stages:

$$3Fe_2O_3(s) + CO \rightarrow 2Fe_3O_4(s) + CO_2(g)$$
$$Fe_3O_4(s) + CO(g) \rightarrow 3FeO(s) + CO_2(g)$$
$$FeO(s) + CO(g) \rightarrow Fe(s) + CO_2(g)$$

For each oxide and reaction temperature a critical CO/CO_2 ratio in the gas mixture must be exceeded for the reaction to proceed. These conditions are illustrated in Fig. 2.1. Typically, the reduction of Fe_2O_3 to Fe_3O_4 occurs between 200–500°C and a minimum CO/CO_2 ratio of approximately $1/10^4$ is required. Fe_3O_4 is reduced to FeO between 500–900°C, and FeO is reduced to iron at temperatures between 900–1300°C.

In all the above equations, the value of $\Delta G°$ (standard free energy change) is negative within the temperature range under consideration. One of the most useful diagrams to identify the stability of the oxide is the *Ellingham diagram* (Fig. 2.2). This shows that below

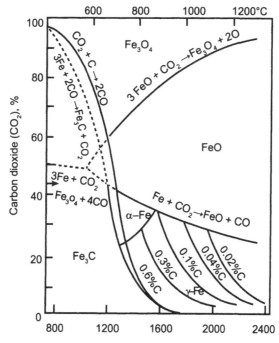

Fig. 2.1 Equilibrium diagram for the system Fe-O-C: Percent of CO_2 in CO/CO_2 mixture versus temperature

POWDER METALLURGY

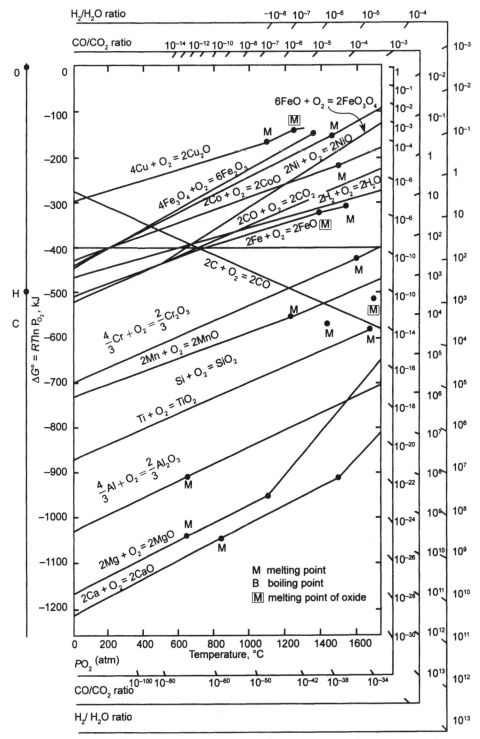

Fig. 2.2 Ellingham diagram of free energy versus temperature for various oxides

any Ellingham line, the metal is stable relative to the oxide. The Ellingham line thus divides the diagram into stability fields. In principle, if the element A can reduce the oxide B_xO_y in the diagram, the Ellingham line for A_xO_y lies below that for B_xO_y.

The noble metals which are easily reduced occur at the top of the diagram and the more reactive metals at the bottom. It is evident that with increase in temperature the oxides become less stable. Most of the ΔG versus T lines therefore slope upwards. An exception to the positive slope of the ΔG versus T line occurs for carbon oxidation to CO or CO_2. In both cases the oxide product is gaseous and thus also has a high free energy. The carbon monoxide reaction $2C + O_2 \rightarrow 2CO$ is favoured at high temperatures, and therefore, carbon is a very effective reducing agent, having a greater affinity for oxygen than most oxides. For many oxides $\Delta G°$ tends to zero at some elevated temperature. This is known as the standard dissociation temperature when the oxide is in equilibrium with the pure element and oxygen at 1 atm pressure (P) according to the expression:

$$\Delta G = \Delta G° + RT \ln P \tag{2.1}$$

At equilibrium $\Delta G = 0$, and the relationship becomes

$$P_o = \exp\frac{\Delta G°}{RT} \tag{2.2}$$

If the pressure is lowered below this value, the oxide will dissociate. On the other hand, if the pressure is raised above this equilibrium value, the oxide is stable.

The reduction of metal oxides using hydrogen gas is also a possibility. By controlling the proportion of hydrogen and steam (H_2O) in the gas mixture, the oxygen potential of the gas can be controlled through the reaction:

$$2H_2(g) + O_2(g) \rightarrow 2H_2O$$

$$K = \frac{P_{H_2O}}{P_{H_2}^2} \times \frac{1}{P_{O_2}} \tag{2.3}$$

At any temperature the effective partial pressure of oxygen is determined by the ratio of the pressures of steam and hydrogen. To obtain low oxygen pressures a high ratio of hydrogen in steam is necessary. To avoid a large number of calculations a monogram was constructed by Richardson (Fig. 2.2), which allows the speedy estimation of the hydrogen/steam ratio required to reduce metal oxide at any reaction temperature. On the left-hand side of the figure, there are two markings 'H' and 'C' representing the hydrogen and carbon. The ruler has to be placed over the point H or C, depending on where we wish to read the H_2/H_2O scale or CO/CO_2 ratio scale. The value so read is the critical value and the temperature of reaction under consideration. Reduction can still be achieved with any mixture containing greater than this proportion of hydrogen or carbon monooxide, as the case may be.

The more stable the metal oxide, the greater the H_2/H_2O or CO/CO_2 ratio necessary to achieve reduction. In the case of Al_2O_3, at 500°C, a minimum H_2/H_2O ratio greater than $10^{18.5}$ is required to achieve reduction—a condition impossible to achieve in practice.

Hydrogen and carbon monoxide do have the advantage that they are readily obtained from either natural gas or coal. The high mobilities of these gaseous reactants reduce the difficulties in supplying reactants to the solid surfaces.

The major inferences drawn from the Ellingham diagram are:

1. The interaction of the free energy change/temperature line with the temperature axis when $\Delta G^\circ = 0$ gives the temperature at which oxygen equilibrium pressure P_{P_2} is equal to the standard pressure of 1 atm. The temperature is known as the decomposition temperature of the oxide.

2. In the range of temperature at which the free energy change line is below the line of $\Delta G^\circ = 0$, the oxide is stable; above this line, it is unstable.

3. When the oxygen pressure in contact with the solid oxide drops to below the value of the equilibrium oxygen pressure of the oxide, the oxide will become unstable and may decompose.

For the reduction of the more stable oxides, either reduction by solid carbon, by electrical methods or by another metal are the only viable techniques. For further discussion on the kinetics of heterogenous reactions, one may refer to Alcock CB, *The Principles of Pyrometallurgy*, Academic Press, 1976.

The reduction of halides by hydrogen can be carried out in a manner similar to oxides and sulphides, e.g., VCl_3, $ZrCl_4$ or $TiCl_4$. One complication is that the solid or liquid chlorides may have high vapour pressures or may react to form volatile halide products. The volatiles would then follow the gas flow and have to be condensed and recycled to avoid product loss.

2.1.2 Hydro-metallurgical reduction

The production of metals directly from aqueous solutions may be carried out by:

- Reduction with another metal
- Gaseous reduction

The process of reduction of metal ions from a solution of another metal is commonly known as *cementation*. The thermodynamic conditions necessary for cementation can be determined by considering the reduction potentials of the species involved in the reaction. For any reaction,

$$M^{n+} + ne \rightarrow M \tag{2.4}$$

the reduction potential is given by

$$E = E^\circ - \frac{RT}{nF} \ln \frac{a_M}{a_{M^{n+}}} \tag{2.5}$$

where E° is the standard reduction potential of metal M, and $a_{M^{n+}}$ and a_M are the activities of the metal ion and metal, respectively. By convention, the activity of the pure metal is unity. Thus,

$$E = E^\circ + 2.303 \frac{RT}{nF} \log a_{M^{n+}} \tag{2.6}$$

The more negative the electrode potential, the more stable the ions are in solution. By comparing the electrode potentials of different metals the more stable metal ion can readily be determined.

In gaseous reduction, the suitable gases which are commonly available in commercial quantities are: H_2S, SO_2, CO and H_2. The most commonly used reagent is hydrogen. The stability of metal ions in solution in the presence of hydrogen can be calculated for ionised metal salt solutions such as sulphate or chloride using a modified form of E_h–pH diagram (Fig. 2.3). The metal potentials are plotted against concentration and the hydrogen potential against pH.

Reduction of metal species in solution by hydrogen can only take place if the hydrogen is at a lower potential than the metal ions at the appropriate metal concentration. The potential of the hydrogen reaction,

$$2H^+(aq) + 2e \rightarrow H_2(g) \tag{2.7}$$

is given by

$$E = E° + \frac{RT}{F} \ln [H^+] \tag{2.8}$$

The hydrogen potential, and thus the ability of this species to act as a reactant, therefore changes considerably with the pH of the solution. The overall reduction reaction of nickel

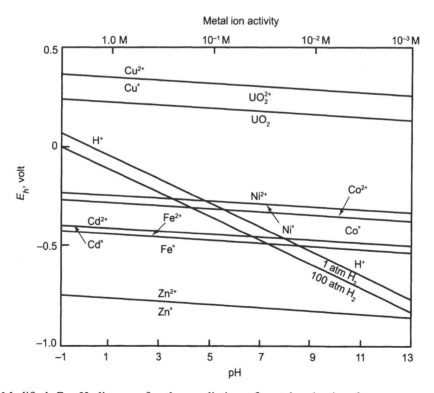

Fig. 2.3 Modified E_h–pH diagram for the prediction of metal reduction from aqueous solutions by hydrogen

ions in the presence of hydrogen can be represented by the equation:

$$Ni^{2+}(aq) + H_2(g) = Ni(s) + 2H^+(aq) \tag{2.9}$$

where, $k = \dfrac{a_{H^+}^2}{a_{Ni^{2+}} p_{H_2}}$.

The major variables in this simplified system are the reaction temperature, the hydrogen ion concentration, the metal ion concentration and the hydrogen pressure.

From the above equilibrium relationship it is seen that increasing hydrogen pressure and decreasing hydrogen ion concentration favour the reduction of metal ions to metal. Hydrogen ions generated by the reduction may be removed by the addition of ammonia (NH_3) to the solution. However, the presence of ammonia also results in the formation of complex ammine ions of the form $[Ni(NH_3)_n]^{2+}$. The reduction of the complex by hydrogen can be represented by the following equations:

$$Ni(NH_3)_n^{2+}(aq) \rightarrow nNH_3(g) + Ni^{2+}(aq) \tag{2.10}$$

$$Ni^{2+}(aq) + H_2(g) \rightarrow Ni(s) + 2H^+(aq) \tag{2.11}$$

$$H^+(aq) + NH_3(g) \rightarrow NH_4^+(aq) \tag{2.12}$$

Increasing the ammonia concentration in the solution results in two opposing effects:

1. The production of nickel metal is favoured because the hydrogen ions generated are removed during reduction by forming ammonium salts instead of free acid.
2. The decrease in free nickel ion concentration in the solution is due to the complexing action of ammonia, making the reduction of Ni^{2+} more difficult.

The stabilities of the ammine complexes as a function of pH and NH_3/Ni ratio are shown in Fig. 2.4. It is clear that the largest driving force for hydrogen reduction of nickel is near a NH_3/Ni ratio of 2. This is the optimum NH_3/Ni^{2+} ratio which minimises the nickel ion concentration remaining in the solution on reduction with hydrogen.

Although the chemical reaction shown in Eq. (2.9) indicates that isolated nickel metal atoms form in the solution, in reality, a critically-sized metal nucleus made up of a number of nickel atoms must be created. Nucleation can occur homogeneously throughout the liquid, or heterogeneously at the existing solid surfaces. The predominant mechanism in any system will depend on the chemical driving force available for metal production, and the availability of suitable surfaces which will initiate metal growth. In many commercial practices the rate of metal production is increased by seeding the solution with metal powder, thus overcoming the problems of nucleation of metal from the solution. As in the case of metal compound formation, the particle size distribution, the density and the shape of the metal powders must be controlled. This can be achieved by the addition of surface-active organic reagents to the solutions. Precipitation of cobalt and nickel from an acid medium takes place homogeneously.

Of late, bacterial leaching, a variant of hydrometallurgical processing, has been found to be an attractive method for preparing nanoparticles of metals. Synthesis of noble-metal nanoparticles using Fe(III)-reducing bacteria has been successfully attempted. The bio-reductive deposit of noble metal is a fast process: 1–2 mol m^{-3} of aqueous Pd(II) ions are completely reduced to crystalline Pd(0) nanoparticles within 60 minutes.

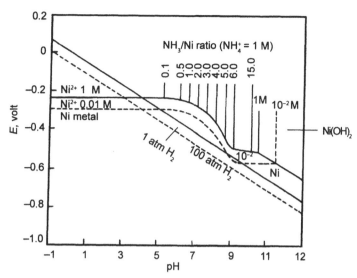

Fig. 2.4 Thermodynamics of nickel ammine solutions

2.1.3 Ion exchange method

The separation of ions from solution is often carried out using the ion exchange method. Metal ions in solution can exchange with specific ions contained in a solid or second liquid phase. Ion exchange reactions are reversible and stoichiometric and can provide a convenient method for:

- removing unwanted ions from aqueous solutions
- increasing the concentration of selected metal ions in the solution.

An ion exchange resin consists of a rigid or semi-rigid three-dimensional framework of atoms or molecules, which must be chemically stable in the solutions to be treated. Functional groups—group of atoms which supply the exchangeable ions—are chemically bound to this framework. The modern synthetic organic resins are based on phenol formaldehyde and amine formaldehyde polymers, which give three-dimensional structures. Specific ion exchangers (*chelating exchangers*) are those in which the functional groups have the properties of a specific reagent. A cation exchange reaction can be expressed as:

$$RH + M^+_{aq} \rightleftarrows RM + H^+_{aq} \tag{2.13}$$

where, R represents the fixed functional groups in the exchange resin and M the metal cation which takes part in the exchange.

Four types of extractants can be used for liquid/liquid extraction:

- Solvating extractants
- Cation exchange extractants
- Chelating extractants
- Anion exchange extractants

Some organic liquids can be used for solvent extraction in an undiluted form. However, most organic reagents are too viscous to be used in this way and must be dissolved in some inert liquid, such as kerosene. In such cases the active reactant is referred to as the *extractant*. The inert organic carrier is called the *diluent* and the total organic phase is referred to as the *solvent*. In liquid ion exchange systems, 5–10 vol% solutions of extractant in the diluent are commonly used. The solubility of the extractant in the diluent should be sufficient to achieve efficient separation.

The solvent extraction of metals from aqueous solutions is carried out in continuous exchange processes. The metal is extracted from the impure aqueous phase into an organic liquid phase. The organic phase, loaded with metal, is physically separated from the aqueous phase and transferred to a separate reactor for stripping. In this second process step the metal transfers back to an aqueous solution to obtain not only a purified solution but also one with a higher metal ion content. Again, the organic and aqueous phases are separated. The barren organic solution is returned to be contacted with incoming impure aqueous feed and extraction of metal ions takes place once more.

2.1.4 Direct synthesis

This method is carried out at high temperature and is adopted for the preparation of pure compound ceramics and intermetallics. The examples are refractory carbides (TiC, WC, etc.), silicon carbide and silicon nitride. Of late, a new method known as 'self propagating high-temperature synthesis' has been adopted, in which a porous compact made from a mixture of the reagents is brought into contact with a hot tungsten coil. This initiates an exothermic reaction which continues as a self-sustaining combustion wave propagating through the porous mass, converting it into the reaction products. This is shown schematically

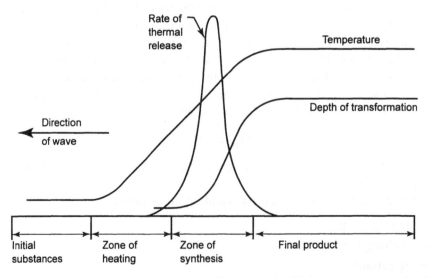

Fig. 2.5 Equilibrium adiabatic structure of a self-propagating high-temperature synthesis wave

Table 2.1 Examples of materials prepared by the combustion synthesis method

Borides	CrB, HfB_2, NbB_2, TaB_2, TiB_2, LaB_6, MoB_2
Carbides	TiC, ZrC, HfC, NbC, SiC, Cr_3C_2, B_4C, WC
Carbonitrides	$TiCN$, $NbCN$, $TaCN$
Cemented carbides	$TiC\text{-}Ni$, $TiC\text{-}(Ni, Mo)$, $WC\text{-}Co$, $C_3C_2\text{-}(Ni, Mo)$
Chalcogenides	$MoSe_2$, $TaSe_2$, $NbSe_2$, WSe_2
Composites	$TiC\text{–}TiB_2$, $TiB_2\text{–}Al_2O_3$, $B_2C\text{–}Al_2O_3$, $TiN\text{–}Al_2O_3$
Hydrides	TiH_2, ZrH_2, NbH
Intermetallics	$NiAl$, $FeAl$, $NbGe$, $TiNi$, $CoTi$, $CuAl$
Nitrides	TiN, ZrN, BN AlN, Si_3N_4, TaN (cubic and hexagonal)
Silicides	$MoSi_2$, $TaSi_2$, $TiSi_2$, $ZrSi_2$

in Fig. 2.5. Depending on the temperature of the reaction, synthesis can be entirely in the solid state, partly in gaseous state or entirely in the gaseous state. The powders used in PM are prepared by gaseous combustion. The synthesis can be designed on the basis of the adiabatic temperature of combustion, the combustion gas velocity and activation energy for the compound formation. With an exothermic reaction, the processing time at high temperatures is short and the purity of the powder can be maintained. This process was first used in Russia for the production of TiC abrasive powders and $MoSi_2$ for heating elements. During direct synthesis, sintering often occurs, and is known as reactive sintering. Table 2.1 shows the examples of materials prepared by combustion synthesis.

2.1.5 Some specific powders produced by chemical method

Iron powder

Sponge iron powder produced by the Höganäs process is a typical example of the chemical method. This is a batch process in which the ground ore does not move during reduction, but is static. This is in contrast to other direct reduction processes which are continuous. The Höganäs process is based on the use of quite pure magnetite (Fe_3O_4) ores found in northern Sweden. The iron ore is reduced with a carbonaceous material. Figure 2.6 shows the steps involved in producing such powders. The ore is ground to a particle size distribution determined by each of the desired iron powders. The ore powder is placed in the centre of cylindrical ceramic containers ('saggers' made of silicon carbide) surrounded on the outside by a concentric layer of a mixture of coke and limestone. The saggers are placed in layers upon cars which are pushed through a fuel-fired tunnel kiln. The carbon monoxide produced from the coke reduces the ore to iron. The total reduction time is of the order of 24 hours at a reduction temperature of 1200°C. The limestone serves to bind any sulphur in the coke and prevents it from contaminating the iron. The sponge iron is mechanically removed from the saggers and ground. The resulting powder is magnetically separated from impurities. In the final reduction step, the powder is carried through a continuous furnace in an atmosphere of hydrogen on a belt made of stainless steel.

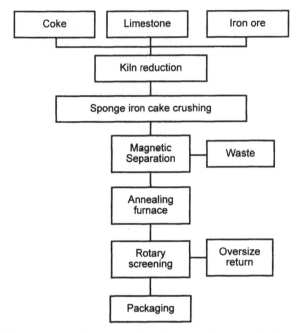

Fig. 2.6 Flowsheet for the production of sponge iron powder from iron ore

Another chemical method for the production of iron powder is the decomposition of its carbonyl. The carbonyls are liquids at normal temperature with a low boiling point. These are formed by reaction of the metal and carbon monoxide gas under pressure. Iron carbonyl $[Fe(CO)_5]$ is formed at 70–200 atm pressure and a temperature of 200–220°C. The carbonyls can now be decomposed by heating the vapour at atmospheric pressure. Care must be taken to conduct the decomposition in the gas phase and not on the surface of the reaction vessel, in order to obtain the metal in the desired powdery form. The usual carbonyl iron powder particles are spherical with an onion skin structure, because the iron powder 'nuclei' first formed catalyse the decomposition of CO into C and CO_2. Carbon deposits on the iron powder nucleus; another layer of iron is deposited on top and so on. This type of iron powder is quite pure with respect to metallic impurities, but contains a considerable amount of carbon and oxygen (fraction of a percent). Powder devoid of such impurities can be produced by adding ammonia during the decomposition of the carbonyl and by a subsequent annealing treatment. But these would naturally enhance the cost. Carbonyl iron powder is usually spherical in shape and very fine (<10 μm). Figure 2.7 illustrates the flowsheet for powder production, while Table 2.2 gives the characteristics.

Copper powder

The major source of raw metal is mill scale from copper rolling or wire drawing operations which typically have an oxygen content of about 10%. An alternate source of raw material involves the air atomisation of molten copper with a subsequent oxidising step. Figure 2.8 illustrates the steps used to produce copper powder by this method. Reduction is commonly

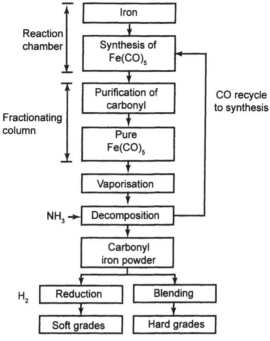

Fig. 2.7 Flowsheet for the production of carbonyl iron powder

Table 2.2 Characteristics of carbonyl iron powders of different grades made by BASF

Carbonyl iron powder	CNS	CCS
Bulk density, kg L^{-1}	~4.2	~4.2
Average particle size, μm	5–6	4–6
Iron, %	99.6	99.6
Carbon, % (by combustion)	0.04	0.05
Oxygen, % (by method of difference)	0.15	0.2
Nitrogen, %	0.01	0.01
Nickel, ppm	<100	<300
Chromium, ppm	<100	<200
Molybdenum	<50	<100

Particle size [Coulter (counter)]	wt%
<2 μm	>2
<4 μm	>30
<6 μm	>85
<8 μm	>99

carried out at temperatures ranging from 425°C to 705°C with a total furnace time of 4 to 8 h. Since hydrogen readily diffuses into copper, the reduction rate is related to hydrogen pressure. As the H_2O formed by reaction with oxide cannot readily escape by diffusion, it must either form internal cavities or escape through fissures. This accounts for the spongy and porous nature of reduced copper powder. The reduction reactions are highly exothermic in nature. The apparent activation energy for the reduction of cuprous oxide is 57.3 kJ mole^{-1}. The reduction of Cu_2O involves the equilibrium adsorption of hydrogen atoms on the oxide

---

Example 2.1: The decomposition of carbonates or hydroxides is commonly known as *calcining*. Describe the characteristics, using thermodynamic data, for calcium carbonate and magnesium hydroxide.

Solution: The reactions for calcining of these compounds are:

$$CaCO_3(s) \rightarrow CaO(s) + CO_2(g) \qquad \Delta H° = 161 \text{ kJ mol}^{-1}$$

$$Mg(OH)_2(s) \rightarrow MgO(s) + H_2O(g) \qquad \Delta H° = 46 \text{ kJ mol}^{-1}$$

Figure Ex. 2.1 shows the decomposition pressures of CO_2 and H_2O for a number of carbonates and magnesium hydroxide with respect to temperature. A number of reaction steps are involved in the decomposition of these materials:

- The decomposition reaction
- Diffusion of the product gas and heat conduction in the product layer
- Heat and mass transfer between the particle surface and the bulk gas.

In the case of limestone, because the reactions are highly endothermic, for spherical particles greater than 40 μm diameter, heat transfer through the product layer is rate limiting. The materials not only undergo changes in chemical composition but also in structure and morphology. For example, after decomposition at 700°C, the surface area of the CaO is 8 m^2 g^{-1} (original 0.07 m^2 g^{-1}), while treatment at 1350°C results in a material with surface area of only 0.2 m^2 g^{-1}.

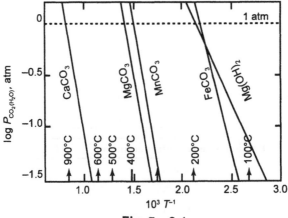

Fig. Ex. 2.1

surface followed by the surface reaction which is the rate controlling step. It has been claimed that while surface oxides do not interfere markedly with the moulding operation or the subsequent green strength of the compact, occluded oxides are highly detrimental and result in a product of low tensile and impact strengths.

Example 2.2: Figure Ex. 2.2 shows the relationship between oxygen partial pressures and compositions for the Ti-O and Zr-O systems at 1000°C. Explain why the halide route is preferred instead of direct reduction.

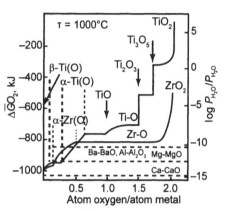

Fig. Ex. 2.2

Solution: Even at low partial pressures of oxygen, considerable concentration of oxygen may be retained in a solid solution. This dissolved oxygen greatly affects the physical properties of the metal, e.g., ductility. From the free energies of formation data, it can be seen that halides are, in general, less stable than the corresponding metal oxides, and hence the halide route is the preferred production route.

Example 2.3: It is desired to prepare barium titanate (an electronic ceramic) from barium carbonate and titania. Write the reaction and discuss any problem faced during the preparation.

Solution: The formation of barium titanate is through the solid-state reation:

$$BaCO_3 + TiO_2 \rightarrow BaTiO_3 + CO_2(g)$$

Some of the problems faced are:

1. A non-equilibrium phase Ba_2TiO_4 initially forms between $BaTiO_3$ and unreacted $BaCO_3$.

$$BaTiO_3 + BaCO_3 \rightarrow Ba_2TiO_4 + CO_2(g)$$

This is undesirable in the calcined product. It is minimised by dispersing agglomerates of TiO_2 and mixing the particle contacts thoroughly, thus, reducing the diffusion path between $BaCO_3$ and TiO_2. The partial pressure of CO_2 in the pores of the product influences the reaction kinetics.

2. During calcination, the partial pressure of oxygen in the air must be controlled to obtain the requisite oxidation states of the titanium metal ions.

The completeness of the reaction and uniformity of the product depend on the particle size and mixedness of the reactants and the time, temperature and atmosphere, and their uniformity during calcination. In application, therefore, different lots of processed materials are blended to maintain a high level of uniformity.

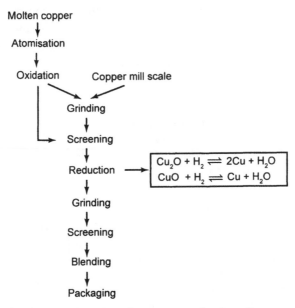

Fig. 2.8 Schematic flowsheet for gaseous reduction of copper oxide

In the atomisation/oxidation process it is necessary to grind the atomised copper powder finely before oxidation, since cuprous oxide forms an impervious layer on the particles and prevents further oxidation from taking place. Thus, the individual particles may show only the spongy structure at the surface and the density reducing effect may be diminished.

Cement copper can be precipitated from copper sulphate solution with iron. Low cost copper powder is produced from the solution obtained by leaching copper ores or copper scrap, where the precipitation of copper powder from an acidified solution of copper sulphate with iron is achieved. Large quantities of this 'cement copper' are produced from the copper sulphate solutions which are a by-product of the copper refinery industry. Most of this cement copper is eventually melted and cast—rather than used as powder—for two reasons:

- The cement copper produced as a by-product is rather impure unless special precautions are taken.
- The powder is quite fluffy, i.e., it has a low apparent density, which is not satisfactory for many copper powder applications.

To make the powder suitable, a furnace treatment would be necessary, which will add to the cost.

It is to be remembered that the copper solution is a dilute one (1–2 kg m⁻³). It is contacted with iron leading to the overall reaction:

$$Cu^{2+}(aq) + Fe^\circ = Fe^{2+}(aq) + Cu^\circ \tag{2.14}$$

and $K = \dfrac{a_{Fe^{2+}}}{a_{Cu^{2+}}}$.

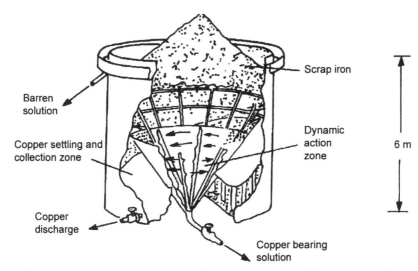

Fig. 2.9 Kennecott cone for precipitating copper from dilute leach solutions

The equilibrium for this reaction at 25°C is achieved when $K = 10^{26}$. In effect this means that almost all the copper should be removed from solution as the system approaches equilibrium.

Figure 2.9 illustrates the Kennecott cone used for copper cementation. The copper bearing solutions are forced up at high velocity through a bed of scrap iron. The copper is cemented onto the iron, but the copper particles get detached from the metal surface and are carried out by the solution through a grid into a region of low liquid velocity. Gravity separation of the copper and the barren solution occurs in this zone. Limitations do exist: in practice the complete recovery of copper is not achieved. The typical analysis of the product is 85%–90% Cu, 0.2%–2% Fe, 0.5% ($SiO_2 + Al_2O_3$), and the balance oxygen.

Nickel/Cobalt powder

These metals are typical examples for the preparation of metal powders by gaseous reduction from aqueous solutions, the basic aspects of which are described in Section 2.1.1.

Nickel powder is produced by the Sherritt–Gordon process. It is a typical hydro-metallurgy method for production of metal powders in which reduction of an ammoniacal solution of nickel sulphate with hydrogen under a pressure of 1.38 MPa (200 psi) and a temperature of 190–200°C in an autoclave is carried out. A nickel salt solution is obtained by leaching complex Cu-Ni-Co ores. Before the nickel is precipitated as metallic powder, the copper is removed from the solution by precipitation as sulphide. For the precipitation of the first nickel powder nuclei from the solution, a catalyst, such as ferrous sulphate, is used. The very fine nickel powder nuclei are allowed to settle in the autoclave; the barren solution is decanted and a new batch of solution is introduced into the autoclave. The nickel powder nuclei are suspended in the solution by agitation and the nickel in the solution are reduced with hydrogen at 1.38 MPa (200 psi) and precipitated on the existing nuclei.

The process—called densification—is repeated many times, say 15–30 times. Finally, the powder is removed from the autoclave, washed and dried. The process permits control of the size and shape of the nickel powder being produced.

The spent solution contains 1 g L^{-1} Ni and 1 g L^{-1} Co. It is treated with H$_2$S at 80°C and atmospheric pressure. The precipitated Ni-Co sulphides are filtered off for recovery, and the solution is evaporated to crystallise ammonium sulphate for use as fertiliser. The mixed sulphides are leached with H$_2$SO$_4$ at 120°C in the presence of air at 7 atm. Acid leaching is used instead of ammonia leaching to avoid the formation of lower oxidation products of sulphur. The solution is purified, removing the last traces of iron by adjusting the pH to 5 and filtering off ferric hydroxide.

Table 2.3 shows the typical characteristics of hydro-metallurgically reduced nickel powder. Figure 2.10 gives the flowsheet of the recovery of nickel and cobalt by the Sherritt–Gordon process.

Another method to produce Ni powders is by the decomposition of the respective carbonyl. This process was originally developed by Mond, primarily as a means of purifying nickel. Crude metal is exposed to carbon monoxide at atmospheric pressure and at temperatures between 40°C and 100°C, to obtain a colourless gas, nickel tetracarbonyl [Ni(Co)$_4$]. This reaction is readily reversible by heating the carbonyl in the 150–300°C range to yield pure nickel and carbon monoxide. Conditions that influence the formation of the self-nucleating nickel particles during carbonyl decomposition can vary greatly, which affects the physical and technical properties of the powders produced. The powder decomposers are steel cylinders, the walls of which are heated with high-capacity electrical resistance heaters. Liquid carbonyl vapour is introduced into the top of the chamber, slightly above atmospheric pressure, where it contacts the heated decomposer walls that are preset at a temperature between 250°C and 350°C. The collected powder is gas purged, stabilised with an oxide coating and transferred to storage, completely free from carbonyl and carbon monoxide. INCO produced three types of nickel powders by this method: single spiky particles, filaments and high-density semi-smooth particles. Table 2.4 gives the data related to chemical composition and physical

Table 2.3 Typical characteristics of hydro-metallurgically reduced nickel powder

Chemical composition	%
Nickel and cobalt	99.9
Cobalt	0.5–0.10
Copper	0.003
Iron	0.005–0.010
Sulphur	0.03
Carbon	0.006

Screen size	%
+100 mesh	0–10
−100 + 150 mesh	5–30
−150 + 200 mesh	20–45
−200 + 250 mesh	10–25
−250 + 325 mesh	10–35
−325 mesh	5–25

Physical properties	
Apparent density, g cm^{-3}	3.4–4.1
Flow rates, 50 g^{-1}	20

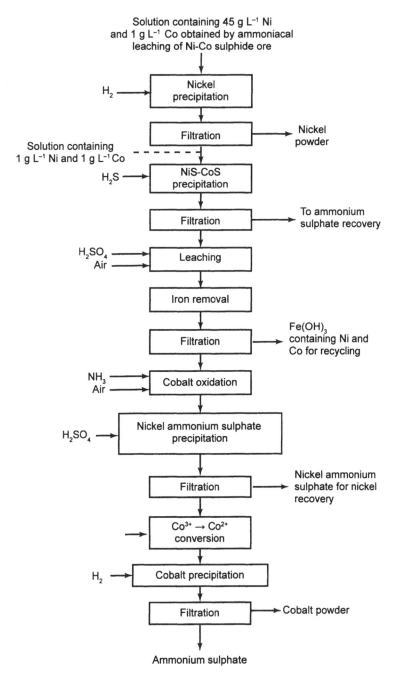

Fig. 2.10 Process flowsheet for the Sherrit–Gordon nickel process for the recovery of nickel and cobalt by hydrogen reduction

Table 2.4 Properties of various grades of carbonyl nickel powder of INCO make

Physical properties	Type 123	Type 255	Type 270	Type 287
Apparent or bulk density, g cm^{-3}	1.6–2.7	0.50–0.65	0.66–0.74	0.75–0.95
Fisher subsieve size, μm	3.0–7.0	2.2–2.8	2.5–3.1	2.6–3.3
BET specific surface area, m^2 g^{-1}	0.4	0.7	0.6	0.6

Typical chemical composition	Typical weight	%
	Type 123	Types 255, 270 and 287
Nickel	Balance	Balance
Carbon	0.06	0.2
Oxygen	0.05	0.07
Iron	0.005	0.005
Sulphur	0.0002	0.0002
Nitrogen	0.003	0.0002
Cobalt	0.0003	0.0003
Copper	0.0001	0.0001
Lead	<0.0001	<0.0001
Other elements	<0.001 total	<0.001 total

properties of four major types. Type 123 powder possesses discrete spiky characters, while the remaining three (Type 255, 270 and 287) belong to low-density filamentary powders, and are mainly used in the electronics industry. The semi-smooth high-density nickel powder (HDNP) is in the 5–10 μm range, with an apparent density of 3.5–4.2 g cm^{-3}.

Titanium powder

Metallothermic reaction is the most common method used for the reduction of titanium compound. In the Kroll process, TiCl$_4$ is reduced by molten magnesium. The schematic process is illustrated in Fig. 2.11. The reactor is made of steel and is heated externally until the magnesium ingots melt. The halide is then introduced as TiCl$_4$. The reduction reaction is exothermic and produces enough heat to maintain a reaction temperature of 800–1000°C.

$$TiCl_4(l) + 2Mg(l) \rightarrow Ti(s) + 2MgCl_2(l) \qquad (2.15)$$

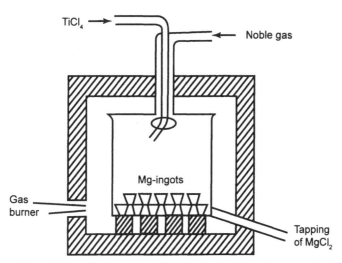

Fig. 2.11 The Kroll reactor used in the production of sponge titanium

After completion of the reduction reaction, most of the product chloride is removed by tapping it off as a liquid. After cooling the reactor to room temperature, the mixture of solid titanium sponge, magnesium chloride and the remaining magnesium metal is removed from the vessel by machining. Separation of the products is then carried out either by vacuum distillation or selective leaching. The fines of unalloyed titanium sponge (<150 μm) are screened out, and can be used for PM processes. Using this elemental powder for the production of PM parts provides cheaper raw material, since the subsequent process and handling operations are eliminated. Such a powder contains small amounts of residual chlorides (~1500 ppm salt) and other interstitials which can be detrimental to the end properties of the product, especially if it has to be used in high performance applications.

Another method of producing titanium powder is by decomposition of TiH_2 in the form of sponge, chip or turning. TiH_2 is formed from titanium in the temperature range 300–500°C. The hydride is quite brittle and can be readily ball-milled into powder of the desired fineness. These may be dehydrided by heating them in a good vacuum at the same temperature at which the hydride was formed. Care must be taken to avoid contamination of O_2, N_2 and C during hydriding or dehydriding.

The powder produced from hydride is active, and special precautions, such as, inert gas absorption on the powder surfaces, are often necessary in order to avoid instantaneous ignition on exposure to air. The process, as such, does not involve any purification of the metal phase. On the contrary, an increase in oxygen and nitrogen can be expected in the final metal powder, depending on the quality of hydrogen and the degree of vacuum used.

Tungsten powder

The important minerals of tungsten are wolframite ($FeWO_4$) and scheelite ($CaWO_4$). They often occur with tin ores. Scheelite ores are leached with HCl to form tungstic acid. Tungstic acid is dissolved and digested in ammonia solution to give rise to ammonium tungstate solution. APT (ammonium paratungstate) is obtained by the crystallisation of ammonium tungstate solution. It is then calcined to give blue oxide.

Tungsten is leached with caustic soda at elevated temperature under pressure to produce sodium tungstate in solution. In either case the solution is purified using solvent extraction and tungsten is finally precipitated as pure WO_3.

The reduction of WO_3 is carried out by hydrogen and the stages can be written as:

$$4WO_3 + H_2 \rightarrow W_4O_{11} + H_2O \tag{2.16}$$

$$1/3\,W_4O_{11} + 4/3\,H_2 \rightarrow WO_2 + H_2O \tag{2.17}$$

$$1/2\,WO_2 + 1/2\,H_2 \rightarrow W + H_2O \tag{2.18}$$

For the last stage of reduction from WO_2 to tungsten at 850°C, there is still a significant concentration of H_2O vapour in the gas phase. The reduction of WO_2 by hydrogen is catalysed by tungsten metal. This is due to the dissociation of hydrogen molecules absorbed at the surface of the metal. This catalytic effect of tungsten is strongly inhibited by the presence of water vapour, with a consequent harmful effect on the reduction kinetics.

In a stationary furnace, the major parameters that affect the reduction rate are the furnace temperature, the amount of oxide loaded in the boat, the speed of boat movement, the rate of hydrogen circulation and the moisture content of the hydrogen. Hydrogen used for reduction is at least 99.5% pure and is usually electrolytically produced from distilled water with an addition of NaOH or KOH as the electrolyte. The preferable dew point range is − 40 to − 43°C. Safety devices are employed to avoid air leaks in the system and to prevent possible explosion.

In the rotary type furnace, the raw material as described above is fed from a hopper by means of screw feed. The material feeds through the reduction tube by gravity; the tube is inclined towards the discharge end and is slowly rotated. The metal powder produced falls into a sealed container. The drawback of the rotary furnace process is that the powder product is less uniform in particle size than that of a stationary furnace. As there is a large difference between the bulk volume of WO_3 and WO_2, for better utilisation of space, two-stage reduction is preferred. The temperature ranges from 500°C to 700°C for the first stage and from 700°C to 850°C for the second stage.

One of the major problems faced in tungsten powder production is the coarsening effect. At high reduction temperature sublimation of the oxide is common. This is a function of total pressure, the hydrogen flow rate and the powder particle size. The vapour pressure of small particles is greater than that of large particles, and hence during reduction under such conditions, the vapour of fine particles will deposit on the surface of coarse tungsten particles, thus giving rise to coarsening.

Tantalum and niobium powder

Tantalum and niobium are very similar in their chemical behaviour, and therefore, the production method is also very similar. Tantalite and columbite, either naturally occurring or synthetically produced as concentrates from tin slags, are digested with hydrofluoric and sulphuric acids at elevated temperatures. The accompanying elements, along with tantalum and niobium are dissolved, forming the complex heptafluorides H_2TaF_7 and H_2NbF_7. After filtering off the insoluble residue (fluorides of alkaline earth and rare earth metals), the aqueous solution of Ta-Nb in hydrofluoric acid is extracted in several continuously operated mixer-settlers or columns with the organic solvent, methylisobutylketone (MIBK). The complex fluorides of niobium and tantalum are extracted by the organic phase and impurities like iron, manganese, titanium, etc., remain in the aqueous phase, the *raffinate*.

In practice, Nb_2O_5 + Ta_2O_5 concentrations of 150–200 g L^{-1} in the organic phase are maintained. The organic phase is commonly washed or scrubbed with 6–15 N sulphuric acid and then extracted with water or dilute sulphuric acid to obtain niobium by selective extraction or stripping. The aqueous phase takes up the complex fluoroniobate and free hydrofluoric acid, while the complex fluorotantalate remains dissolved in the organic phase. The aqueous niobium solution is contacted with a small amount of MIBK to remove traces of co-extracted tantalum. The resulting organic phase is returned to the combined Ta-Nb extraction stage. Gaseous or aqueous ammonia is added to the aqueous niobium solution to precipitate niobium oxide hydrate; tantalum is extracted or stripped from the organic phase with steam, water or dilute ammonium hydroxide. Tantalum oxide hydrate is precipitated by ammonia, or alternatively, potassium salts are added to produce the salt, K_2TaF.

The precipitation of the oxides as well as the crystallisation can be done in a batch process or in a continuously operated facility. The oxide hydrates are collected by filtration, dried and calcined at temperatures up to 1100°C. The conditions of precipitation, drying and calcination are varied to produce different particle size distributions needed for oxides for various applications. Depending on quality requirements, the calcination is carried out in a directly- or indirectly-heated chamber or rotary furnace. The nature of the furnace lining has considerable influence on the purity of the oxide: for example, a lining made with Inconel will elevate the Ni and Cr impurities. Sophisticated process control and optimisation enable niobium and tantalum to be produced with high yield (> 95%) and purity (> 99.9%).

Figure 2.12 illustrates the flow diagram of the hydro-metallurgical processing of tantalum and niobium compounds.

Alumina powder

The most common method of production of alumina powder is the Bayer process, which consists of five stages: preparation of raw material, digestion, classification, precipitation and calcination. The general reaction is:

$$Al_2O(OH)_4 + NaOH + H_2O \xrightarrow[160-170\ C]{4\ atm} \text{sodium aluminate in solution} + \text{waste}$$

The separated sodium aluminate liquid is concentrated to saturation and after seeding with a charge of fine gibbsite, a heavy precipitation of $Al(OH)_3$ is produced by hydrolysis for subsequent calcining. Final calcination is achieved at a temperature around 1200°C to convert gibbsite into ceramic grade alumina. The principal impurities removed at the digestion stage are iron, silicon and titanium.

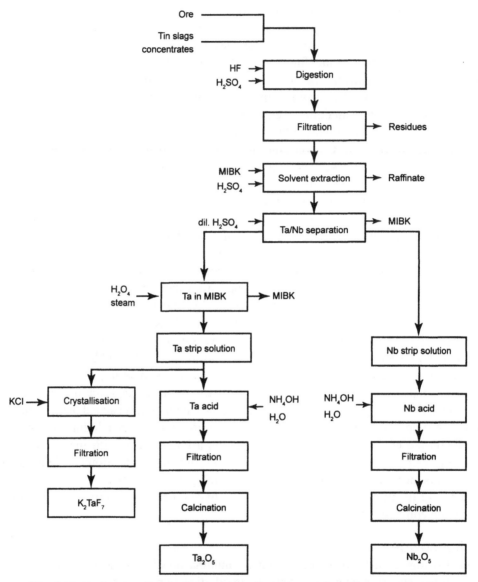

Fig. 2.12 Hydro-metallurgical processing of tantalum and niobium compounds

As advanced ceramic applications require high purity and very fine alumina powders, various synthetic methods have been adopted. They are:

- Sol–gel process (SG)
- Controlled precipitation (CP)
- Pyrolysis of salts/salt decomposition (SD).

As regards chemical purity, the SG process produces the most, followed by the SD, CP and Gibbsite process (G) (Table 2.5). Maximum surface area is obtained in SG (9.6 $m^2\,g^{-1}$)

Table 2.5 Chemical composition of Al_2O_3 powders prepared by different methods

	Sol–gel	Pyrolysis	Controlled precipitation	Bayer
Al_2O_3 (%)	99.93	99.9	99.73	99.52
SiO_2	335	365	1250	300
Fe_2O_3	65	160	840	270
Na_2O	110	195	210	4000
K_2O	30	110	100	60
Mn_2O_3	10	15	10	15
TiO_2	40	50	30	45
CaO	20	25	150	60
MgO	40	70	10	35
CuO	25	25	10	15

Note: All impurities in ppm.

followed by SD, G and CP. The response of these powders is also different for different binders. The green strength with polyvinyl alcohol (2%) is good.

There is another type of alumina-related compound known as β-alumina. This covers a wide range of non-stoichiometric compounds of soda and alumina ($NaO.xAl_2O_3$) where x falls in the range 5–11. The crystal structure is known to be layered. The most important application of β-alumina is as solid electrolytes (e.g., in oxygen meters, batteries, etc.).

Whisker production

Whiskers are short fibres, usually single crystals, with an aspect ratio of 10/1 or greater. They can be considered as a special type of *acicular powder*. They possess high strength and are used as random reinforcement: for example, SiC whiskers in alumina cutting tools. SiC whiskers are produced by the following methods:

- Plasma arc based on the reaction of SiO(g) and CO(g),
- Carbothermal reduction of silica with the addition of a halide as an auxiliary bath, forming β-SiC,
- Vapour–liquid–solid process using an iron catalyst,
- Thermal decomposition of rice hulls, a waste product of rice milling.

A typical SiC whisker has a diameter of 0.7–1.2 μm, an aspect ratio of 10–25 and a density of 3.26 g cm^{-3}. Its surface is smooth and metallic impurities are less than 1000 ppm.

2.2 ELECTROLYTIC METHOD

Basic aspects: In electrodeposition of metal, dissolution of the impure metal (anode) producing metal ions in solution and electrons takes place.

$$M^o_{impure} \rightarrow M_{n+} + ne \qquad (2.19)$$

The electrons produced in the reaction are conducted through an external circuit and power supply towards the cathode. The electrons and the metal ions recombine at the cathode surface to produce the pure metal.

$$M^{n+} + ne \rightarrow M \tag{2.20}$$

The overall electrochemical reaction is:

$$M^{\circ}_{impure} \rightarrow M^{\circ}_{pure} \qquad E^{\circ} = 0 \tag{2.21}$$

This suggests that no power is required to transfer the metal from the impure to the pure metal electrode. However, in order to drive the reaction at a certain current density, an overvoltage is necessary to overcome the electrical resistance in the circuit. Figure 2.13 shows the schematics of a electrolytic cell, while Table 2.6 illustrates the standard electrode potential series relative to the standard hydrogen electrode.

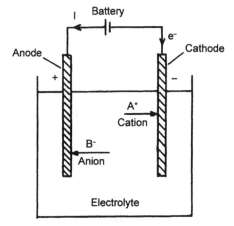

Fig. 2.13 Schematics of an electrolytic cell

Adjustment of the chemical and physical conditions during electrodeposition makes it possible for the metal to deposit loosely on the cathode of the cell, either as a light cake or as flakes. Both are readily crushed to a powder. The method yields a high-purity metal with excellent properties for conventional powder metallurgy processing. The process involves control and manipulation of many variables and in some cases it is significantly more costly than other techniques.

The following factors promote powdery deposits:

- High current density
- Weak metal concentration
- Additions of colloids and acids
- Low temperature
- High viscosity
- Avoidance of agitation
- Suppression of convection

The general shape of the electrolytic metal powders is dendritic; the full discussion is given in Section 2.3.

Diaphragms are sometimes used either to prevent deposition of undesirable solids or metal ions on the cathode, or to maintain substantially different conditions. The classical example is the use of canvas diaphragms during silver refining to prevent slimes from anodes containing the refined silver.

Table 2.6 Standard electrode potential series relative to SHE ($T = 298.15$ K)

Electrode	E_a, volt	Reaction
Au^{++}/Au	+ 1.35	$Au^{++} + 2e^- \rightarrow Au$
Ag^+/Ag	+ 0.80	$Ag^+ + e^- \rightarrow Ag$
Cu^{++}/Cu	+ 0.34	$Cu^{++} + 2e^- \rightarrow Cu$
H^+/H_2	0.00	$H^+ + e^- \rightarrow 1/2 H_2$
Pb^{++}/Pb	− 0.12	$Pb^{++} + 2e^- \rightarrow Pb$
Sn^{++}/Sn	− 0.14	$Sn^{++} + 2e^- \rightarrow Sn$
Ni^{++}/Ni	− 0.24	$Ni^{++} + 2e^- \rightarrow Ni$
Fe^{++}/Fe	− 0.44	$Fe^{++} + 2e^- \rightarrow Fe$
Cr^{+++}/Cr	− 0.60	$Cr^{+++} + 3e^- \rightarrow Cr$
Zn^{++}/Zn	− 0.76	$Zn^{++} + 2e^- \rightarrow Zn$
Al^{+++}/Al	− 1.67	$Al^{+++} + 3e^- \rightarrow Al$
Mg^{++}/Mg	− 2.39	$Mg^{++} + 2e^- \rightarrow Mg$
Na^+/Na	− 2.71	$Na^+ + e^- \rightarrow Na$

Electrodeposition of an alloy is also possible. Examples are Cu-Zn and Zn-Ni alloys. In this method it is essential that two or more elements should co-deposit at the same electrode potential. The simplest case is one where the polarisation curves of the single element can be added to give the polarisation curve for alloy deposition (Fig. 2.14). However, sometimes it is required to use selective complexants in order to bring the deposition potential of the elements closer to each other, or to influence the polarisation of the elements at the operating deposition current. As a result, in practice the relative positions of the polarisation curves for the single element and the alloy differ greatly from the situation shown in Fig. 2.14.

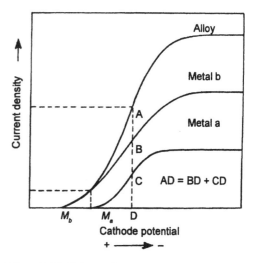

Fig. 2.14 Schematic representation of the polarisation curve of an alloy considered as the sum of the partial polarisation curves of the base elements

2.2.1 Some specific powders produced by electrolysis

Copper powder

The electrolytic process for copper powder production is similar to that used in copper refinery. However, instead of using impure cast copper anodes, electrolytically refined copper anodes are used. Generally, the cathode is cast antimonial lead. The electrolyte concentration is 50 g L^{-1} of $CuSO_4 \cdot 5H_2O$ (in contrast to 150 g L^{-1} for refining), while the current density is

Example 2.4: Write the cell reaction for electrodeposition of copper powder over an electrode of pure copper (the other electrode being of impure copper), the electrolyte being an aqueous solution of copper sulphate with some sulphuric acid. How would you calculate the deposition rate?

Solution

$$2Cu^{2+} + 4e^- \rightarrow 2Cu(s) \quad \text{Cathode reaction} \qquad (2.22a)$$

$$2H_2O \rightarrow 4H^+ + (O_2)(g) + 4e^- \quad \text{Anode reaction} \qquad (2.22b)$$

The overall cell reaction is:

$$2Cu^{2+} + 2H_2O \rightarrow 2Cu(s) + 4H^+ + O_2(g)$$

The rate of copper deposition equals the rate of the overall cell reaction, which can be calculated in terms of the cell current:

$$\text{Deposition rate (g s}^{-1}) = 1(A) \times \text{atomic weight (g mol}^{-1} \text{ n}^{-1} \text{ F}^{-1})$$

where n is the number of faradays to deposit one mole and F is the Faraday constant or $96{,}490$ C mol^{-1}

A faraday is equal to the magnitude of charge on one mole (i.e., 6.022×10^{23}) of electrons. In the case of copper, $n = 2$, and atomic weight = 63.54 g mol^{-1}; therefore,

$$\text{Copper deposition rate (g s}^{-1}) \approx (3.3 \times 10^{-4}) \times I(A)$$

In case we wish to determine the deposition rate per unit cathode area, i in the above equation is to be substituted by current density $i = i/\text{area of cathode surface}$.

Example 2.5: Why does the aqueous electrolyte used during electrolytic copper powder production contain sulphuric acid?

Solution: The reasons are:

1. The E_n-pH diagram of the system Cu-S-H$_2$O confirms that it is necessary to maintain an acidic solution to prevent the formation of cupric oxide or hydroxide at the anode surface at high copper concentration.

2. The electrical conductivity of the electrolyte increases with increasing acid content.

535 A m^{-2} (in contrast to 107–215 A m^{-2} for refining). The copper powder is washed and filtered and finally given an annealing and reducing treatment at temperatures between 500°C and 800°C in a belt furnace with an atmosphere of partially combusted hydrocarbon gas. Powder properties, particularly the apparent density, are primarily controlled by the reducing treatment after electrolytic deposition. The reduced powder forms a cake which must be broken up into powder by a hammer mill.

The addition of chlorine ions prevents formation of highly branched individual particles and promotes greater agglomeration. The impurities in the deposited powder are of particular significance from the electrical conductivity viewpoint. Sulphur is picked up from the electrolyte and calcium and sodium from the wash water. The increase in lead is caused by the lead lining of the electrolyte storage tanks. Silver, which is deposited as extremely small particles of anode slime, falls to the bottom of the cell and mixes with the powder already lying

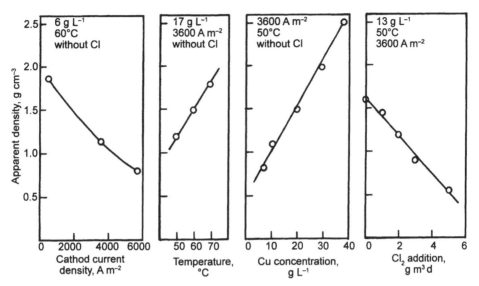

Fig. 2.15 Apparent density of electrolytic copper powder in relation to various parameters: cathode current density, temperature, copper concentration and chlorine addition

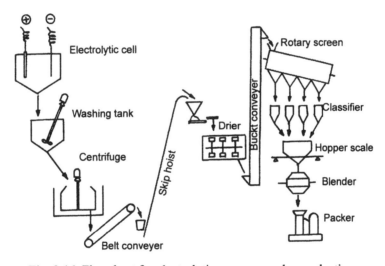

Fig. 2.16 Flowsheet for electrolytic copper powder production

there. Figure 2.15 illustrates some of the effects of electrolysis parameters—cathode current density, temperature, copper concentration and chlorine addition—on the apparent density of the copper powder. In this method, the disposal of effluent is an environmental problem. In addition, the process is both electric power and labour intensive. Figure 2.16 shows the flowsheet for electrolytic copper powder production.

Figure 2.17 illustrates the microstructures of electrolytic copper powder of different apparent densities.

(a)   (b)

Fig. 2.17 Stereo views of electrolytic copper powder of varying apparent densities: (a) 0.8 g cm^{-3} and (b) 2.4 g cm^{-3}

Iron powder

This method was used to produce a highly pure sinterable iron powder but the cost is high: about three times greater than that for water-atomised powder. In the past, many countries, particularly developing ones, met their indigenous requirements through this powder, but now the process is obsolete as it is uneconomical. Iron powder is prepared from aqueous electrolytes, such as, ferrous chloride, and ammonium chloride is added to increase conductivity. The disadvantages of using the sulphate electrolyte are its low conductivity and the need for additional cleaning of the powder to remove any residues after electrolysis. In contrast to copper powder, electrolytic iron is not deposited as a powder but as a brittle, lightly adhering sheet-like deposit on stainless steel cathodes. The anodes are usually Armco iron or low-carbon steel. The brittle deposits can be readily milled into powder in ball mills and the powder must be reduction annealed to make it soft and pure.

Titanium powder

The electrolytic process for titanium powder production is prevalent in Russia. The process is based on fused-salt electrolysis, using soluble anodes of sponge or scrap. The process can also electrolyse TiCl$_4$. Investigations have shown that impurities such as iron, nitrogen and chlorine were lower in titanium powder produced in chloride–fluoride electrolytes compared to chloride electrolytes only. The extent of the reduction in impurity levels is determined by the fluorine ion concentration in the electrolyte and the particle sizes of the powder. The oxygen content in these powders is slightly higher. The most marked changes in powder particle size occur at an atomic fluorine–titanium ratio of 6.1 in the electrolyte. It has also been established that the presence of fluorine (F:Ti = 1.5:1) in the electrolyte decreases the apparent density and flowability of the powder. The process is useful for recycling zircaloy scrap into powder form.

2.3 ATOMISATION METHOD

Any material that can be melted can be made into powder by disintegration of the liquid. Apart from chemical reactivity, which may necessitate specific atmosphere or materials, the atomisation process is independent of the normal physical and mechanical properties associated with the solid material. The method is being widely adopted, especially because of the relative ease of making highly pure metals and pre-alloyed powders directly from the melt. The basic procedure employed is to force the melt fluid through an orifice, possibly at the bottom of a crucible, and impinge the gas or liquid stream on the emerging melt. A great deal depends on the exact design of the orifice. It may induce turbulence in the melt, which atomises the material directly and allows the impinging gas or liquid to reduce the size of the particle even faster.

Figure 2.18 illustrates the main variables in the atomising process.

Fine particle size is favoured by:

- Low metal viscosity
- Low metal surface tension
- Superheated metal
- Small nozzle diameter, i.e., low metal feed rate
- High atomising pressure
- High atomising agent volume
- High atomising agent velocity
- High atomising agent viscosity
- Short metal stream
- Short jet length
- Optimum apex angle

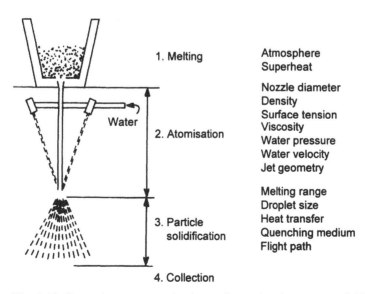

Fig. 2.18 Stages in water atomisation and associated process variables

Particle shapes of atomised powders can be modified from almost perfectly spherical to highly irregular, by controlling the processes which take place in the interval between disintegration of the liquid metal stream and the solidification of the drop. Sphericity of a metal powder is favoured by:

- High metal surface tension
- Narrow melting range
- High pouring temperature
- Gas atomisation, especially inert gas
- Low jet velocity
- Long apex angles in water atomisation
- Long flight paths

In addition to water and gas atomisation processes, there are other methods, which may be categorised as mechanical atomisation. One of the typical example of this is *roller atomisation*, where the stream of molten metal is fed between rapidly rotating twin rollers. The primary operating variables are roll speed, roll gap, metal flow rate, metal stream velocity and metal superheat. A wide range of particle shapes and sizes can be achieved by roller atomisation. In centrifugal atomisation, the molten metal is ejected from a rapidly spinning container, plate or disc.

Basic aspects

In case of water atomisation, it was proposed that the liquid metal stream breaks up under impact from the water droplets, rather than shear mechanism. It is illustrated in Fig. 2.19. It is seen that the metal droplets travel in the same direction as the atomising water droplet. It was confirmed that the velocity component of the water normal to the liquid metal stream, rather than the velocity component parallel to the liquid metal stream, is the dominant parameter in controlling the mean particle size d_m. The mode gives a relation between d_m and V_w as:

$$D_m = \frac{B}{V_w} \sin \alpha \qquad (2.23)$$

where, d_m is in μm, V_w is in m s^{-1}, α is the angle (in degrees) between the axis of the metal stream and the water jets, and B is constant. By considering the units of various parameters and calculating, the value of B is found to be 2750.

In general the mean particle size falls with an increase in the atomising water pressure. The data (Fig. 2.20) are plotted on log–log coordinates and the relationship between P and d_m can be expressed as:

$$d_m = KP^{-n} \qquad (2.24)$$

where, K and n are constants that are specific to each atomisation facility, d_m is in μm, n is typically in the range 0.6–0.8 for water pressures from 0.1 to 20 MPa. The value of K is also a function of the molten metal or alloy that is being atomised. It is determined by the physical and chemical properties of the molten metal/alloy, in particular viscosity and surface tension.

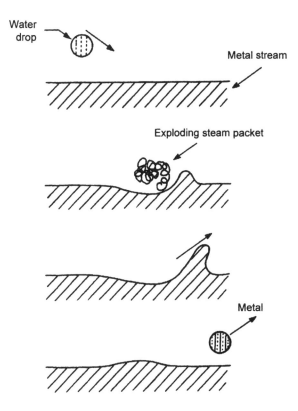

Fig. 2.19 Proposed mechanism of metal droplet formation in water atomisation

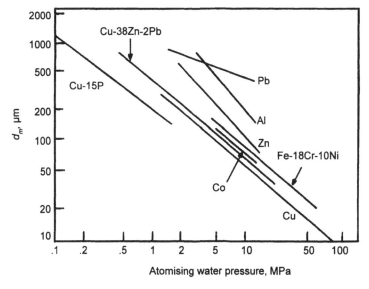

Fig. 2.20 Mass median particle size as a function of water pressure

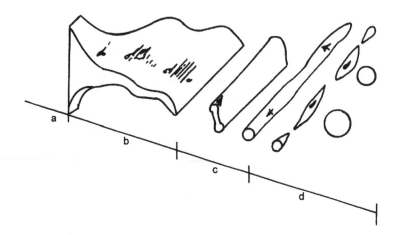

Fig. 2.21 Disintegration of a sheet of liquid during gas atomisation: (a) Stable sheet, (b) Growth of waves sheet, (c) Ligament formation and (d) Ligament breakdown

In gas atomisation, the mechanism of melt droplet formation involves:

- Initiation of a sinuus wave which rapidly increases in amplitude,
- Detachment of the wave from the bulk of the liquid to produce a ligament, the dimensions of which depend on the wavelength at disintegration,
- Breakup of the ligament into spherical droplets.

Figure 2.21 illustrates the above features.

The liquid metal stream has a velocity V given by:

$$V = A \left[\frac{2g(P_1 - P_2)}{\rho} \right]^{0.5} \tag{2.25}$$

where, A is a geometric constant, g is the acceleration due to gravity, P_1 is the injection pressure of the liquid metal, P_2 is the pressure of the atomising gas and ρ is the density of the liquid metal. The steady forces in atomisation that lead to the formation of ligaments from the liquid metal depend on Reynold's number, which in turn is related to the size and velocity of the stream, and the density and viscosity of the liquid metal.

There are a number of empirical equations for the mean particle size d_m. One of which is:

$$\frac{d_m}{d_\ell} = K \left[\frac{\eta_\ell}{W\eta_g} \left(1 - \frac{f_\ell}{f_g} \right) \right]^{0.5} \tag{2.26}$$

where, W is the Weber number of the metal stream, d_ℓ the diameter of the liquid metal stream, η_ℓ and η_g are the kinemetic viscosity of the liquid metal and gas, f_ℓ and f_a are the flow rates of liquid and gas, and K is a constant.

In another empirical relationship (after H Schmitt), the maximum particle size diameter is expressed as:

$$d_{max} = \left(\frac{10^5}{4.8} \times \frac{\sigma}{\rho_M V^2}\right)^{12/13} \left(10^6 \times \frac{\eta_M^2}{\sigma \rho_M}\right)^{1/13} \tag{2.27}$$

where, σ is the surface tension of the melt in N m^{-1}, V is the velocity in m s^{-1}, η_M is dynamic viscosity of the melt in N s m^{-2} and ρ_M is the density of the melt in kg m^{-3}.

Solidification and microstructural evolution

During atomisation, solidification of melt droplets occurs. Hence all the basic aspects of solidification of metals and alloys are applicable here too. In contrast to casting of melt, which is carried out in a sand or metal mould, in atomisation each particle might be considered as a free-standing micro-ingot.

When a liquid metal freezes, the heat of fusion must be withdrawn from the liquid, and the manner in which this is done affects the solidification microstructure. The process of solidification occurs by the mechanism of nucleation and growth, i.e., minute nuclei or seed crystals are formed in various parts of the melt stream, which then grow at the expense of the surrounding liquid. During the freezing process the nuclei grow more rapidly along certain crystallographic directions, and this results in the formation of long branch-like crystals, known as dendrites (Fig. 2.22). Eventually, the outward growth of a dendrite is halted when contact is made with neighbouring growths, and then the remaining liquid freezes in the interstices between the dendrite arms.

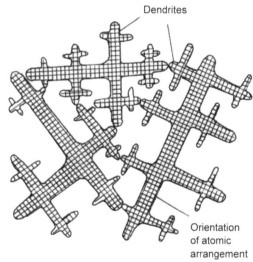

Fig. 2.22 A typical view of dendrites

The shape of the solid–liquid interface is considerably influenced by the local temperature gradient which itself is determined by the way in which the heat of crystallisation is dissipated. It can be assumed that the temperature of the interface is always T_m (equilibrium melting temperature). The velocity of the interface is then determined by the rate of heat loss. This can take place in two ways (Fig. 2.23).

1. The heat flows from liquid to solid and the heat of crystallisation are dissipated through the solid. The interface is, therefore, stable.
2. The heat is dissipated through the liquid as well as through the solid, and this proceeds more efficiently the farther the bulge projects into the liquid. This instability of the flat interface is the cause of the dendritic growth.

In almost all practical cases, heterogenous nucleation occurs below, but very near to the equilibrium point, since inhomogeneities exist in a melt.

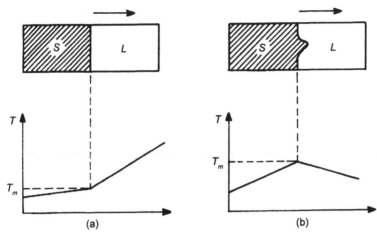

Fig. 2.23 Solidification at an advancing solid–liquid interface for different temperature profiles: S – solid L – liquid

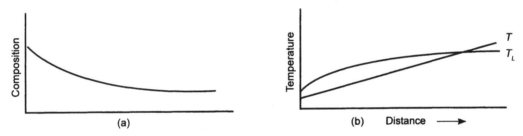

Fig. 2.24 Variation with distance from the solid–liquid interface of (a) composition and (b) actual temperature T and freezing temperature T_L.

Even a small amount of solute element in the melt, either as an intentional alloying or as an impurity, greatly affects the solidification process. Here, apart from the thermal undercooling described, there is an additional undercooling. On freezing, the solute atoms tend to remain in the liquid, rather than solidify with the solvent atoms at the crystal–liquid interface. The solute rejected by the crystals becomes concentrated in the remaining liquid, as a result of which the freezing point of the liquid decreases. However, the liquid just ahead of a freezing solid–liquid interface is more enriched than liquid further away. This means that the composition of the liquid varies with distance from the interface (Fig. 2.24). The temperature at which the liquid will freeze varies with the distance from the interface. This is due to the fact that the liquid which is close to the interface will have its freezing temperature lowered more than liquid further away. For a positive temperature gradient as shown in Fig. 2.24(b), there is a layer of liquid in which the actual temperature T is below the freezing temperature T_L. Such an undercooling effect is known as *constitutional undercooling*. This also causes dendritic growth.

The degree of cooling rates helps in evaluating the relative efficiencies of different cooling/quenching techniques for powder production. The magnitude of the cooling rate and its variation help in interpreting the microstructures of the particles. It has been highlighted earlier that under conditions of non-equilibrium cooling many solid solution alloys solidify in a dendritic fashion. Relationships can be developed between the cooling rate (T) and the primary or secondary dendrite arm spacing. However, the latter has been widely used as a convenient method to assess the cooling rate. A simple relation between the secondary dendrite arm spacing (λ) and cooling rate (T) is given by:

$$\lambda = KT^n \qquad (2.28)$$

where K and n ($\sim 1/3$) are constants of a given material.

Figure 2.25a shows a typical dendritic microstructure of an etched section of rapidly solidified susperalloy, 1N-100 powder, produced by rotating electrode process, while Fig. 2.25b shows a plot relating the secondary dendrite arm spacing to the cooling rate for some metallic alloys. It is to be noted that the slopes are different for different alloys.

A brief mention of the type of alloys which can be rapidly solidified into amorphous (Met Glass) form is called for. There are three main classes:

1. $T_{80}M_{20}$ *glasses,* where 80% of a transition metal (like Fe, Co, Ni, Pd) and 20% of a non-metal (B, C, Si or P) are present. Examples: $Fe_{40}N_{40}B_{20}$ and $Pd_{80}Si_{20}$.

2. $T^I T^{II}$ *glasses* between a 'late' transition metal as above and an 'early' one (like Ti, Zr, Nb). Examples: $Ni_{60}Nb_{40}$ and CuZr.

3. *AB glasses* between Mg, Ca on the one side and Al, Zn, Cu on the other. Example: $Mg_{70}Zn_{30}$ glass.

Many of the phase diagrams of glass forming alloys show deep eutectics as in Fig. 2.26. The compositions of the glasses are marked in the hatched rectangle. A deep eutectic indicates a tendency towards short-range order or compound formation already in the melt, as also expressed by a large negative heat of mixing. The deep eutectic is also favourable for maintaining the molten state at low temperatures. The question why deep eutectics and easy glass formation often appear at a T:M ratio of 4:1 has been the topic of research of many materials scientists.

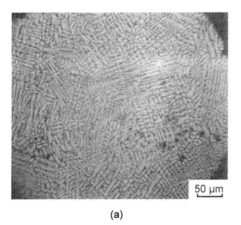

(a)

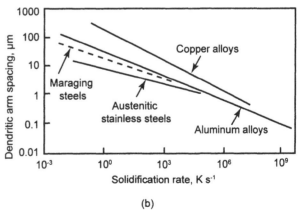

(b)

Fig. 2.25 (a) Polished and etched section of 1N-100 rapid solidified powder showing uniform dendritic structure; (b) Secondary dendrite arm spacing as a function of cooling rate for steels, copper and aluminium alloys.

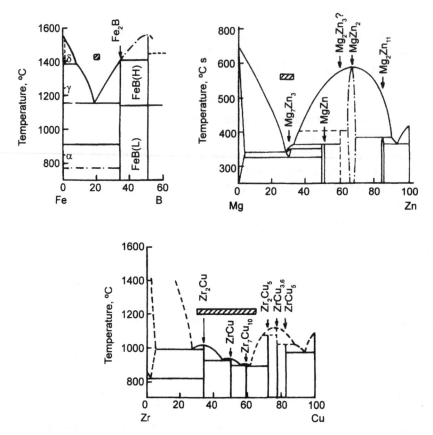

Fig. 2.26 Phase diagram of some glass-forming alloys. The compositions of the existing glasses are hatched.

2.3.1 Gas atomisation

The common atomising media are nitrogen, argon or air. Various atomisation geometries are used in commercial practice. In what is known as 'external mixing' (Fig. 2.27), contact between the atomising medium and melt takes place outside the respective nozzles. This type of mixing is used exclusively for the atomisation of metals. 'Internal mixing' is quite common for the atomisation of materials which are liquid at room temperature. The axes of the gas jets are equally inclined to the melt stream axis and

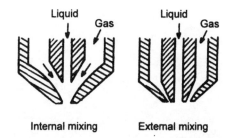

Fig. 2.27 Two fluid atomisation designs

intersect this axis at the geometrical impingement point. The process is governed by a number of interrelated operating parameters. Controllable variables include jet distance, jet pressure, nozzle geometry, velocity of gas and metal, and melt superheat.

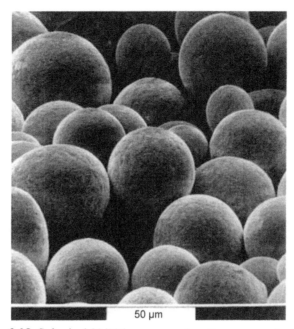

Fig. 2.28 Spherical 90/10 bronze powder after gas atomisation

Gas-atomised powders are typically spherical, with relatively smooth surfaces. Higher pressure/smaller jet distance produce finer powder. Gas atomisation pressures are typically in the range 14×10^5 Pa to 42×10^5 Pa at gas velocities from 50 m s^{-1} to 150 m s^{-1}; under these conditions, the particle quench rate is -10^2 K s^{-1}. This production method is used for preparing powders of superalloys, titanium, high-speed steel and other reactive metals. The method suffers from a very low overall energy efficiency ($\sim 3\%$) and is expensive if inert gases, other than nitrogen, have to be used.

Gas atomiser equipments are of two types: horizontal atomiser and vertical atomiser. Low-temperature atomisers are based on horizontal design. The high-velocity gas emerging from the nozzle creates a siphon, pulling molten metal into the gas expansion area. For high-temperature metals, a closed, vertical, inert gas–filled chamber is used to prevent oxidation.

Figure 2.28 shows the picture of spherical 90/10 pre-alloyed bronze powders, obtained after gas atomisation.

2.3.2 Water atomisation

In water atomisation, a high-pressure water stream is forced through nozzles to form a disperse phase of droplets which then impact the metal stream. In this method, large quantities of energy are required to supply water at high pressure. It is estimated that the overall energy process efficiency is $\leq 4\%$. This production method is significant for low- and high-alloy steels, including stainless steel. Because of oxide formation, water atomisation is not likely to be used in the atomisation of highly reactive metals such as titanium and the

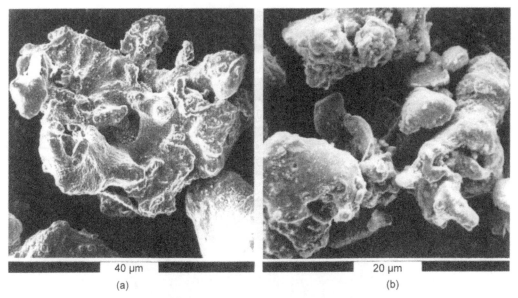

40 µm   20 µm
(a)   (b)

Fig. 2.29 Sterio views of water-atomised 90/10 bronze powder at different magnifications

super alloys. In general, water-atomised powders are irregular in shape, with rough oxidised surfaces.

In water atomisation, water pressures are commonly in the range of 35×10^5 Pa to 210×10^5 Pa with associated water velocities from 40 m s^{-1} to 15 m s^{-1}. The particle cooling rate is ~10^3 K s^{-1} to 10^4 K s^{-1}.

The surface tension of liquid metals is high and a droplet once formed tends to assume the spherical shape. Higher the viscosity of the atomising medium, greater the deformation of the droplet; higher the cooling rate, shorter the time during which the surface tensional forces can operate to spheroidise the droplet, and therefore, the more irregular the particle shape. Impurities and alloying elements in the metal, or reactions on the surface of the droplets that decrease the surface energy, will promote irregular particle shapes. Small quantities of phosphorus in copper lead to the formation of a P_2O_5 film on the surface of particles, which increases the surface energy and results in the formation of spherical droplets. The existence of a solid oxide film such as ZnO acts in the opposite fashion, tending to give less-rounded particles. The addition of silicon is a well-known method of influencing the particle shape of atomised stainless steel powder.

The particle shape is not appreciably influenced by metal pouring temperatures in gas atomisation, but it is in water-atomised powders. At higher pouring temperatures, there remains enough superheat after atomisation to allow surface tension forces to create spheroids. Higher water pressure results in more irregular particle shapes, due to greater impact forces and larger volumes of water, resulting in more rapid quench.

During atomisation of multiphase alloy systems, it is also the aim to eliminate gross metal segregations. The particle structure is therefore, a function of the solidification rate. Fine

Table 2.7 Typical operating conditions for gas and water atomisation processes

Parameter	Gas atomisation	Water atomisation
Pressure, MPa	1.4–4.2	3.5–21
Velocity, m s^{-1}	50–150	40–150
Superheat, °C	100–200	100–250
Angle of impingement	15–90°	$\leq 30°$
Particle size, μm	50–150	50–200

Particle shape	Smooth and Spherical	Irregular and rough surface
Yield	40% at −325 mesh	60% at −35 mesh

microstructure particles are promoted with water atomisation as opposed to gas atomisation, by lower metal pouring temperatures, higher atomising agent pressure, flow rate and viscosity, and by shorter particle flight paths.

Figure 2.29 shows water-atomised 90/10 pre-alloyed bronze powders. In contrast to gas-atomised powder of the same composition (Fig. 2.28), the shape is irregular.

Table 2.7 summarises the differences in various operating conditdions for gas and water atomisation processes.

2.3.3 Liquid gas atomisation

The Krupp Company (Germany) introduced a novel version of atomisation in which the melt is atomised with cryogenic liquid gas (argon or nitrogen) at −200°C. During the process, the pressure of the liquid gas is increased up to 300 bar, while a recooling unit prevents the temperature from rising, in spite of compression, and prevents the cryogenic liquid from vaporising instantaneously at the jet opening. An even stream is generated, which atomises the melt comparably to water atomisation and cools rapidly. Since the atomisation liquid vaporises completely, gas and powder can be easily separated in the cyclone. The resulting powder has the following properties:

- It is much purer than the powder atomised with water and can be compared to the quality of the gas atomised powder.

- The cooling rate is ten times higher than in gas atomisation and almost reaches the quality of water atomisation. For example, particles of 100 μm in diameter are quenched for approximately 10^6 K s^{-1}.

- As in gas atomisation the powders are spherical and have an average size of 6–125 μm. Due to the distinctive presence of satellites, mainly low-alloy powders give satisfactory results in cold forming, and also have good flowability. Gas-atomised powders, on the other hand, have poor green strength.

Example 2.6: Figure Ex. 2.6 shows two types of nozzles for gas atomisation of a molten metal. What are the differences?

Solution: The left-hand side is the free fall version, while the one on the right-hand side is for confined atomisation. The metal stream energy for the melt bath in a free fall is impinged on by the atomising gas stream energy at an angle *a* for the nozzle. In the confined version, the gas stream emerges tangentially and the melt stream is atomised by it. The degree of turbulence is much greater in the confined type of nozzle design and the energy is efficiently utilised.

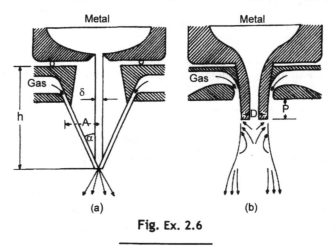

Fig. Ex. 2.6

Example 2.7: The indirect method for estimating the cooling rate of a melt during atomisation (e.g., secondary dendrite arm spacing) has many limitations. Elaborate.

Solution: The limitations are:

1. It is not applicable to glass forming alloy compositions.
2. It can only give an average cooling rate.
3. Each family of alloys requires a separate calibration.
4. The dendrites may coarsen after solidification and so yield misleading results.

2.3.4 Centrifugal atomisation

The basis of centrifugal atomisation is the ejection of molten metal from a rapidly spinning container, plate or disc. The rotating electrode process (REP) is a further example of centrifugal atomisation. The material in the form of a rod electrode is rotated rapidly while being melted at one end by an electric arc. Molten metal spins off the bar and solidifies before hitting the walls of the inert gas–filled outer container (Fig. 2.30).

Tungsten contamination from the stationary electrode is a limitation of REP powders. To eliminate this, the PREP (plasma rotating electrode process) method has been commercialised (Fig. 2.31). It is important that the electrodes are precisely dimensioned and straight. This can be achieved by subjecting the cold drawn rod to cross roll straightening.

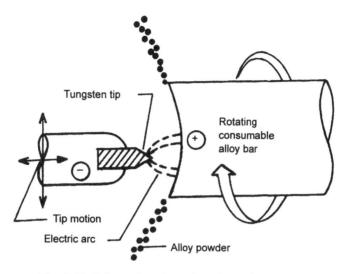

Fig. 2.30 Schematics of rotating electrode process

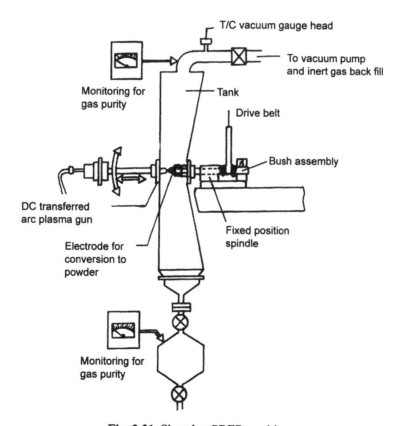

Fig. 2.31 Short bar PREP machine

Resonance effects are experienced in REP. The natural frequency (*f*) experienced is a function of the modulus of elasticity (*E*), the sectional moment of inertia (*I*), the beam length (*l*) and the beam weight (*W*) or weight per unit length (*m*), and is given by the relationship:

$$F = C_n \sqrt{\frac{EI_g}{Wl^3}} \ or \ C_n \sqrt{\frac{EI_g}{ml^4}} \tag{2.29}$$

where, C_n is a constant depending on beam support conditions and mode number of vibration, and *g* is the gravitational constant (9.81 m s^{-2}).

In order to obtain a desired range of particle size distribution, accurately controlled rotation of the anode is important. The diameter of the molten droplet in a given material is determined by such parameters as the surface tension of the liquid metal, centrifugal forces (related to rotation speed) and to some extent by the 'aerodynamics' of the droplet's trajectory through the inert cover gas.

The design of REP or PREP equipment includes sufficient damping to suppress or withstand resonant vibrations of moderate amplitude. However, when the electrodes are marked by out of tolerance for straightness, severe loading is imposed on the spindle and seal mechanism.

Spherical metal powders made by either REP or gas atomisation are not well suited for cold pressing into green compacts to be followed by sintering. They are used in more specialised applications where consolidation is achieved by hot isostatic pressing (HIP) or some other high-temperature method in which interparticle voids are more readily closed.

2.3.5 Vacuum atomisation

Vacuum or soluble gas atomisation is a commercial batch process based on the principle that when a molten metal supersaturated with gas under pressure is suddenly exposed to vacuum, the gas expands, comes out of solution, and causes the liquid metal to be atomised. Alloy powders based on nickel, copper, cobalt, iron and aluminium can be vacuum atomised with hydrogen. The powders are spherical, clean and of a high purity, compared to powders produced by other processing methods. The process was developed and patented by Homogeneous Metal, Inc. Figure 2.32 illustrates a schematic of the equipment used for the atomisation process. The principal use of powder made by vacuum atomisation has been in the production of gas turbine discs and intricate parts by injection moulding.

The configuration of the vacuum atomiser is invariably such that the nozzle points upward. The molten stream is sprayed up and then the particles fall down. This shortens the size of the atomiser. It should be noted that the rate of heat extraction is very slow in vacuum.

Figure 2.33 illustrates the representative particle size distributions for metal powders prepared by inert gas/rotating electrode or vacuum atomisation processes.

2.3.6 Ultrasonic gas atomisation

In ultrasonic gas atomisation, the charge—generally aluminium alloy—is inductively melted in an inert atmosphere. When the desired melt temperature, typically 750°C, is attained, atomisation

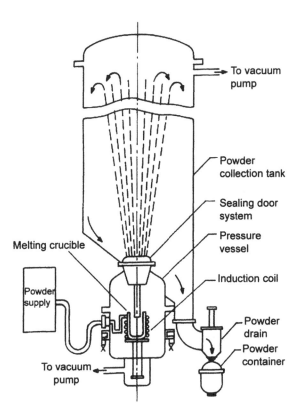

Fig. 2.32 Vacuum atomisation equipment

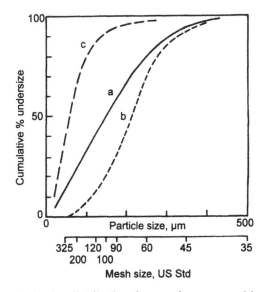

Fig. 2.33 Representative particle size distribution for powders prepared by (a) inert gas atomisation, (b) rotating electrode process and (c) vacuum atomisation.

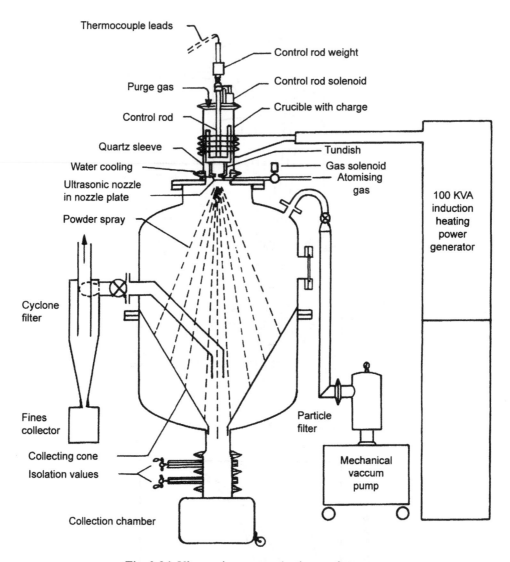

Fig. 2.34 Ultrasonic gas atomisation equipment

is initiated by simultaneous activation of solenoids which open the gas line—an orifice in the flow of the melting crucible. Molten metal is gravity fed through an unheated tundish and pour tube and atomised and solidified by a pulsed, ultrasonic stream of inert gas emanating from an annular nozzle ring. The resultant powder is spherical, with fewer satellites than typical vacuum atomisation powder. Figure 2.34 illustrates the details of the ultrasonic gas atomisation equipment.

2.3.7 Chill methods

Chill methods are common for producing amorphous particulates of metallic alloys. In this method the melt is brought into contact with a chill substrate. The various methods include injecting the melt into a die cavity, or forming the melt into a thin section by forging between a hammer or piston and anvil, or extruding the melt on a chill surface, or extracting the melt by contact with a rotating disc. These amorphous particulates are not further consolidated by press and sinter technique. Non-equilibrium processes are involved in this process. A detailed description of such chill methods can be had from a review by C Suryanarayana (Suryanarayana, 1991).

2.3.8 Some specific atomised metal and alloy powders

Low- and high-alloy ferrous systems

As iron and steel powders are used extensively, there has been much interest in various atomisation methods. The first method was a hybrid air atomisation of cast iron melt followed by decarburisation to obtain RZ grade iron powder, originally patented by Mannesmann of Germany during the Second World War. The method is now obsolete. Later, AO Smith Corporation, USA, pioneered water-atomisation of ferrous powders in the mid-1960s. In 1985, to meet the demand for highly compressible iron powder, Höganäs, Sweden, developed ABC 100.30 grade water-atomised powder. In Canada, Quebec Metal Powders Ltd, established a hybrid process of atomisation milling to produce iron powders from a pig iron (4% C) melt obtained by electric smelting. Of all the steps, decarburisation is the most energy consuming part of the QMP process. A total of 5.3 MBtu per ton is used in heating the furnaces, dissociating the ammonia and transporting the powder from the ball mill circuit to the exits of the furnace. In Japan, much work has been done in developing oil-atomised low-alloy steel powders, and a 4100 grade steel powder with oxygen as low as 0.09% has been produced. This method is very attractive for obtaining pre-alloyed steel powders containing elements with high oxygen affinity, such as manganese and chromium.

In case of high-speed steels (a high-alloy steel), water atomisation, although a much cheaper powder production method, did pose problems of oxygen impurity in the powder, in the form of inclusions. This inhibits diffusion during sintering and has a marked deleterious effect on the ductility and impact strength. Oxide inclusions are reduced by vacuum or hydrogen annealing treatment. In contrast to water atomisation, in nitrogen or argon atomisation, the oxygen content in the powder is as low as 200 ppm and the powder is characterised by good flowability and high apparent density.

Powders of stainless steels—another group of high-alloy steels—are also extensively produced by the atomisation process. In case of water atomisation, the melt chemistry during air or vacuum induction melting is controlled such that low manganese concentrations (<0.3%), and deoxidation with ferrosilicon to achieve 0.7%–1% Si are accomplished. A typical water pressure for producing a predominantly –80 mesh powder is about 14 MPa. This powder has an apparent density ranging from 2.5–3.2 g cm^{-3}, adequate green strength and reasonably good compressibility. Gas(N_2, Ar)-atomised stainless steel powders have spherical particles, high apparent densities of about 5 g cm^{-3} and excellent flow rates. Oxygen content is less than 200 ppm.

Aluminium

Water atomisation of aluminium melt is feasible, but the problems of avoiding corrosion in drying are severe. Oxidation is about 0.2% before drying, and the particles obtained are coarse(~100 µm) and irregular. The dominant process for the production of aluminium powder is still air atomisation, as it is cheap. The different designs include vertically upward, horizontal and horizontal atomisation of a vertically downward-falling stream. Both closed- and open-nozzle systems are used, with closed systems giving finer powders; but open systems have certain operational advantages. As there is a large market for aluminium powder, for different applications, the powder is sieved or classified. The oxide film on the most carefully prepared inert gas–atomised powders is still about half as thick as that on air-atomised powders (~50 nm). The oxide surface is inert and cannot be reduced by hydrogen, carbon monoxide or carbon during sintering. The aluminium surface is hygroscopic and contains physically absorbed water and hydrated aluminium oxide.

In a variation of the air atomising process, nitrogen gas, which is less reactive, is used. The gas which carries the powder through the system to the collection stages is air. Typically, this gas is ten times the volume of the atomising gas. The use of nitrogen for atomising in this way renders the particles near-spherical and probably helps to maintain a thinner skin of oxide on the surface. The cooling rate of nitrogen is no greater than that of air.

Two atomisation processes described earlier, namely, vacuum atomisation and ultrasonic gas atomisation, are used as rapid solidification processing methods. In case of vacuum atomisation, hydrogen is not used as the additional soluble pressurising gas, in contrast to vacuum atomisation of superalloys, in order to avoid contamination of the powder. Table 2.8 illustrates the distinctive features of aluminium powders obtained through these two atomisation methods.

Copper

The atomisation process is most versatile for making pre-alloyed powders and is extensively used for the manufacture of brass, bronze, Cu-Pb, Cu-Pb-Sn and Cu-Ni-Sn powders. It is particularly suitable for lead-containing alloys as it forms islands of lead in the copper matrix.

Water atomisation is preferred to air atomisation for producing moulding grades of copper powder. Water's greater viscosity and superior heat removal property produce higher yields of finer, more irregular particles with lower apparent density and greater compressibility. Since the surface tension of copper is rather high and its oxides are soluble in the molten metal, air-atomised copper consists of round, non-mouldable particles. When alloying zinc with copper, the surface tension changes; a surface oxide forms and the particle shape becomes quite irregular. Phosphorus increases the surface tension of liquid copper and copper-base alloys. It 'boils off' profusely (BP 445°C) and continuously disrupts the oxide film forming on the liquid. Phosphorus is intentionally used to produce spherical copper-base powders for filters.

As the rate of cooling in some of the atomisation methods is far from equilibrium cooling, it may be possible to have other hard intermediate phases in the copper particles. In particular,

Table 2.8 Characteristics of aluminium alloy powders (rapid solidification) obtained under two atomisation methods

Characteristics	Vacuum atomisation	Ultrasonic atomisation
Quench medium	Vacuum (expansion of pressurising gas)	Argon at 8.3 MPa flowing at Mach 2, pulsed at ~ 80,000 Hz.
Particulate shape	Spherical powder	Spherical powder
Particulate dimensions	Geometric mean dia. = 46 μm, Standard deviation = 1.58	Geometric mean dia. = 59 μm, Standard deviation = 1.64
Dendritie arm spacing (DAS)	0.7–4 μm at average particle dia. of 2.4 μm	0.7–5.5 μm at average particle dia. of 2.9 μm
Average solidification rate	~ 10^4 K s^{-1}	~ 10^4 K s^{-1}

it is possible to retain δ-phase at room temperature in water-atomised bronze powders, which can have a significant effect on the performance of sintered bearing.

2.4 EVAPORATION METHODS

This method is very useful in preparing nano-scale particles. The material is vaporised in low-pressure argon at about 10% of atmospheric pressure. From the vapour the particles are nucleated homogeneously. Such fine particles are collected on a cold substrate. The heating source may be electron beam, lasers, plasma flames or induction fields. In case compounds are required, the powders so obtained may be reacted with process atmosphere: for example, oxygen for oxides and nitrogen for nitrides. The shape of the powders is generally agglomerated. The powders so produced are expensive. In addition, the handling of such small-sized particles is very difficult, due to their high reactivity. This limitation necessitates glove-box handling. Such small-sized particles are also a health hazard.

The vapour phase technique for preparation of ceramic powders is well suited to the preparation of powders which are much less aggregated than those prepared by solution techniques. On the other hand, it is considerably more difficult to make homogeneous powders of complex compositions by vapour phase techniques. Another variation is to decompose the vapour, yielding solid particulates. Examples include the thermal decomposition of alkoxide vapours to yield oxide powders and decomposition of $(CH_3)_2SiO_{12}$ to form SiC powder. Another variation of this method is thermal plasma synthesis, which involves the quenching of vapours produced by either a physical transformation or a chemical reaction.

2.5 MECHANICAL METHODS

Mechanical methods are not much used as primary methods for the production of metal powders. Mechanical comminution is possible by methods such as impact, attrition, shear and compression. The formation of metal powders by mechanical methods relies on various combinations of these four basic mechanisms. Such methods have been used as the primary process in the following cases:

- Materials which are relatively easy to fracture such as pure antimony and bismuth, relatively hard and brittle metal alloys and ceramics,
- Reactive materials such as beryllium and metal hydrides,
- Common metals such as aluminium and iron which are required sometimes in the form of flake powder.

In ceramic systems the starting materials may be beneficiated physically using operations such as crushing and milling. Other supplementary operations like washing, chemical dissolving, settling, magnetic separation, dispersion, mixing, classification, de-airing, filtration and spray drying form an essential part of the powder preparation cycle. The description of some of these is given in Chapter 4. β-alumina is an example of a powder produced by milling the commercially available fused-cast bricks.

Machining is another mechanical method to produce particulates. Sometimes, it is done intentionally for very specific cases obtained by ingot metallurgy route, for example, beryllium. The resultant product is *swarf*, which can be used as such or can be further milled as per the requirement. On the other hand, these swarfs may be the by-product in conventional metal machining shops. The use of swarf of value-added materials, say copper, obtained in electric conductor industries, is much sought after.

Basic aspects: The main causes for different stresses during crushing/milling may be categorised as:

- *Impact/compression*—by falling or vibrating media
- *Shear*—by seizure between two surfaces moving with different velocities
- *Attrition*—by frictional stresses.

In real practice, a combination of factors may prevail. For example, the role of attrition increases as the media size and impact force decrease, and as the frequency of rubbing contacts increases.

The grinding energy produced during milling is proportional to the mass (m) and the change in velocity (v) of the media on impact is:

$$\text{Energy} = \Delta \frac{mv^2}{2} \qquad (2.30)$$

The mass is increased by using media of larger size or density. Table 2.9 offers density and hardness data of major industrial grinding media.

Table 2.9 Some grinding element materials

Material	Abrasion behaviour	Specific weight, g cm^{-3}	Mohs hardness	No chemical resistance in
Agate (SiO$_2$)	Abrasion-proof; approx. 200 times more resistant than hard porcelain	2.65	7	Hydrofluoric acid (HF)
Zirconium oxide (ZrO$_2$)	Abrasion-proof; approx. 10 times more resistant than sintered corundum	5.7	8.5	Sulphuric acid (H$_2$SO$_4$), hydrofluoric acid (HF)
Alumina (Al$_2$O$_3$)	Good abrasion resistance	4.0	9	Conc. acids
Hard porcelain	Sufficiently abrasion proof	3.1	8	Conc. acids
Hardened chromium steel 2080	Good abrasion resistance; better than CrNi steel	7.9		Acids
Hardened steel CK45	Good abrasion resistance	7.9		Acids
Stainless steel 4301	Average abrasion resistance	7.9		Acids
Hardmetal (Cemented tungsten carbide)	Extremely abrasion proof; approx. 200 times more resistant than agate	14.75	8.5	Nitric acid (HNO$_3$), hydrochloric acid (HCl)

Defects in the binding of particles are the major cause for their weakening. Figure 2.35 illustrates some microstructural defects in a particle. From solid-state mechanics we know that the reduction of the fracture strength (σ_f) of a material due to the presence of stress-raising flaws of depth (c) is given by the expression:

$$\sigma_f = \frac{K_k}{Y\sqrt{c}} \qquad (2.31)$$

where, Y is a constant which depends on the flow geometry and K_{lc} is the fracture toughness parameter that includes the work required to propagate the crack. Fracture fragments containing defects that produce smaller stress intensification are more resistant to grinding. Dense particles with a finer grain size and higher fracture toughness are more resistant to attrition. Abrasion grinding may increase the number of edge and surface flaws which promote fracture. Attrition is important for the

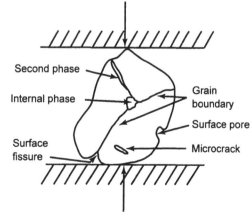

Fig. 2.35 Various types of microstructural defects causing a fall in the strength of a particle

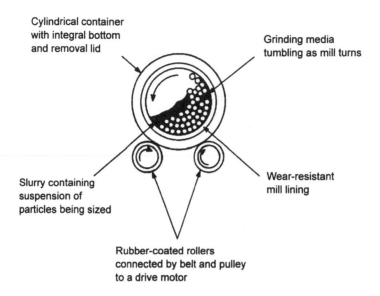

Cylindrical container
with integral bottom
and removal lid

Grinding media
tumbling as mill turns

Slurry containing
suspension of
particles being sized

Wear-resistant
mill lining

Rubber-coated rollers
connected by belt and pulley
to a drive motor

Fig. 2.36 Schematics of the action in a jar mill

grinding of inelastic particles and particles finer than a few micrometres. Microstructures, intergrannular fracture paths, and material defects all have a cumulative effect on apparent reduction ratio during milling.

During wet milling, additional features come into play. In wet milling, the impact stresses are lower. Chemical leaching of a less resistant phase by the milling liquid may significantly reduce the milling time. Water may be absorbed in the cracks, thus reducing the strength of the material. On the other hand, water vapour often causes agglomeration during dry milling. Surfactants such as alcohols, oleic acid, glycols and silicones are added (<1%) to minimise agglomeration. In wet milling, a deflocculant is added to disperse agglomerates and to increase the solids loading in the slurry.

Figure 2.36 shows the general view of a jar mill. The height of the media before cascading is a function of media charge, solids loading, angular speed and the viscosity of the suspension in case of wet milling. The critical angular frequency W_{cr} (Hz) causing centrifuging is expressed as:

$$W_{cr} = 0.5R^{-1/2} \qquad (2.32)$$

where, R is the radius of the mill in metres. Adhesion produced by a viscous slurry that wets the mill effectively reduces W_{cr} and the mill normally operates at $0.65–0.85W_{cr}$ to generate a maximum media height of 50–60° measured from the horizontal through the centre of rotation. Typically, media charge is about 50% of the volume of the mill. The critical factor is the speed of the drum's rotation. A very high speed will cause the material and the ball to be pressed against the walls of the drum, because of the centrifugal forces and prevent relative motion between the material and the balls. Too low a speed will result in an insignificant amount of movement in the lower part of the drum. The optimum speed corresponds to a situation in which some amount of ball and material is lifted up to the top of the drum and falls down on the remaining material.

A reactive ball milling, where a reagent is added, offers unique properties in the end product. For example, after reactive milling, $LaCoO_3$ displays a better stability of perovskite structure under a reducing atmosphere. Such stability is of significance in catalytic reactions such as perovskite-type oxide catalysed hydrogenation reactions.

Equipment for reducing particle size

Before selecting an equipment for size reduction, the following aspects should be looked into:

Type of material: What are the physical and chemical properties of the material? Is the size reduction process affected by, for example, hardness or toughness or by chemical reactions?

Final size: To what particle size should the material be reduced? Should as narrow a particle size distribution as possible be achieved?

Abrasion: Which impurities caused by abrasion of the grinding elements must be avoided? What quantity of impurities is still permissible in other cases?

Versatility: If necessary, should wet size reduction be possible? Should the size reduction be carried out in an inert atmosphere or in a vacuum?

Figure 2.37 schematically shows the product size versus feed size relations during various types of size reductions.

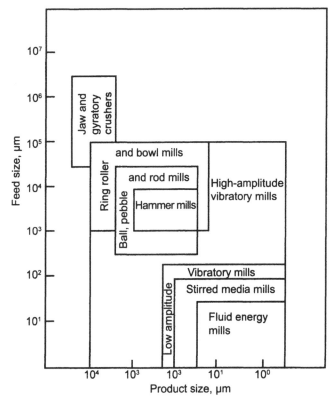

Fig. 2.37 Nominal feed and product mean size capabilities of various industrial mills

Crusher: The main equipments are mortar and pestle, heavy drop hammer and jaw crushers. In jaw crushing the crushing chamber is formed by one fixed jaw and one movable jaw between stationary side support walls. The movable jaw is mounted eccentrically at the top and is driven by a motor via a flywheel. During the crushing process, the material is drawn down into the tapered crushing chamber as a result of the eccentric jaw motion and is reduced in size by compression and frictional forces. In addition, the angle between the jaws can be altered to give optimum adjustment to the crushing of the material. The final crushed product particle size can be adjusted to values between 15 and 1 mm by adjusting the gap at the base of the jaws. The standard jaws and side walls are manufactured from hardened steel. If the feed material is very abrasive or particular impurities are to be avoided, these components are also available in stainless steel or hard tungsten carbide. If the fixed jaw is removed, access is given to the crushing chamber and the equipment can be cleared easily. In addition, the connection of an exhaust device allows work to be carried out in dust free conditions.

Figure 2.38 shows a schematic of jaw crusher, rotary crusher, roll crusher and hammer mill.

Ball mill: The ball mills can be divided into two types: centrifugal and planetary mills. In a *centrifugal ball mill*, a single bowl fastener is moved horizontally and eccentrically driven while not rotating itself. In spite of this, the velocity of the grinding balls in this case is still six times that of the grinding balls in the gravity ball mills.

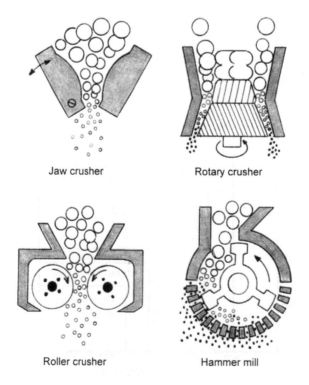

Jaw crusher   Rotary crusher

Roller crusher   Hammer mill

Fig. 2.38 Schematic diagrams of (a) jaw crusher, (b) rotary crusher, (c) roller crusher and (d) hammer mill

In *planetary ball mills,* two or four bowl fasteners, each of which accommodates one grinding bowl, are attached to a supporting disc. During grinding, the grinding bowls and supporting disc rotate in opposite directions, so that two different centrifugal forces act on the bowl contents. The grinding material is reduced in size as a result of both attrition and impact. The grinding balls have a velocity which is twelve times that of the grinding balls in a grinding ball mill. Each grinding medium influences the grinding process via the specific weight of balls. For example, if a grinding medium of agate (specific weight 2.6 g cm^{-3}) is used, the corresponding balls have a smaller impact energy than those of tungsten carbide (specific weight 14.75 g cm^{-3}). The more mixing/grinding intended, the smaller is the ball size selected. The grinding time in case of the centrifugal ball mill is increased by smaller ball acceleration.

The chemical composition of the grinding medium qualitatively determines the type of contamination which can occur, whilst the abrasion behaviour determines the quantity thereof. Other details like specific weight (ball weight in ball mills), Mohs hardness, compression and breaking strength (in the crushers) and chemical resistance should also be considered.

Disc grinder: The disc grinders are suited for processing hard brittle materials. Wet grinding in a closed grinding vessel which is driven horizontally and eccentrically and which contains grinding elements (disc or ring) is used to reduce material to a particle size of a few micrometres. Impact and friction between the grinding elements and the grinding vessel cause such grinding energy that the process is generally concluded within a few minutes.

Attritor mill: Historically, attritor milling or mechanical alloying was discovered first for high-energy ball milling under conditions such that powders are not only fragmented but also re-welded together. Later, with the advent of nanoparticulates, its significance has been enhanced for straight high-energy mechanical milling. In attritor milling, the ball charge is stirred vigorously with rotating paddles. In mechanical alloying, the charge is a blend of elemental powders, at least one of which is a ductile material. To provide a dispersed phase in superalloys, fine inert oxides can be included in the charge, usually Y_2O_3. The mechanical alloying process was first invented by JS Benjamin at INCO, New York, in the early 1970s.

Figure 2.39 illustrates the effect of a single high-energy collision between two balls or the powder trapped between them. The ductile elemental metal powders are flattened, and where they overlap, the atomically clean surfaces just created weld together, building up layers of composite powder, between which are trapped fragments of the brittle powder and the dispersoid. At the same time, work-hardened elemental or composite powders fracture. These competing processes of cold welding and fracture occur repeatedly throughout milling, gradually kneading the composites so that their structure is continually refined and homogenised.

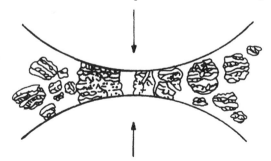

Fig. 2.39 Effect of single collision between two balls on trapped powder

When the change comprises malleable powder particles, the particles are subjected to repetitive deformation, fracture and welding when entrapped between colliding grinding media. These events are necessary for microstructural refinement. Even when brittle materials are milled, there is evidence of particle coalescence, as well as their anticipated fracture. In fact, an appropriate balance between the fracture and welding frequencies of the powder particles is needed for successful processing. If fracture unduly predominates, submicrometre size particles—many of them pyrophoric and some of them potentially explosive—are produced. Conversely, excessive welding eventuates in large particles and coating of the grinding media container walls by the powder. Powder welding can be reduced through process control agents (PCAs), which are particularly effective for this purpose when ductile materials are milled. PCAs are usually organic fluids, but welding frequencies can also be changed by varying the mill atmosphere (oxygen generally reduces welding frequencies), mill temperature (cold temperatures promote fracture) and by addition of a small percentage of a hard dispersoid which reduces the metal–metal contact required for powder particle welding. The degree to which particles deform, and the frequency with which they fracture and weld, are material dependent; for example, ceramics fracture more and plastically deform less than metals. Nonetheless, nominally brittle materials manifest plasticity under appropriate milling conditions and, as noted above, their coalescence is requisite for successful processing.

Mechanical alloying is not just mixing on a fine scale; true alloying occurs. The progress of alloying can be monitored by X-ray diffraction studies. The microstructure of the alloy so processed is an extremely fine one. It is suggested that localised melting may occur during high-energy milling due to the extensive plastic deformation. Since 10^{-2} to 10^{-3} J per impact is dissipated during collision, the average bulk temperature might be raised to about 600 K. Alloying is associated with an increase in temperature by enhanced diffusion due to formation of lattice defects, and due to shorter diffusion paths as the grains become smaller and smaller. Another example of an inorganic mechanically alloyed compound is zinc ferrite $ZnFe_2O_4$ from ZnO and Fe_2O_3. A distinctive feature of such a processed ferrite is that it has partially inversed spinel structure. This establishes that atomic size effects may override preferential chemical binding under extremely high localised pressures. Another interesting aspect of attritor milling is the amorphisation of the crystalline powder. A typical example is glass forming Ni-Nb alloy system through high-energy ball milling.

Milling temperatures affect the rate at which the nanocrystalline structure develops. Milling at temperatures lower than ambient can bias the defect accumulation induced by plastic deformation. The very first use of milling at cryogenic temperatures has been for introducing fine nano-scale nitrides or oxy-nitrides into aluminium. To achieve this, liquid nitrogen is introduced into the milling vial along with powders and milling balls. The chemical reaction of nitrogen with metal powder produces fine nitride particles which help stabilise the nano-scale grain size during subsequent thermal/mechanical powder consolidation as well as act as a dispersion hardening agent.

Recently, investigations have also been made in developing nano-scale networked structure on the surface of metal and alloy powders, particularly, austenitic stainless steel, by applying mechanical milling, such that the core structure of coarse grains remain unaffected. One of the limitations of using mechanical milling/alloying has been the lack of predictiveness. An urgent need is to put forward reliable predictive models that will shorten the time and labour of the investigators. Another anamoly has been in confusing nanocrystallites present in the bulk of coarse grains with nanoparticles. A powder metallurgist is more interested in nanoparticle assemblies, where the advantage of large surface area available is exploited in developing high-density parts.

Cold stream process: This process is another mechanical method which uses high velocity gas jets to collide the powder with a cold target. The initial input powder is a rather coarse one. The gas pressure is around 7 MPa. The resultant powder is generally above 10 µm in size with a rounded, but irregular shape. The low temperature enhances metal brittleness and thus facilitates impact attritioning. The method has been extensively used in preparing flame spray powders and stainless steel powders for filters. Figure 2.40 schematically illustrates the flowsheet of the cold stream process.

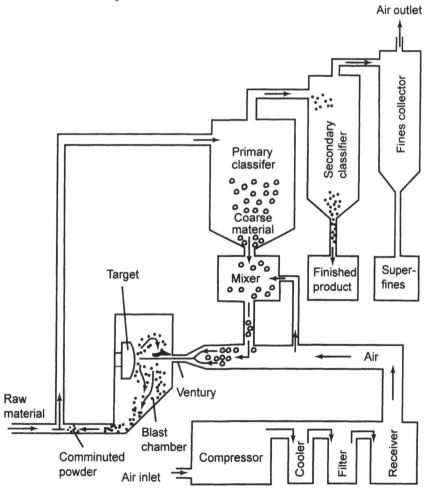

Fig. 2.40 Schematics of the cold stream process

Example 2.8: Describe with the help of a flow diagram the production of PM grade powder from the machining swarf of mild or medium carbon steel. 15% of the swarf is the cutting fluid.

Solution: The first step is to crush the swarf to get particles up to 12 mm length. The next operation is to remove the cutting fluid. The untreated swarf is loaded into the crusher (1) and fed via conveyers (2, 3) to the hopper (4) of the swarf centrifuge (5), where most of the oil is removed (Fig. Ex. 2.8). If the cutting fluid is reclaimable cutting oil, it is cleaned and treated to make it suitable for subsequent reuse. The chips are then washed in an alkaline chamber to remove the remainder of the cutting fluid, and then dried. The non-metallic particles are separated magnetically (6). Subsequently, the chips are pulverised either at cryogenic temperatures or at room temperature in a hammer mill (9), and the coarser particles are separated (10) and recycled (11) until swarf powder of an acceptable size range is obtained. The final operation of annealing (12) is carried out to ensure that particle properties are favourable for compaction.

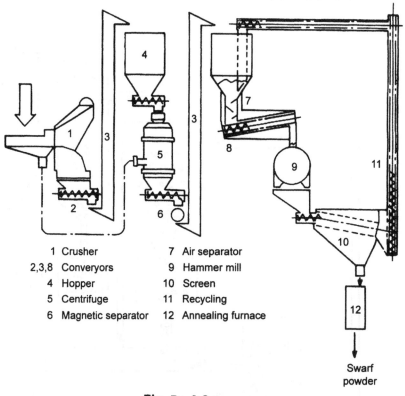

1	Crusher	7	Air separator
2,3,8	Converyors	9	Hammer mill
4	Hopper	10	Screen
5	Centrifuge	11	Recycling
6	Magnetic separator	12	Annealing furnace

Swarf
powder

Fig. Ex. 2.8

SUMMARY

- Chemical reduction involves chemical compounds, most frequently an oxide, but sometimes a halide or other salts of metals. The operation may be carried out in solid state, gaseous state or aqueous solution state.

- Metal powder production from aqueous solutions can be carried out by reduction with another metal or gaseous reduction.

- In ion exchange method, the metal ions in solution can exchange with specific ions contained in a solid or second liquid phase. The metal is extracted from the impure aqueous phase into an organic liquid phase.

- Direct synthesis is a common high temperature process for pure compounds, e.g., for preparation of ceramics or intermetallics.

- Adjustment of chemical and physical conditions during electodeposition produces a loose deposit of the metal on the cathode. The purity of such powders is high.

- Virtually any material that can be melted can be made into powder by disintegration of the liquid by gas or water. The atomisation process is independent of normal physical and mechanical properties associated with the solid metal.

Further reading

Bockris JO'M and Reddy AKN, *Modern Electrochemistry*, Plenum Press, New York, 1970.

Bose A, *Advances in Particulate Materials*, Butterworth-Heinemann, London, 1995.

Hydrometallurgy: Theory and Practice, First Technical Symposium on Hydrometallurgy, May 19–20, 1972, The University of Denver, Denver, Colorado, USA.

Exner HE and Danninger H, in *Metallurgy of Iron*, Vol. 10, Springer-Verlag, Berlin, 1991.

Hayes PC, *Process Selection in Extractive Metallurgy*, SBA Publications, Calcutta, 1987.

Geiger GH and Poirier DR, *Transport Phenomena in Metallurgy*, Addison-Wesley, New York, 1973.

Koch CC, (ed), *Nanostructured Materials*, 2nd edn, William Andrew Publishing, Norwich, NY, 2007.

Kubaschewski O, Evans EL and Alcock CB, *Metallurgical Thermodynamics*, 5th edn, Pergamon Press, Oxford, 1979.

Lawley A, *Atomisation*, Metal Powder Industries Federation, Princeton, 1992.

Lu L and Lai O, *Mechanical Alloying*, Kluwer Academic Publishers, Boston, MA, 1998.

Munir ZA and Holt JH, (eds), *Combustion and Plasma Synthesis of High Temperature Materials*, VCH, Weinheim, 1990.

Otooni MA, (ed), *Science and Technology of Rapid Solididfication and Processing*, Kluwer Academic Publishers, Dordrecht, The Netherlands, 1995.

Ring TA, *Fundamentals of Ceramic Powder Processing and Synthesis*, Academic Press, San Diego, CA, 1996.

Rosenquist T, *Principles of Extractive Metallurgy*, McGraw Hill, New York, 1974.

Ryan W, *Properties of Ceramic Raw Materials*, 2nd edn, Pergamon Press, Oxford, 1978.

Segal D, *Chemical Synthesis of Advanced Ceramic Materials*, Cambridge University Press, Cambridge, 1989.

Suryanarayana C, *Processing of Metals and Alloys*, Vol. 15, Cahn RW, Haasen P and Kramer EJ, (eds), *Materials Science and Technology*, VCH, Weinkeim, 1991.

Van Bogdandy L and Engell HJ, *The Reduction of Iron Ores*, Springer-Verlag, Berlin, 1971.

Worral WE, *Clays and Ceramic Raw Materials*, Halsted Press Div., Wiley-Interscience, New York, 1975.

Yih SW and Wang CT, *Tungsten*, Plenum Press, New York, 1979.

Yule AJ and Dunkley JJ, *Atomisation of Melts for Powder Production and Spray Deposition*, Clarendon Press, Oxford, 1994.

EXERCISES

2.1 Höganäs sponge iron powder production from iron oxide (magnetite) is a two-stage process, i.e., reduction to sponge followed by pulverisation. Suggest a single-stage operation to produce the reduced powder.

2.2 Figure P. 2.2, illustrates the stability range of various copper oxides. Explain how it is useful in air atomisation/reduction process for producing copper powder.

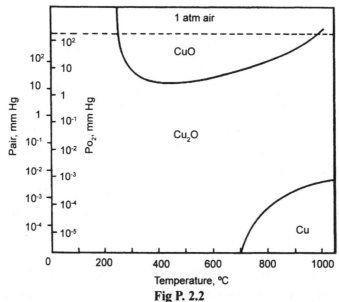

Fig P. 2.2

2.3 It is established that at 500°C the chemical reduction of Al_2O_3 is not possible in practice. What if the temperature is increased to 1000°C? Using Ellingham diagram determine the critical H_2/H_2O and CO/CO_2 ratios for this temperature.

2.4 It is observed that fine aluminium powder exposed to air oxidises rapidly at 600°C, whereas a solid block of aluminium oxidises slowly at the same temperature. Explain.

2.5 Why is the reduction of tungsten oxide generally done in two stages?

2.6 The common titanium mineral is ilmenite ($FeTiO_3$). To produce titanium sponge, why should the mineral be first partially reduced to TiO_2?

2.7 $TiCl_4$ cannot be obtained by direct chlorination; the presence of carbon makes it necessary at 500°C. Why? (Hint: Write down the reaction and obtain the ΔG_{500} values.)

2.8 A spinel, $MgO.Al_2O_3$, can be made by direct synthesis from MgO and Al_2O_3 powders. If you are asked to prepare a pure spinel, what method will you follow?

2.9 In a pure ceramic powder, a trace of undesirable titanium oxide (TiO_2) is present. Indicate one of the possible ways to remove TiO_2? (Hint: TiO_2 is weakly magnetic.)

2.10 During calcination of a pure ceramic, it was noticed that the ceramic got contaminated. What remedy do you suggest?

2.11 If calcium carbonate powder is milled, will there be any change in decomposition temperature with respect to milling period? Will the mean particle size be identical after decomposition for a particular milled powder?

2.12 An oxide ceramic is calcined in a rotary kiln. On which side of the kiln should the burner be installed?

2.13 Two specimens of zinc aluminate spinel, nominally $ZnO.Al_2O_3$, are prepared by reacting ZnO and Al_2O_3. The weights of ZnO and Al_2O_3 are 80 g and 100 g in batch 1 and 81.38 g and 101.94 g in batch 2. Calculate the stoichiometry of each spinel, assuming the components react completely.

2.14 A copper base alloy containing Al and Ni is to be prepared by PM route for shape memory applications. Can you prepare the alloy by an exothermic reaction? Comment.

2.15 It is required to prepare Mo(C,N) powder from the molybdenum powder. What possible methods of synthesis should be adopted? Will the temperature of synthesis be lower or greater than that for W(C,N) powder synthesis?

2.16 Using the E_h–pH diagram (Fig. 2.3), determine the pH of the following aqueous solutions to reduce them at 100 atm: (a) 1 M nickel salt solution, (b) 1 M Cd^{2+} solution and (c) 1 M Fe^{2+} solution. Can zinc salt solution be reduced by hydrogen?

2.17 Construct a processing flow diagram for the formation of a $BaTiO_3$ powder.

2.18 What are the technical factors that will favour the production of a ceramic powder using special chemical methods rather than the solid-state reaction of powders?

2.19 In processing ceramic raw materials, which of the following steps consumes maximum energy?

(a) Mining

(b) Crushing and grinding

(c) Drying

(d) Calcination

(e) Separation

(f) Transportation.

2.20 How many grams of iron are deposited in 17.3 min with a current flow of 4.38 A and a current efficiency of 97.4%? The solution contains $FeCl_2$.

2.21 What problems might you face in electrodepositing a 70/30 brass powder electrolytically?

2.22 In an under-cooled copper melt, which solid plane, {111} or {100}, will grow more rapidly? Why?

2.23 Why is solute enriching present in the melt and not in the solid during non-equilibrium solidification of an alloy?

2.24 In case the supercooled layer in advance of the solid–liquid interface is thin, will you get a dendritic growth or some other type of microstructure in the atomised metal powder?

2.25 What is splat cooling and what are typical cooling rates observed in splat cooling? What kinds of structures result from splat cooling of metals?

2.26 In case there is some additional uniform agitation facility in the melt bath, will it have the same effect on the morphology of the atomised metal powder?

2.27 Gas-atomised nickel powder is produced in the following gaseous media: (a) Ar, (b) He and (c) N_2. Arrange the gaseous media in the descending order of the secondary dendrite arm spacing. Is the spacing dependent on particle size?

2.28 The table below gives the properties of rapidly solidified Al-20Si-2Fe-2Ni powders, prepared under air or argon atomising medium. Identify the atmosphere and justify your answer.

Atomising medium	Particle size, μm	Specific surface area, $m^2\,g^{-1}$	O_2 combined wt%	Oxide thickness, nm	H_2 content, ppm
A	−125	0.22	0.21	33.36	3.55
B	−125	0.16	0.10	10–13	0.55

2.29 Inert gas atomisation is an ideal process to produce spherical powder particles. However, the process is costly. In case it is required to produce similar shaped powder by much cheaper water atomisation, what special care will have to be taken?

2.30 Gold and copper are miscible in melt and form a continuous solid solution. Indicate what alloy will be more convenient for powder production by atomisation process. (Hint: Refer to the binary Au-Cu phase diagram).

2.31 Table 2.8 shows the particle behaviour of rapid solidified Al-alloy powders prepared by vacuum and ultrasonic atomisations. If the powder is to be prepared by Twin-roller quench method, complete the above table as best as possible.

2.32 Why is secondary dendrite arm spacing measured through quantitative metallography considered to be more reliable than primary dendrite arm spacing for measuring the cooling rate?

2.33 Why is metal powder produced by gas atomisation generally used for powder injection moulding purpose? Why is the powder produced by PIM costly on weight basis?

2.34 Figure 2.33 illustrates particle size distribution of metal powders produced by three types of atomisation process. Explain the reasons for the distinctive features.

2.35 Nitrogen added high-speed steels have better wear resistance (HRC 65 to 72)—both abrasive and adhesive. Which methods should be applied to produce powders of such steels?

2.36 Cu-10Sn bronze powders were prepared by the following routes: (i) Premix electrolytic Cu, (ii) Partial pre-alloyed electrolytic Cu and (iii) Pre-alloyed water atomised. List the response of these powders for the following properties:

(a) Oxidation response

(b) Flowability

(c) Compressibility

(d) Green strength.

(Rank: 1 – good, 2 – moderate, 3 – poor).

2.37 Copper forms a β-intermetallic phase with other metals such as Si, Ga, Sn, Al, Zn and In as shown in fig. P. 2.37 below at a fixed electron/atom rate known as Hume-Rothery electron compounds. For preparing these β-phase powders as master alloys of reasonable purity, what possible methods do you envisage? Arrange these intermetallic powders in ascending order of hardness. (Notice the degree of stability of these β-phases in the diagram). Convert the *e/a* ratios into weight percent of solutes.

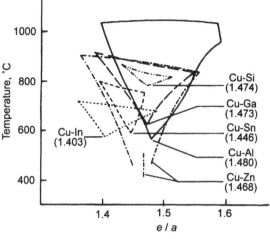

Fig P. 2.37

2.38 The table below gives the typical properties of copper powder produced by atomisation, reduction, electrolytic and hydro-metallurgical processes. Identify the production method.

	A	**B**	**C**	**D**
Copper, %	98–99.5	99–99.5	99–99.5	98.5–99.5
H loss, %	0.1–0.75	0.1–0.75	0.1–0.5	0.3–1.0
Insoluble, %	0.3 max.	0.02 max.	0.1 max.	0.05 max.
Apparent density, g cm^{-3}	2–4	1–4	2–5.5	1–4
Flow, s (50 g)$^{-1}$	20–35	25–40	15–35	25–0
–325 mesh, %	15–90	15–90	25–90	35–90
Green strength, MPa	2.8–13.8	2.8–20.7	1.4–15.8	2.8–20.7

2.39 A ceramic powder is much easier to mill than a metal or intermetallic powder. Justify the fact with scientific reasoning.

2.40 Write down an equation inferring that the energy for size reduction during crushing increases rapidly as the product size decreases. What is the assumption in the equation?

2.41 In a ball mill a lining with baffle bars is used sometimes. Why?

2.42 Which one of the following processes would most likely result in the lowest pick up of impurities?

(a) Precipitation

(b) Ball milling

(c) Vibratory milling

(d) Attrition milling.

2.43 Why are smaller size media used during ball milling for achieving micron size particles or dispersing agglomerates?

2.44 Explain why the temperature of the slurry in wet milling should be monitored and controlled.

2.45 What factors are used for correct viscosity selection during slurry milling?

2.46 What types of media are used when the viscosity of the suspension during wet milling is high?

2.47 Why do milling media of mixed size wear more rapidly? In which cases can they be decreased?

2.48 A W-Ni premix powder is mechanically alloyed for 4 hours in an attritor mill at 140 RPM. What should be the rotational rate in order to achieve the same degree of alloying in 1 hour?

2.49 A dispersion-strengthened Al-alloy, DISPAL powder, was prepared through two routes: ball milling and attritor milling. The table below shows some of the operating conditions. Identify the routes and justify your answer.

Route	A	B
Milling time, h	1.5–3	12–20
Throughput capacity, kg h^{-1}	0.5–5	5–40
Temperature, °C	125–250	50–70
Energy density, kW vol^{-1}	10	1

2.50 Nano-spherical particles of copper are produced by RF induction thermal plasma process. Can the same be produced by attritor milling? Comment. What should be the atmosphere in the plasma process?

2.51 How will the following vary with the powder feed rate for producing submicron copper spherical powder by RF induction thermal plasma method?

(a) Particle size

(b) Evaporation ratio

(c) Amount of vaporisation.

2.52 Since both manganese and silicon are inexpensive, it is proposed to prepare an iron-based master alloy powder with these elements. The method used is chill casting followed by milling. With the help of the ternary phase diagram, suggest possible compositions, keeping in view the low melting point and ease of friability during milling.

2.53 In PM high-speed steel production plants heat sizes are generally large (5–8 tons) for commercial efficiency. Consequently, the gas atomisation process runs for many hours without interruption. For a high degree of stability against variation, suggest the exact atomisation detail. To achieve fine powder, what modifications are necessary? For obtaining clean powder, what additional processing should be done before atomisation?

2.54 Write the various reactions involved in synthesis of uranium carbide from urania and carbon. (Note: Uranium carbide has more than one composition).

2.55 Based on thermodynamic free energy data, which oxide will be least susceptible to reduction at 1500°C: (a) Y_2O_3, (b) SiO_2, (c) ZrO_2 and (d) Al_2O_3?

3

Powder Characterisation

LEARNING OBJECTIVES

- Principles of chemical characterisation and structure of powders
- Experimental methods for measuring particle size
- Significance of true, apparent and tap densities of powders
- Flow rate of powders and its significance
- Mechanical properties of powder green compacts
- Relationship between method of powder production and powder characteristics

The success of any powder metallurgical process depends to a great extent on the complete characterisation and control of the metal powders. The method of powder production influences particle chemistry and structure, apart from the precise nature of particle size distribution. These properties also influence the behaviour of the powder during compaction and sintering, and the composition, structure and properties of the sintered material. In this chapter, various characteristics of the powder—some interrelated—are considered. Table 3.1 gives a brief summary of principles involved in the characterisation of various powder properties which will be described in subsequent images.

3.1 POWDER SAMPLING

Since many tests are performed using very small amounts of powder, it is important that the test portion used be representative of the whole. The practice is to take into account the possibility of segregation of the powder during filling of containers. Three steps are involved in the process.

1. Receive the powder in a rectangular receptacle, which is moved across the stream of flowing powder at a constant speed.

2. Blend the gross sample of at least 5 kg in a blender for 10 to 15 revolutions.

3. Pass the blended powder sample through the sample splitter to form a number of test samples depending on the design of the splitter (Fig. 3.1).

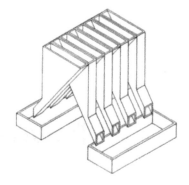

These test samples should be kept carefully and supplied as test portions for subsequent powder characterisations. Complete details are available in ASTM Standard B215.

Fig. 3.1 Schematic illustration of splitter used for powder sampling

3.2 CHEMICAL COMPOSITION AND STRUCTURE

The levels of impurity elements in metal powders are significant for both the processing and properties of the final product. It is necessary to know whether such elements are present in their elemental form or in the form of a chemical compound. For example, in reduced iron powder, silicon is present as an impurity in the form of silica. Other ceramic and inert compounds may exist; they can be reported in terms of an acid insoluble figure. The effect of impurity elements on the hardness of the particles and the degree of chemical reactivity during sintering will differ widely, depending on the actual form they are in.

Hydrogen loss is a common parameter to know the level of oxygen impurities in those metal powders, the oxides of which are easily reducible by hydrogen, e.g., iron, tungsten,

Table 3.1 Powder properties and methods of measuring them

Property	Method
Particle size and size distribution	Sieve analysis; permeability; sedimentation; electrical resistance; light obscuration; light scattering; microscopy; surface area
Particle shape (external)	SEM; shape parameters; morphological analysis; fractals
Particle shape (external and internal)	Stereology; mercury porosimetry; gas adsorption
Particle density	Pycnometry; mercury porosimetry
Specific surface area	Gas absorption; permeametry
Surface chemistry	X-ray photoelectron spectroscopy (ESCA); Auger electrón spectroscopy; secondary ion mass spectroscopy; ion scattering spectroscopy
Alloy phases and phase distribution	Optical metallography; stereology; electron microscopy; EDAX; X-ray diffraction
Quality of mixing (segregation)	Macroregion: Variability coefficient (by chemical analysis); Microregion: Variability coefficient (2nd comp. >5%); Homogeneity coefficient (2nd comp. <5%); by metallography

copper, nickel, etc. However, this value can be in error due to incomplete reduction of oxides, and some oxides may not be reduced at all. The annealing of the powder in a reducing atmosphere is an effective way of reducing oxygen content. The standards of Metal Powder Industries Federation (MPIF) and the American Society for Testing and Materials (ASTM) contain the details of procedures for determining these parameters.

Any metal powder adsorbs significant quantities of gases and water vapour from the atmosphere during storage. Such adsorption can lead to the formation of surface oxides on the metal which may interfere with compaction and sintering and possibly remain in the sintered material. The amount of such contamination increases with decreasing particle size and increasing chemical activity of the surface.

Chemical composition of a crystalline powder can be determined by X-ray or spectrographic techniques, apart from conventional wet analysis. The details of X-ray methods are given in Chapter 11. Spectrographic analysis is often performed when a rapid determination of low concentrations of elements is required. A small pulverised specimen, highly excited in an electric arc, emits a spectrum of radiation characteristic of its chemical composition. Analysis of the spectrum is performed by wavelength (element) and intensity (concentration) measurements. Most of the wavelengths used in metallurgical analyses lie in the ultraviolet region. Photoelectric intensity measurement lends itself to the construction of almost completely automatic equipment, with which the analysis of a sample for several elements can be completed in two or three minutes. The metallurgical condition of the sample does not affect the results unless marked heterogeneity is present.

The particle shape of a metal powder indirectly depends on the production method (Fig. 3.2).

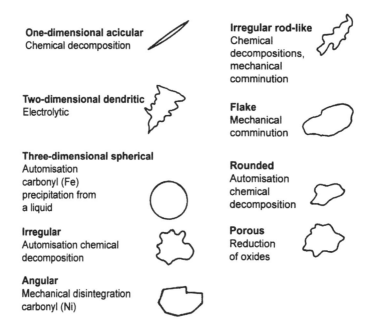

Fig. 3.2 Effect of method of metal powder production on particle shape

Most powder particles are three-dimensional in nature and they can be considered to be somewhat equi-axed. Spherical particles represent the simplest and ideal example of this shape. Porous particles differ from irregular ones, because the porosity itself may be irregular in both size and shape. A large amount of porosity makes any shape characterisation very difficult.

The microstructure of the crystalline powder has significant influence on the behaviour of the powder during compaction and sintering, and on the properties of the final product.

Pre-alloyed powders may contain various phases depending on their exact composition, the appropriate phase diagram, their thermal history and the method of powder production. Multiphase microstructures can result from alloying such as in steels, cast irons and superalloys. In the case of water-atomised solid solution–type alloys, the microstructure, like the chilled structure, consists of a cored structure. Microporosity associated with entrapped gases is also common. A cold worked powder, e.g., ball milled, exhibits high dislocation density which could be lowered by annealing. Such imperfections influence the compaction and sintering response of the concerned powders.

The detailed experimental methods of the study of microstructure are given in Chapter 11.

3.3 PARTICLE SIZE

In a real mass of powder, all prepared in the same manner, all the particles will not have exactly the same size, although the shape may be essentially the same. Consequently, we must deal with size distribution when accurately describing powders. Various methods are used to calculate average diameters. Particle size, therefore, is not a concise quantity but for any given non-spherical particle it may have several values with different meanings, depending on the sizing method used.

Distribution curves for particle size relate particle size to the corresponding fraction of powder with that size. Figure 3.3 illustrates various size distributions. In unimodal distribution, there is one high point or maximum amount of a certain critical size. The polymodal distribution consists of two or more narrow bands of particle sizes, each with a maximum, with virtually no particles between such bands. The broad band distribution simply corresponds to a uniform concentration of particle sizes over a rather broad size interval with virtually no particle sizes occuring outside this range. The irregular distribution represents a continuous and finite variation of particle sizes within a relatively broad range. It can be thus concluded that particle size distribution is necessary for complete characterisation, instead of an average, maximum or minimum value.

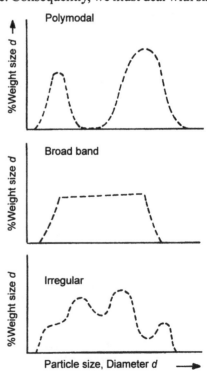

Fig. 3.3 Various types of possible powder size distributions

Table 3.2 Common methods for determining particle size and their limits of applicability

Class	Method	Approximate useful size range, μm
Sieving	Sieving using mechanical agitation or ultrasonic induced agitation and screens	44–800
Microscopy	Micromesh screens	5–50
	Visible light	0.2–100
	Electron microscopy	0.001–5
Sedimentation	Gravitational	1–250
	Centrifugal	0.05–60
Turbidimetry	Turbidimetry (light intensity attenuation measurements)	0.05–500
Elutriation	Elutriation	5–50
Electrolytic resistivity	Coulter counter	0.5–800
Permeability	Fisher subsieve sizer	0.2–50
Surface area	Adsorption from gas phase	0.01–20
	Adsorption from liquid phase	0.01–50

A number of techniques for measurement of particle size are available in powder metallurgy, each with its own limitations. Table 3.2 classifies some of the common methods for determination of particle size and their limits of applicability. It is worth noting that the results obtained by various methods using different physical principles may disagree. The results are strongly influenced by the physical principles employed by each method of particle size analysis. The results of any particle sizing method should be used only in a relative sense; they should not be regarded as absolute when comparing results obtained by other methods.

3.3.1 Sieving method

Sieving is technologically the most satisfactory method for reporting and plotting particle size distribution, in which successive sizes form a geometrical series. The reference point for their scale is 75 μm, which is the opening of the 200-mesh woven wire screen standardised by the National Bureau of Standards. Because of the widespread use of this method, one distinguishes between particles which are larger than 45 μm and fines or subsieve powder that are smaller than 45 μm. Table 3.3 gives the most pertinent data for both the Tyler standard and United States Sieve Series. The new US Series standard sieves adopted in 1970 are preferred. The old US Series and equivalent sieves manufactured by other companies, such as Tyler, may also be used if the new US Series is not available. Care should be taken to make sure that sieve opening sizes are correct when standardisation work is performed. Micromesh sieves are also available with openings down to at least 5 μm, but are rather difficult to use

Table 3.3 US standard series test sieves and equivalent Tyler standard sieves

Mesh designation number	Sieve opening, μm		
	New US Series	Old US Series	Tyler Series
20	850	841	833
35	-	-	417
40	425	420	-
60	250	250	295
80	180	177	175
100	150	149	147
140	106	105	-
150	-	-	104
200	75	74	74
230	63	63	-
250	-	-	63
325	45	44	44

and maintain; they are fragile and have low load capacities. These sieves are produced by electrodeposition of nickel or copper on photo-sensitised machine-ruled lines.

Another advantage of screening is that any desired distribution, according to the type of blend required by the manufacturer, can be synthesised. Conventional sieving requires a sample of 50 g for accurate analysis and this becomes difficult in analysis of expensive metal powder. The development of small sieves has overcome this problem. A Ro-tap type machine,which is a common sieve shaker, is used in size analysis. The choice of Sieve shaker and shaking mechanism depends on the shape of powder. A swirling motion is effective for round particles but is extremely slow for elongated particles; a jumping motion is effective for elongated particles because it throws them into the air, allows them to rotate and perhaps land point first to pass through the hole. Thus consistent shaking motion is needed.

3.3.2 Permeability method

Another important industrial method for measuring the subsieve particle size is by what is known as the *Fisher subsieve sizer*. This is a common method used in refractory metal powder and cemented carbide industries. The measured surface area is converted into an equivalent spherical surface diameter, which is only an approximate measurement tool. The technique does not measure the surface-connected porosity. A pre-weighed amount of powder is exposed to a known flow rate and the pressure drop is measured to determine permeability. From a knowledge of powder porosity and theoretical density, the surface area is calculated. The advantage of this method is that it is a direct reading one, which is convenient for quick industrial quality control of fine powders. It must be clearly recognised that the value of specific surface obtained from a permeability experiment is representative of the 'friction' surface presented by powder mass to the flowing fluid.

In the Fisher subsieve sizer, values related to porosity and the air permeability of a bed of powder are determined. A value related to porosity is determined by shifting the calculator chart laterally. The value related to permeability is determined by reading the liquid level of a manometer tube. From the porosity and permeability of the powder bed, the specific surface area of the powder is derived using Kozeny–Carman relationship as follows:

$$S_v = \sqrt{\frac{\varepsilon^3}{\kappa(1-\varepsilon)^2 p}} \qquad (3.1)$$

where, S_v is the volume specific surface area m^2/m^3 or m^{-1}, ε is the porosity of the bed equal to $1 - M/AL\rho$, where M is the mass in kg, A the cross-sectional area in m^2, L the length of the bed in m, and ρ the density of powder in kg m^{-3}; κ is the the Kozeny-Carman number, taken as 5.0, and p is the the permeability of the bed in m^2 equal to $Q.\eta.L/A\Delta P$ with Q the volume rate flow in m^2 s and η the viscosity in Pa s, and ΔP the pressure drop in Pa.

For a powder in which all particles are spherical and of uniform size, the particle size d in micrometres may be calculated from the volume specific surface area, S_v:

$$d = \frac{6 \times 10^6}{S_v} \qquad (3.2)$$

Calculation of the 'average' particle size based on this equation is performed automatically by the calculator chart of the Fisher subsieve sizer from the values related to porosity and permeability of the powder bed measured by the instrument. In other words, what is determined with the instrument is the specific surface of the powder. If desired, the average particle size may be converted into a volume specific surface area (S_v) in m^{-1} or a mass specific surface (S_m) in $m^2 kg^{-1}$ using the following equations:

$$S_v = \frac{6 \times 10^6}{d} \qquad (3.3)$$

$$S_m = \frac{6 \times 10^6}{\rho d} \qquad (3.4)$$

where, d is particle size in μm and ρ is true density of the powder in kg m^{-3}.

3.3.3 Light scattering method

A prepared sample of particulate material is dispersed in water or a compatible organic liquid, and circulated through the path of a light beam or some other suitable light source. A dry sample may be aspirated through the light in a carrier gas. The particles pass through the light beam and scatter it. Photodetector arrays collect the scattered light that is converted to electrical signals, which are then analysed in a microprocessor. The signal is converted to a size distribution using Fraunhofer diffraction or Mie scattering, or a combination of both. Scattering information is analysed assuming a spherical model. Calculated particle sizes are therefore presented as equivalent spherical diameters. The MicroTrac particle size analyser uses the scattering of a laser light beam by particles in a flowing stream. Fraunhofer diffraction—which uses low-angle forward light scattering—is used for measuring the particle sizes ranging from 2 to 2000 µm. Other scattering techniques can be used for determining the particle size distributions down to 0.1 µm.

3.3.4 Sedimentation method

A carefully dispersed homogeneous suspension of the powder is permitted to settle in a cell that is electronically programmed to move downwards with respect to a fixed, well-collimated beam of X-rays of constant intensity. The net X-ray signal is inversely proportional to the sample concentration in the dispersing medium, and the particle diameter is related to the cell position. Cumulative mass per cent versus equivalent spherical diameter is simultaneously and instantaneously plotted on an x–y recorder to yield a particle size distribution curve.

3.3.5 Turbidimetry method

A uniform suspension of the powder in a liquid medium is allowed to settle in a glass cell. A beam of light is passed through the cell at a level with a known vertical distance from the liquid level. The intensity of the light beam is determined using a photo cell. This intensity increases with time as sedimentation of the dispersion takes place.

The times at which all particles of a given size settle below the level of the transmitted light beam are calculated from Stoke's law for the series of sizes chosen for the particle size analysis.

The intensity of the light beam at these times is measured as per cent of the light transmitted through the cell with clear liquid medium. The size distribution in the powder can be calculated from these relative intensities using the Lambert–Beer law in the modified form:

$$\Delta W_{1-2} = d_m (\log I_{d1} - \log I_{d2}) \tag{3.5}$$

where, I_{d1} and I_{d2} are the intensities measured at times when all particles with diameters larger than d_1 and d_2, respectively, have settled below the level of the light beam, d_m is the arithmetic mean of particle sizes d_1 and d_2. ΔW is determined for each of the particle size ranges chosen. The sum of these values is $T\Delta W$. The weight per cent of particles in the size range d_1 to d_2 can then be calculated as:

$$\text{weight\%} = \left(\frac{\Delta W_{1-2}}{\sum \Delta W} \right) \tag{3.6}$$

Table 3.4 summarises the relative advantages and disadvantages of different particle size analysers.

Table 3.4 Advantages and disadvantages of particle size analysers

Method (size range)	Advantages	Disadvantages
Particle counting (nm to mm)	• extremely accurate • reproducible • direct measurement • inexpensive • can be automated	• time consuming if done manually • sample size may not be representative of powder lot
Sieve analysis (44 μm to several mm)	• widely used • inexpensive • can handle large samples • easy to use	• no information on particle sizes between cuts • particle shape can greatly skew results • problem with agglomeration and adhesion to screen • must be calibrated periodically to check wear/damage
Fisher subsieve analyser (0.2 to 25 μm)	• widely used in certain industries • easy to use • inexpensive	• does not give true particle size • assumes a spherical particle shape • no information on particle size distribution
Laser diffraction (0.1 μm to several mm)	• fast • reproducible • easy to use • many instruments commercially available • sample does not have to be run in conjunction with a known standard to interpret results • effective for statistical process control	• expensive • care must be taken with powder dispersion • complicated algorithms involved in interpreting the diffracted signal
Traditional sedimentation devices (non-centrifuge; 1 μm to 50 μm)	• widely used in certain industries • does not need to be calibrated against a standard sample to interpret data • relatively easy to use	• slow • limited particle size range • care must be taken with powder dispersion • accuracy
Centrifugal sedimentation (10 nm to 300 μm)	• can measure submicron particles • analysis time is greatly reduced from non-centrifugal methods • does not need to be calibrated against a standard sample to interpret data • easy to use	• expensive • limited number of commercial instruments • very complicated algorithms • difficult to verify results

3.3.6 Crystallite size measurement

Fine size powders, particularly of brittle solids, are produced by mechanical milling methods (Chapter 2). The fine particulate material contains small size crystallites in it. The crystallite size can be measured through optical/electron microscopic methods. It can also be estimated by X-ray diffraction studies. The details of this method are given in Chapter 11.

3.4 PARTICLE SURFACE TOPOGRAPHY

The nature of the surface of individual particles is also an important powder characteristic. A spherical particle may appear smooth, but on closer examination at high magnification the surface may actually consist of many protuberances. Reduced metal powder has a highly roughened surface. Atomised/metal powders, on the other hand, have finer degree of surface roughness, which is rounded rather than sharp and irregular. The scanning electron microscope is a powerful tool for examining surface topography. Surface contamination of particles and agglomeration of fine particles can also be studied by this technique.

The exact nature of surface topography will influence the frictional forces between particles. These are important in the case of bulk movement of the particles, when the powder is flowing, settling or during compaction. The extent of actual particle-to-particle contact during sintering will also be affected by the nature of surface roughness. Chemical reactivity of the powder also tends to increase with increasing surface roughness, especially the irregular type.

3.5 SURFACE AREA

The actual amount of surface area per unit mass of powder is of great significance. Any reaction between the particles or between the powder and its environment starts at these surfaces. This affects sinterability. For highly irregular-shaped particles with a high degree of surface roughness, the specific surface area can be very high.

The surface area of a given powder is measured by the BET method, in which adsorption of a species in solution is used to obtain a value of specific surface (S_w) if the surface is completely covered by a monomolecular layer of the solute. From a knowledge of the area occupied by one molecule, the total area of the powder sample, and finally, S_w can be obtained. The amount of gas adsorbed in a monomolecular layer in square metres is calculated from an adsorption isotherm, i.e., a series of measurements of the volume of gas adsorbed as a function of pressure.

The BET method of determining the specific surface is widely used for catalysts. It is used primarily for very fine powders, particularly those of the refractory metals, and for characterising the total surface area of porous powders.

3.6 TRUE, APPARENT AND TAP DENSITY

The *true density* of a given powder is also expressed as *pycnometric density*, since the true volume is determined by a gas pycnometer. The schematic is shown in Fig. 3.4. Two precision

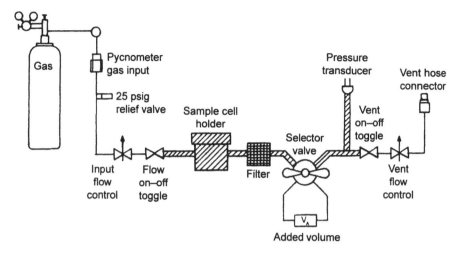

Fig. 3.4 Schematic showing the functioning of the gas pycnometer

cells of known volume are used. There is a pressure release valve between the cells. A weighed sample of powder is placed in the first cell, with the second at ambient pressure. The pressure in the first cell is increased to about 0.1 MPa (15 psi) over ambient, and the relief valve is opened, equalising the pressure between the cells. The ideal gas law is assumed to hold, and this permits the true value of the powder (V_t) to be calculated as:

$$V_t = \frac{V_C - V_A}{(1 - P_2/P_3)} \tag{3.7}$$

where, V_c is the volume of the empty powder test cell, V_A is the volume of the pressure release cell, P_2 is the increased test cell pressure, usually 0.1 MPa (15 psi) and P_3 is the final pressure in the system after the relief valve has been opened.

The accuracy of determination is within 0.1% to 0.2%. Helium is generally used as the pressuring gas to approximate ideal gas behaviour. However, other gases such as nitrogen may be used in most cases without causing major error.

The *apparent density* of a powder refers to the mass of unit volume of loose powder usually expressed in g cm^{-3}. It is one of the critical characteristics of a powder for the following reasons:

- It determines the size of the compaction tooling and the magnitude of press motions necessary to compact and densify the loose powder.
- It determines the selection of equipment used to transport and treat the initial powder.
- It influences the behaviour of the powder during sintering.

Other characteristics which have direct bearing on apparent density are the density of the solid material, particle size and shape, surface area, topography and its distribution.

Apparent density is determined by the *Hall flowmeter*, where a container of known volume (25 cm^3) is completely filled by flowing metal powder through a Hall funnel (Fig. 3.5).

Another method to determine apparent density of a powder is by *Arnold meter* (ASTM Standard B703). The method consists of sliding a bushing filled with powder over a hole in a steel block, filling the hole with powder, collecting and weighing the powder, and calculating its apparent density. The steel block has a hole in the centre with diameter 31.664 ± 0.0025 mm and thickness 25.4 ± 0.0025 mm that corresponds to a volume of 20 cm³. The bushing is approximately 38 mm long with an inside diameter of 38 mm and outside diameter of 45 mm. The density is reported nearest to 0.01 g cm⁻³.

Often a mass of loose powder is mechanically vibrated or tapped. The density of the loose powder increases due to this treatment and is always higher than the apparent density. The greatest increase in density occurs during the

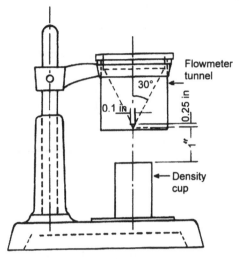

Fig. 3.5 Schematic of Hall flowmeter used for apparent density measurement

initial tapping period and eventually the density becomes constant. The final stable density is the value reported as the *tap density* (Fig. 3.6). The amount of increase in density due to tapping depends on the extent of original frictional forces between the particles: greater the frictional conditions in the original powder (small sizes, irregular shapes and roughened surface), greater the increase in density due to tapping.

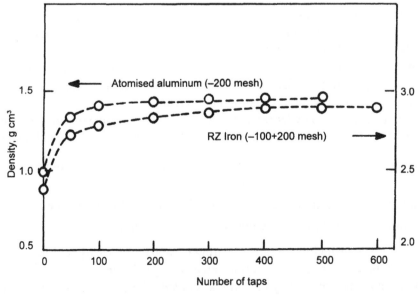

Fig. 3.6 Effect of number of tappings on the density of atomised aluminium and reduced iron powders

For fine powders (in the submicron range down to 0.2 μm), the measurement of tap density by constant mass can yield unreliable results. Some powder is lost during transfer from the weighing dish to the graduated cylinder. In addition, some of the powder may adhere to the cylinder wall and not be included in the measured volume. The relatively high aspect ratio of the graduated cylinder increases particle-to-cylinder wall friction, inhibiting migration of the particles to their final packed position. Since these fine powders are not 'free-flowing', they do not pack evenly, often resulting in an uneven top surface of the packed column. Thus, visual interpretation of the volume is required. To address these problems, ASTM International Subcommittee B09.02 is developing a proposed standard (WK 13023) for the determination of the tap density of very fine powders, based on a constant volume measuring method. This approach reduces or eliminates the problems cited for constant mass. The low aspect ratio of the density cup reduces friction. The mass of the sample is weighed after tapping with no loss of powder. Because the tapped height of the column is constant, no visual interpretation is necessary.

Table 3.5 gives typical data of apparent and tap densities of iron powder produced by different methods.

In this section, it is also worthwhile to talk about the density of a *slip*. The description of slip casting will be given in Chapter 6. In case one intends to measure the density of a previously prepared slip, the simplest method is to use a hydrometer—a closed, weighted tube of fixed displacement that will partially submerge in a fluid. Lower the density of the slip, deeper the level of submergence of the hydrometer. The depth of submergence is read on the stem of the hydrometer, which is graduated directly in specific gravity units. In case the slip is prepared

Table 3.5 Apparent and tap densities of various powders

Material	Apparent density, g cm^{-3}	Tap density, g cm^{-3}	Per cent increase
Copper*			
spherical	4.5	5.3	18
irregular	2.3	3.14	35
flake	0.4	0.7	75
Iron (−100 + 200 mesh)			
electrolytic	3.31	3.75	13
atomised	2.66	3.26	23
sponge	2.29	2.73	19
Aluminium (−200 mesh)			
atomised	0.98	1.46	49

*All copper powders with same size distribution;

Source: Poster AR, (ed.), *Handbook of Metal Powders*, Reinhold, New York, 1966.

by mixing solid and liquid, the density of the slurry is given by the expression:

$$\text{Density of slurry} = \frac{W_S + W_L}{W_S/\rho_S + W_L/\rho_L} \qquad (3.8)$$

where, W_S and W_L are the respective weights of insoluble powder and the liquid, and ρ_S and ρ_L are the densities of these two materials.

3.7 FLOW RATE

A rapid rate of production of PM parts requires a relatively rapid flow of powder from storage containers to dies. The standard method of determination is by the Hall flowmeter, where the time in seconds necessary for 50 g of powder to flow through a prescribed small orifice is measured. The test offers only a means of comparison and evaluation, because, in the majority of operating conditions the powder does not have to flow through a small orifice. Therefore, flow times are proportional to the reciprocal of the flow rates. Very fine powders do not flow through a small orifice. This is a result of the drastic increase in the specific surface area as the size becomes very small. For a given powder, higher the apparent density, lower the flow time. When a fine size powder is mixed in a coarse powder, because of the increase in the apparent density, the flow time is decreased, irrespective of whether the particles are irregular-shaped or spherical. However, in case of addition of irregular-shaped powder, an amount is reached for which no flow behaviour is observed. This corresponds to the presence of an excessive amount of frictional surface area.

3.8 COMPRESSIBILITY

Compressibility is a measure of how much a powder will compress or densify on application of external pressure. Compressibility is reported as the density in g cm^{-3}, rounded to the nearest 0.01 g m^{-3}, at a specified compaction pressure, or as the pressure needed to reach a specified density. Typically, a cylinder or rectangular test piece is made by pressing powder in a die, with pressure applied simultaneously from above and below.

Compressibility of the powder is influenced by factors such as, inherent hardness of the concerned metal or alloy, particle shape, internal porosity, particle size distribution, presence of non-metallics, addition of alloying elements or solid lubricants.

Alternatively, compressibility is defined in terms of the densification parameter.

$$\text{Densification parameter} = \frac{\text{green density} - \text{apparent density}}{\text{theoretical density} - \text{apparent density}} \qquad (3.9)$$

In general, compressibility increases with increasing apparent density. A rather large amount of densification occurs at relatively low compaction pressure. Another term, which is very important for tooling design, is the compression ratio. It is the ratio of the volume of loose powder to the volume of the compact made from it. A low compression ratio is desirable for the following reasons:

- Size of the die cavity and tooling can be reduced
- Breakage and wear of tooling is reduced
- Press motion can be reduced
- A faster die fill and thus a higher production rate can be achieved.

3.9 GREEN STRENGTH

Green strength is the mechanical strength of a green, i.e., unsintered powder compact. This characteristic is very important, as it determines the ability of a green compact to maintain its size and shape during handling prior to sintering.

Green strength is promoted by:

- increasing particle surface roughness, since more sites are available for mechanical interlocking;
- increasing the powder surface area by increasing the irregularity and reducing the particle size;
- decreasing the powder apparent density—a consequence of the first two factors;
- decreasing particle surface oxidation and contamination;
- increasing green density (or compaction pressure);
- decreasing the amount of certain interfering additives, e.g., the addition of small alloying elements, such as soft graphite to iron and lubricant, which prevents mechanical interlocking.

The standard green strength test is a transverse bend test of a 12.7 mm by 31.7 mm (0.50 by 1.25 inch) rectangular specimen of thickness 6.35 mm (0.25 inch). It is the stress calculated from the flexure formula, required to break the specimen.

$$\text{Green strength} = \frac{3PL}{2wt^2} \ (\text{N/mm}^2) \tag{3.10}$$

where, P is the breaking load in N, L is the distance between the supporting rods in mm, t is the specimen thickness in mm, w is the width of the specimen in mm.

Table 3.6 shows the relationship among green strength, apparent density, compacting pressure and green density for several types of iron powders.

3.10 PYROPHORICITY AND TOXICITY

Pyrophoricity is a potential danger for many metals—including the more common types—when they are in a finely divided form with large surface area-to-volume ratios. The toxicity of powder is normally related to inhalation or ingestion of the material and the resulting toxic effect. The chemical reactivity of a material increases as the surface area-to-volume ratio

Table 3.6 Green density and green strength for various types of iron powders

Powder	Apparent density, g cm^{-3}	Compaction pressure		Green density, g cm^{-3}	Green strength	
		MPa	tsi		MPa	psi
Sponge*	2.4	415	30	6.2	14.41	2100
		550	40	6.6	22.05	3200
		690	50	6.8	28.25	4100
Atomised sponge**	2.5	414	30	6.55	13.09	1900
		550	40	6.8	18.80	2700
		690	50	7.0		
Reduced*	2.5	415	30	6.5	15.85	2300
		550	40	6.7	20.67	3000
		690	50	6.9	24.11	3500
Sponge*	2.6	415	30	6.6	18.60	2700
		550	40	6.8	24.80	3600
		690	50	7.0	26.87	3900
Electro†	2.6	415	30	6.3	31.69	4600
		550	40	6.7	42.72	6200
		690	50	6.95	53.74	7800

* powders containing 1% zinc stearate

** powders containing 0.75% zinc stearate

† unlike the other powders, this one was isostatically pressed

Source: Buren CE and Hirsch HH, In: *Powder Metallurgy*, Interscience, New York.

increases. For this reason, fine particles of many materials combine with oxygen, ignite and result in explosive conditions.

Powder production methods and characteristics

After going through Chapter 2 (Powder Production) and the current chapter, readers will be able to appreciate the intimate relationship between production methods and powder characteristics. Such a close relationship is not so prevalent in metal production methods, other than powder production. Table 3.7 summarises different types of characteristics, already described in the earlier sections, along with different metal powder production routes.

Table 3.7 Some characteristics of metal powders made by various commercial methods

Method of production	Typical purity (est.)	Particle characteristics		Compressibility	Apparent density	Green strength
		Shape	Meshes available			
Atomisation	High 99.5+	Irregular to smooth, rounded dense particles	Coarse shot to 325 mesh	Low to high	Generally high	Generally low
Gaseous reduction of oxides	Medium 98.5 to 99.0+	Irregular, spongy	Usually 100 mesh and finer	Medium	Low to medium	High to medium
Gaseous reduction of solutions	High 99.2 to 99.8	Irregular, spongy	Usually 100 mesh and finer	Medium	Low to medium	High
Reduction with carbon	Medium 98.5 to 99.0+	Irregular, spongy	Most meshes from 8 down	Medium	Medium	Medium to high
Electrolytic	High 99.5+	Irregular, flaky to dense	All mesh sizes	High	Medium to high	Medium
Carbonyl decomposition	High 99.5+	Spherical	Usually in low micron ranges	Medium	Medium to high	Low
Grinding	Medium 99.0+	Flaky and dense	All mesh sizes	Medium	Medium to high	Low

SUMMARY

- Powder sampling is essential before any characterisation.
- Impurities in metal powders are present in elemental or compound form.
- The chemical composition of a crystalline powder is determined by X-ray or spectrographic technique, apart from conventional wet analysis.
- The shape of the powder particles is dependent on the production method.
- The microstructure of the powder has significant influence on the behaviour of the powder during compaction or sintering and on the end properties of the product.
- Particle size is not a concise quantity for a given non-spherical particle.
- Particle size of a powder is influenced by the physical principles employed by each method of particle size analysis.
- Surface topography of the powder influences frictional forces between particles.
- BET method is a common method to measure the surface area of a given powder.
- The apparent density of a powder refers to the mass of unit volume of loose powder. It influences the behaviour of the powder during sintering.
- The compressibility of a powder increases with increasing apparent density.

Further Reading

Allen T, *Particle Size Measurement*, Wiley-Interscience, New York, 1981.

Gregg SJ and Sing KSW, *Adsorption, Surface Area and Porosity*, Academic Press, New York, 1987.

Wachtman JB, *Characterization of Materials*, Butterworth-Heinemann, Oxford, 1973.

EXERCISES

3.1 What are the distinctive features of any chemical analysis by X-ray and spectrographic method?

3.2 A specimen of an unknown Al-alloy powder premix was subjected to spectrographic analysis. Observation of the spectrogram revealed lines at the following wavelengths: 643.8, 518.4, 517.3, 481, 472.2, 468, 383.8, 382.9, 361.1, 346.6 and 340.3 nm, plus many lines of aluminium. Besides aluminium, what other elements are present?

3.3 Nitrogen gas is adsorbed on an alumina powder at 77.4 K. The area occupied by an adsorbate molecule is 0.163 nm^2. At STP, the volume of a mole of gas is 2.24×10^4 cm^3 and the volume of gas required for monolayer adsorption on 1 g of powder is 2.52 cm^3. Calculate the specific surface area of the powder.

3.4 Calculate the mean density of a powder composed of 90 wt% alumina ($D_a = 3.98$ g cm^{-3}) and 10 wt% zirconia (6.03 g cm^{-3}).

3.5 Two grades of tungsten powder gave specific surface areas as 0.3 and 0.14 m^2 g^{-1}. What are the equivalent surface diameters?

3.6 Calculate the mass and number of formula units for alumina particles with diameters 10^{-2}, 10^{0} and 10^{2} μm.

3.7 Give reasons why the flow time in the Hall flowmeter is observed to increase with the ratio of the tap density to apparent density? Is the statement true for all types of free-flowing powders?

3.8 The sieve analysis of a sponge iron powder carried out by two laboratories was found to differ widely. What could be the possible reasons for such differences?

3.9 Under what conditions would the surface areas of a powder measured by gas permeability and gas adsorption methods be similar?

3.10 A vendor supplied fine metal powder to a MIM production unit. The particle size of the powder was tested by the vendor using laser diffractometry. Describe what the test report must contain.

3.11 A nano-scale metal powder is to be tested for its particle size. Which one of the following methods will be preferred for more accurate results? Give reasons.

(a) X-ray diffraction

(b) Laser diffraction

(c) BET surface area.

4

Powder Treatment

LEARNING OBJECTIVES

- The need to anneal metal powders
- How to blend powders
- Role of lubricant/binder additives in powder
- When to coat a metal powder
- How to coat using electrolysis and hydro-metallurgical routes
- Why a metal should be degassed
- How to degas a metal powder

Powder treatment is an essential part of PM processing, before pressing or forming operation. The main purpose is to make the powder amenable to forming. For example, if the powder is hard, it should be softened by annealing. In a multi-component metallic system, concurrent diffusion alloying of different particle species may be permitted during annealing. Sometimes, particles in a particular range of size are required, for which screening is adopted. Impurity particles in raw materials and those introduced during processing—from the abrasion of materials handling and processing machinery—are sometimes removed, based on the difference in size, density, surface behaviour or electromagnetic properties. Mixing is another treatment, where more than one type of powder is mixed with solid lubricant to improve the chemical and physical uniformity of the mixture. Often, the bulk powder containing admixed binder has agglomerates varying in size, shape and density. They do not flow well, pack densely or compact into a uniform microstructure. During granulation treatment, satisfactory feed material consisting of controlled agglomerates called *granules* is produced. Often, coating of a metal or ceramic powder becomes necessary. In some metals, particularly in the case of aluminium powder, degassing of the powder is also desirable.

4.1 PARTICLE SEPARATION

Sizing involves the separation of particles to achieve:

- separation in a relatively narrow or discrete range of sizes
- elimination of extremely coarse or fine sizes.

Screens are commonly used for the continuous separation of particles up to 25 μm in size. Preliminary details of the standard sieve sizes and sieving is given in Chapter 3. Both dry forced air screening and sonic screening are used for dry separation. Production models may be configured in series to obtain multiple size cuts.

Cyclone and centrifugal techniques are commonly used for particle sizing in the subsieve range of size. Cyclones, which are inertial separators, are used for particle separation in ceramic processing. They are relatively simple in construction and have no moving parts. The feed suspension enters the cyclone tangentially under pressure and circulates in a spiral path. The centrifugal force on a particle is opposed by the inward directed drag force of the fluid.

Hydro-cycloning, filtration and centrifuging are used to concentrate particles in slurry. Washing of powder mass is carried out to remove soluble substances. This is commonly done for metal powders obtained through electrolysis. Magnetic separation and other separation processes are used to remove impurity particles and to concentrate particles of different density and surface behaviour.

4.2 ANNEALING AND DIFFUSION ALLOYING

The powder is delivered to the fabricator in a state ready for mixing with lubricant and alloying additions, if any. The aims of annealing are:

- to soften the powder
- to reduce the residual amount of oxygen, carbon and nitrogen in the powder.

The annealing operation is carried out at elevated temperatures in an atmosphere furnace or a vacuum furnace. The former may be of batch or continuous type. The furnace construction is similar to that given in Chapter 8, which deals with sintering technology. Annealing temperatures are kept as low as possible to minimise sintering of powder.

In diffusion alloying or bonding the additive element is diffused at the surface in the parent metal particles at a temperature depending on the characteristics of the latter. In case of diffusion bonding over iron powder, the temperature is in the range 600–1000°C. At increasing temperature, bridges are formed between the particles of host and additive elements. This prevents segregation in the powder premix during handling. Partial pre-alloying of iron powder is ideal for those elements, e.g., copper, nickel and molybdenum, which unlike carbon and phosphorus, have relatively low diffusion rates. Höganäs Company, Sweden, where such iron powders originated, called them 'Distaloy'. Since fully pre-alloyed powders are less compressible due to solid solution strengthening, the partially pre-alloyed powders ensure reasonable compressibility. Figure 4.1 schematically illustrates premix, partially pre-alloyed and pre-alloyed powders. Coating is another way of alloying, which will be described in a later section of this chapter.

Sometimes, the conventional diffusion alloying is accomplished by starting with additive metals that are in the form of their oxides. Such powders were found to have better sintering activity and

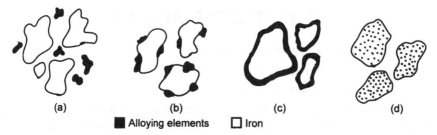

■ Alloying elements   □ Iron

Fig. 4.1 Schematic illustration of powders: (a) Mixed powders, (b) Partially pre-alloyed powder, (c) Coated powders and (d) Completely pre-alloyed powder

extremely fine distribution. In this process, heat treatment combines the operation of alloying as well as reduction. Sometimes, the original powder is a pre-alloyed powder (e.g., Fe-Ni) over which diffusion bonding with high diffusivity alloying addition, i.e., molybdenum or copper, is accomplished.

4.3 POWDER MIXING

The term 'blending' is strictly applied to a one-component operation; mixing involves more than one type of powder, e.g., mixing of solid lubricant with a metal powder or powders of several other metals. Sometimes the additive acts as a lubricant as well as an alloying addition, e.g., graphite in iron powder.

The variables in the powder mixing process are:

- Type of mixer
- Volume of the mixer
- Geometry of the mixer
- Inner surface area of the mixer
- Constructional material and surface finish of the mixer
- Volume of the powder in the mixer before mixing
- Volume of the powder in the mixer after mixing
- Volume ratio of mixer to powder
- Characteristics of component powders
- Type, location and number of loading and emptying devices
- Rotational speed of mixer
- Mixing time
- Mixing temperature
- Mixing medium (gaseous or liquid)
- Humidity, when mixing in air.

Example 4.1: Refer Metals Handbook for the following Ni-based binary phase diagrams: (a) Cu-Ni, (b) W-Ni and (c) Mo-Ni. Select an alloy on the Ni-rich side of each system and subject it to solid-state homogenisation treatment. Describe the distinctive features and the resultant microstructures.

Solution

The binary phase diagram of the three systems is shown in Figs. Ex. 4.1a, 4.1b and 4.1c.

1. Ni-Cu phase diagram exhibits complete solid solubility of the components Cu and Ni and is therefore a single phase system. The homogenisation proceeds by levelling the concentration–distance profiles. With increase in time, the solid solution will be generated in higher volume fraction.

2. Ni-W phase diagram contains two terminal solid solutions, separated by a two-phase field. The Ni and W premix undergoes interdiffusion during homogenisation, but the two phases always exist. As the selected composition lies in the Ni-rich field, the W-rich phase would dissolve completely, with subsequent homogenisation taking place by levelling of the concentration gradients in the Ni-rich phase.

3. The Ni-Mo phase diagram consists of the Ni-rich solid solution, the Mo-rich solid solution and the Mo-Ni intermetallic phase. In the initial stage of homogenisation, Mo-Ni phase layer falls at the interface between the dispersed Mo particles and the Ni matrix. The subsequent stage would depend on the composition. Since, in the present case, the composition lies in the Ni-rich solid solution field, the Mo-rich phase will dissolve, creating a two phase homogenisation situation.

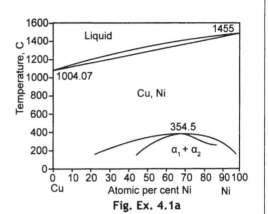

Fig. Ex. 4.1a

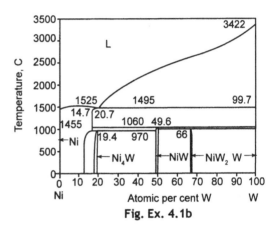

Fig. Ex. 4.1b

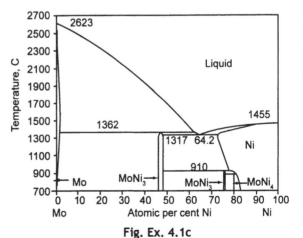

Fig. Ex. 4.1c

Mixing efficiency is best when the powder volume is about 50%–60% of the mixer volume. Optimum mixing time may be between 5 to 30 min, but this can be determined only by experience with a given mixture in a particular mixer. The aim is to mix the powders only as long as it is necessary to achieve a thorough mix and to fix a uniform apparent density of the mix from batch to batch. The apparent density of the mix tends to increase with mixing time.

During mixing, segregation of particles may occur, if they are of different sizes and densities. Particle segregation can occur by one or more of five primary mechanisms in powder mixing. These mechanisms include sifting, entrainment of air (fluidisation), particle velocities along a surface, entrainment of a particle in an air stream, and, to a lesser extent, dynamic effects such as electrostatic separation. The occurrence of these mechanisms depends on the physical properties of the powder, the flow rate of the powder, and the type of handling equipment and surfaces being used.

At times, alloying through the premix route becomes advantageous. Firstly, the alloying can be affected from readily available powders and the compositional adjustment can be readily achieved by varying the proportions of powders in the blend. Secondly, there is a better control of microstructure. For example, through proper control the formation of any intermetallic—which may adversely affect powder compaction—can be avoided.

4.3.1 Lubricant additive

The main function of a lubricant is to minimise friction between powder particles, and between particles and die wall; which would otherwise adversely affect powder compressibility. The lubricant is added during the mixing stage. Stearates of Al, Zn, Li, Mg, Cr and Ca are the common lubricants, which in metal powders are about 0.5–1.5 wt% of the mix. The mean particle size is in the range 10–30 μm. Besides stearates, other lubricants are waxes and cellulose additives. Table 4.1 gives the properties of some stearate lubricants. In ceramics, apart from stearates, wax and talc, sometimes even clay is used as a lubricant.

Mixing time plays an important role in the resultant powder mass: larger the mixing time, greater the apparent density, an effect which is often desired. On the other hand, the fine branches on the powder surface are damaged during the mixing process. The result is often an unexpectedly great drop in green strength. Figure 4.2 illustrates the variation in the apparent densities for premix, partially pre-alloyed and pre-alloyed bronze (90/10) powders. All the powders had 0.5% solid lubricant. The green strength variation of the compacts out of similar powders is shown in Fig. 4.3.

Table 4.1 Properties of some common stearate lubricants

Common type	Oxide	Per cent oxide	Softening temp, °C	Melting temp, °C	Density, g cm^{-3}
Zinc stearate	ZnO	14	100–120	130	1.09
Calcium stearate	CaO	9	115–120	160	1.03
Lithium stearate	Li$_2$O	5	195–200	220	1.01

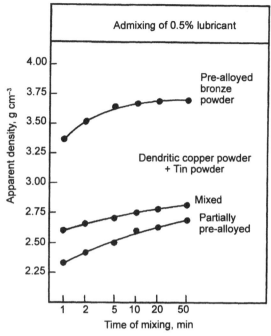

Fig. 4.2 Effect of mixing time on the apparent density of premix, partially pre-alloyed and pre-alloyed 90/10 bronze with 0.5% admixed lubricant

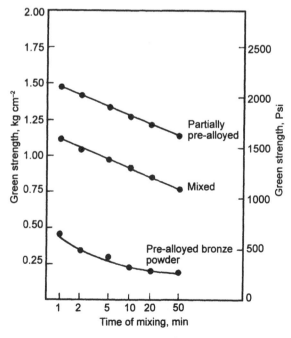

Fig. 4.3 Effect of mixing time on the green strength of compacts prepared out of premix, partially pre-alloyed and pre-alloyed 90/10 bronze

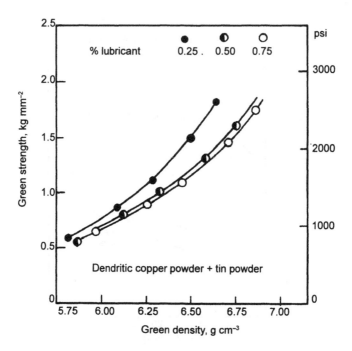

Fig. 4.4 Effect of different quantities of lubricants on the green strength of compacts prepared out of premix 90/10 bronze powders

The effect of different quantities of lubricant on the green strength of 90/10 bronze premix compacts is shown in Fig. 4.4. A relatively large quantity of lubricant hampers the interlocking between particles and hence a drop in green strength occurs.

Sometimes, milling is also carried out to increase the homogeneity of non-uniform mixtures. In addition, this also increases the chemical reaction rates during the subsequent processing operation. The actual requirements of a suitable size reduction process are extremely varied and depend on several parameters. A detailed description of size reduction was given in Chapter 2.

4.3.2 Binder additive

The main purpose of binders is to provide strength to the moulded green body. Various polymers and waxes are used as binders. These are polyvinyl alcohol, polyethylene glycol, paraffin wax or water soluble waxes. All except paraffin wax are water soluble. Paraffin wax requires hydrocarbon solvent and is used extensively in processing cemented carbides. Table 4.2 lists some of the binders used in processing metal/ceramic powders.

The bonding efficiency of powder mixes is evaluated by the dusting test. 25 g of powder mix is pound into a 25 mm cylindrical tube and an air stream is allowed to flow into the tube at a rate of 6 L min^{-1} for 5 min. The flow of air is strong enough to partially fluidise the powder. The powder is analysed before and after the test and the dusting resistance of a specific component

Table 4.2 Composition of some binders used in processing metal/ceramic powders

Binder application	Composition, wt %
Extrusion	15% ammonium polyacrylate
	15% ammonium stearate
	45% methyl cellulose
	25% glycerine
	56% water
	25% methyl cellulose
	13% glycerine
	6% boric acid
Injection moulding	69% paraffin wax
	20% polypropylene
	10% carnauba wax
	1% stearic acid
	75% peanut oil
	25% polyethylene
	55% paraffin wax
	35% polyethylene
	10% stearic acid
Slip casting	4% sodium lignosulfonate
	95% water
	1% calcium nitrate
	93% water
	4% agar
	3% glycerine
Tape casting	3% ammonium polyacrylate
	75% water
	11% polyacrylate emulsion
	11% glycerine
	80% toluene
	13% polyethylene glycol
	7% polyvinyl alcohol

powder is determined by means of the relation:

$$\text{Dusting resistance} = \frac{\text{Weight after test}}{\text{Weight before test}} \times 100 \tag{4.1}$$

In Chapter 1, a brief introduction to powder injection moulding was given. Here the thermoplastic binder system used must be stable and repeatable. Because the green part undergoes considerable shrinkage, the volume ratio of binder to fine powder must be held within a tight tolerance. The goals of mixing are:

- to coat the particles with the binder
- to break up agglomerates
- to attain uniform distributions of binder and particle size throughout the feedstock.

For thermoplastic binders, mixing is performed at an intermediate temperature, where shearing is predominant.

In general procedure, the binder is placed in a temperature-controlled mixer and heated to its melting temperature. The blended metal powder is added to the molten binder and mixed for a prescribed period to achieve a uniform blend. The mixture is then cooled and removed from the mixer. The feedstock mass can be extruded and cut into pellets for easy feeding into the moulding machine. Throughout mixing, special care is needed to maintain a repeatable mixture. Weighing and temperature control must be precise, since, even a small change gives rise to variations in final dimensions after sintering. The common mixers used for mixing dry powders are not useful in the preparation of PIM feedstock, as the binder components require shear to cause molecular scale thinning and dispersal between the particles.

4.3.3 Equipments

Most metal or ceramic powder mixing and blending is carried out using rotating containers as shown in Fig. 4.5. The interior of such mixing devices is important in achieving the mixing efficiency. Though various types of mixers are available, the following are the most common for metal powders.

Double cone mixer: This consists of vertical cylinders with conical ends which rotate about a horizontal axis. This rotation imparts a continuous rolling motion which spreads and folds the powders as they move in and out of the conical area. This action thoroughly mixes the powders with little or no change in the size and shape of the individual particles. Figure 4.6 shows the double cone mixer flow pattern.

V Mixer: This is constructed by joining two cylinders of equal length into a 'V'. As the 'V' rotates about its horizontal axis, the powder charge splits and refolds. Figure 4.6 illustrates the V mixer flow pattern. Baffles, spinning blades and dividers are additional provisions to promote intermixing.

For the best mixing centrifugal forces must be low, but not that low that turbulence does not occur. A desirable rotational speed is one which balances gravitational and centrifugal forces. The optimum rotational speed N_o (in RPM) for a cylindrical-type mixer is given by the expression:

$$N_o = \frac{32}{\sqrt{d}} \tag{4.2}$$

where d is the drum diameter in metres.

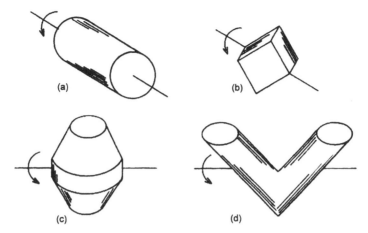

Fig. 4.5 Some common equipment geometries for mixing or blending powders: (a) Rotating cylindrical, (b) Rotating cube, (c) Double cone and (d) Twin shell

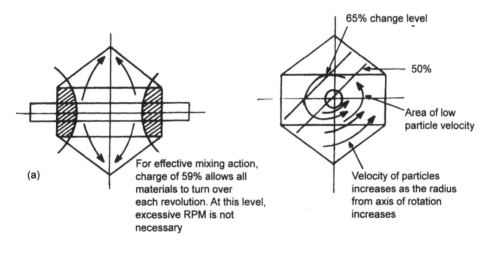

65% change level

50%

Area of low particle velocity

(a)

For effective mixing action, charge of 59% allows all materials to turn over each revolution. At this level, excessive RPM is not necessary

Velocity of particles increases as the radius from axis of rotation increases

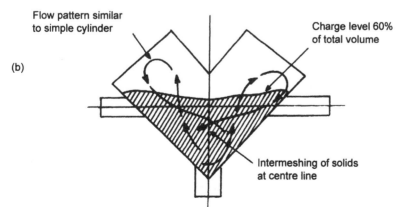

Flow pattern similar to simple cylinder

Charge level 60% of total volume

(b)

Intermeshing of solids at centre line

Fig. 4.6 Flow pattern in (a) double cone and (b) V-type mixers

The mechanism of powder mixing varies from mixer to mixer. For example, in the rotating mixer it is diffusional, in the screw mixer it is convective, and in the blade mixer it is shear mechanism. Figure 4.7 schematically illustrates these three mechanisms.

Figure 4.8 illustrates a continuous twin screw mixer for the PIM cycle. They have two overlapping screws that rotate in a figure eight–shaped barrel. These screws intermesh to constantly compress, shear, separate and smear the feedstock. Depending on the production rate, barrels of different diameters are used. Viscosity is the direct and convenient measure for the homogeneity. For example, strainless injection moulding feedstock has the following approximate features: powder 67 vol%, mixture density 5.6 g cm^{-3}, viscosity at 130°C 87 Pa s, and room temperature green strength 20 MPa.

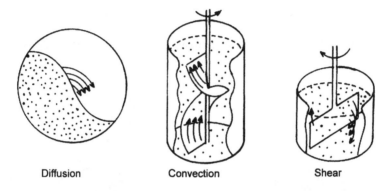

Diffusion   Convection   Shear

Fig. 4.7 Various modes of powder mixing (Reproduced with permission from German RM, *Powder Metallurgy Science*, 2nd edn, Metal Powder Industries Federation, Princeton, USA, 1994)

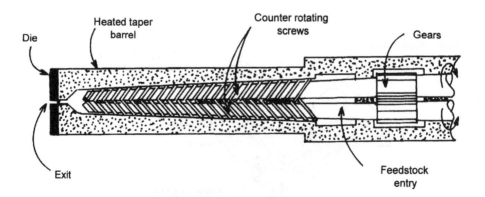

Fig. 4.8 A continuous twin screw mixer (Reproduced with permission from German RM, *Powder Metallurgy and Particulate Materials Processing*, Metal Powder Industries Federation, Princeton, USA, 2005).

Example 4.2: Compare the advantages of dry and wet ball milling.

Solution

The advantages of dry milling are:

- Avoids reaction of the powder with liquid medium
- Less media and lining wear than wet milling
- Can be started/stopped any time
- Easier to optimise.

The advantages of wet milling are:

- Low power requirement
- No dust problems
- Higher rotational speeds
- Good homogenisation
- Smaller particle size than dry milling
- Narrower particle size distribution than dry milling
- Compatible with spray drying and casting processes.

4.4 GRANULATION

Fine, hard particles of materials such as tungsten, molybdenum and WC-Co are not free flowing and are therefore difficult to press. The handling of such fine particles is also difficult. Consequently, large agglomerates are formed by granulation method. In order to achieve this, the continuous stirring of powder–organic slurry is used, while the volatile agent is removed by heating. Such a process is better suited for small batch sizes.

One of the better forms of processing the slurry is known as *spray drying*. The slurry is sprayed into a heated free fall chamber where due to surface tension spherical agglomerates are formed. Heating the agglomerate during free fall causes vaporisation of the volatile agent, producing a hard dense-packed agglomerate.

Three standard techniques are used to atomise slurry for spray drying:

1. Single-fluid nozzle atomisation
2. Centrifugal (rotating disc) atomisation
3. Two-fluid nozzle atomisation.

The largest agglomerate sizes (600 μm) are achieved by the single-fluid nozzle method. The centrifugal atomiser yields agglomerate sizes up to 300 μm, and the two-fluid nozzle produces agglomerates only up to about 200 μm in size.

Suitable binder materials must be homogenously dispersable (preferably soluble) in the liquid used to form the slurry. Plasticisers, e.g., ethylene glycol, may be used with binding materials that are hard or brittle and that tend to crack during drying. Suspending agents, e.g., sodium carboxymethyl cellulose, are needed to prevent solids from settling within the slurry. Deflocculating agents, e.g., sodium hexametaphosphate, aid in the formation of slurry by preventing the agglomeration of fine particles. Wetting agents, e.g., synthetic detergents, can also be used to maintain solids in suspension.

Figure 4.9 illustrates the schematics of centrifugal and single-fluid nozzle atomisers. In the former, the atomiser and the inlet for drying air are positioned at the top of the dryer. The atomised slurry has maximum liquid content when it encounters the laminar flow of hot incoming air. The maximum product temperature is relatively low and the evaporation time is relatively short. However, the product exists with moist air, and the exit temperature must be relatively high to obtain a dry product. In the nozzle atomiser, the slurry and drying air inlet are at opposite ends of the drying chamber. Partially dried droplets encounter the incoming hot air, and the product heating is greater. In the two-fluid nozzle atomisation, the residence time of the spray is increased and it provides a means for reducing the size of the chamber needed for drying.

The majority of the spray dried product is discharged through a rotary valve into interchangeable containers attached at the base. Product fines entrained in the exhaust air are separated using a cyclone or bag filter. Product fines may be used in some pressing operations, but are often recycled into the feed slurry. Periodic cleaning of the dryer using a liquid spray is required to remove the granulate crust. Abnormally large granules and flakes of granulate crust from the wall of the dryer are undesirable and are commonly removed by screening.

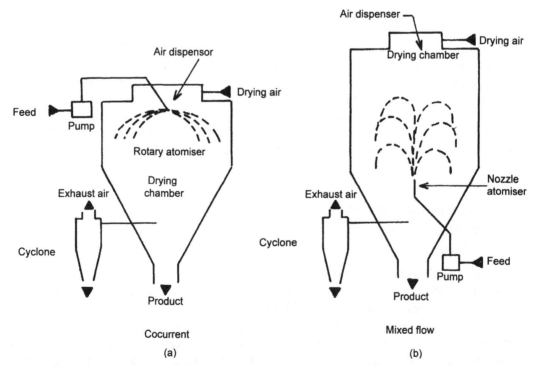

Fig. 4.9 Spray dryers: (a) Cocurrent spray-dryer with centrifugal atomiser and (b) Mixer flow spray-dryer with a nozzle atomizer.

Air bubbles present in the feed slurry or occluded during atomisation using a two-fluid nozzle may persist as relatively large pores in the granule. Granules with a large crater are often noticed in industrial spray-dried material. The tendency for these shapes is higher when the inlet temperature is relatively high, the binder content or molecular weight is relatively high and the weights of solids loading in the slurry are relatively low. Rapid surface drying and the formation of a surface of low permeability cause bursting of the granule due to gas pressure.

A major precaution to be taken in spray drying is that granules should not be so strong that they lose their identity during compaction. The popularity of spray drying as a means of using fine powder sizes in high productivity equipment is increasing. A disadvantage is that the organic binder must be removed in the sintering cycle.

Figures 4.10a to 4.10c show typical scanning electron microstructures of a spray dried W-15Cu (wt%) powder. The low magnification picture (Fig. 4.10a) indicates that the powders are coarse and spherical. This enhances the flowability of the powders into the die during compaction. The individual agglomerated granules consist of submicron-sized powders (Fig. 4.10b). It is also interesting to note that the agglomerated granules are hollow (Fig. 4.10c).

Granulation or pelletising of powder injection moulding feedstock is an operation which may be kept optimal. There are two reasons for such operation: to prepare easily transported clusters of powder and binder, and to incorporate recycled material back into the moulding process. In some cases, the mixing and moulding equipment are combined and the pelletisation (granulation) step can be avoided. The

(a)

(b)

(c)

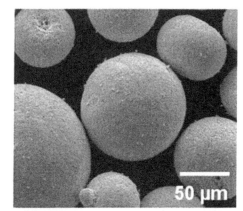
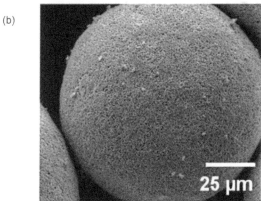
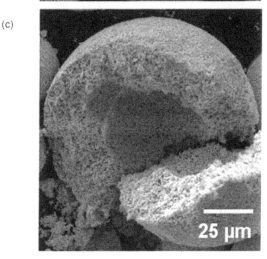

Fig. 4.10 Scanning electron micrographs of the spray dried W-15Cu powder at (a) lower, and (b) and (c) higher magnifications (Photograph courtesy: Avijit Mondal, Indian Institute of Technology, Kanpur)

powder/binder mixture can be mixed with a solvent to lower the viscosity. This slurry can be used to form agglomerates using rotary dryers, spray dryers or disc pelletisers. The sprues, runners and recycled compacts are usually granulated using a cut-and-grind process.

4.5 COATING ON METAL POWDERS

Often, the base metal powders may be coated by another chemical species. The purpose may be to produce homogeneous mixing, e.g., W-Cu, where a soft and relatively low-melting point metal is coated on the hard substrate to impart better compressibility during compaction. The simplest method is the mechanical method, for example, ball milling of W-Cu powder premix. However, the uniformity of coating in such cases is questionable.

4.5.1 Electroplating

In this method, a second dispersed phase of controlled particle size may be introduced into a plating solution. This is referred to as dispersion, inclusion, occlusion, composite or electrophoretic plating, deposition or coating. The requisites are:

- The particles must be insoluble (or only slightly soluble) in the solution.
- The particles must be compatible with the solution, i.e., they should not produce any detrimental effect.
- The particles must be dispersed either naturally (as colloid size particles) or mechanically (stirring, agitation) in order to physically come into contact with the surface being coated.

Electrophoretic coating is used to coat a conducting substrate from a dispersion of colloidal particles. The powder to be coated is immersed in an aqueous dispersion which dissociates into negatively-charged colloidal particles and positive cations. An electric field is applied with the powder mass as anode; the colloidal particles are transported to the anode, where they are discharged and form a film. The coating is air dried and baked to remove the solvent medium. The coating is non-adherent and must be processed faster by compaction and sintering. The coating thickness can be varied by controlling voltage, electrode spacing, suspension concentration and time.

4.5.2 Electroless deposition

The difference between electroless plating process and electroplating is that in the former no external current source is required. Metal coatings are produced by chemical reduction with the necessary electrons supplied by a reducing agent (RA) present in the solution:

$$M^{n+} + ne^- \text{ (supplied by RA)} \xrightarrow{\text{catalytic surface}} M^0 \text{ (+ reaction products)} \qquad (4.3)$$

The significance of the process is that the reaction is catalysed by certain metals immersed in the solution and proceeds in a controlled manner on the substrate surface.

The deposit itself continues to catalyse the reduction reaction so that the deposition process becomes self-sustaining or auto-catalytic. These features permit the deposition of relatively thick deposits. This process is different from other types of chemical reduction:

- **Simple immersion or displacement reactions** in which deposition ceases when equilibrium between the coating and the solution is established (e.g., copper on steel from copper sulphate solution)
- **Homogeneous reduction** where deposition occurs over all surfaces in contact with the solution.

The reducing agents most widely used are sodium hypophosphate (for Ni, Co), sodium borohydride (for Ni, Au), dimethyl amine boron (for Ni, Co, Au, Cu, Ag), hydrazine (for Ni, Au, Pd) and formaldehyde (for Cu). To ensure spontaneous reduction other chemicals are added. These are generally organic complexing agents and buffering agents. Other additives—additional stabilisers, brighteners, stress relievers—provide special functions as in electroplating solutions.

The advantages of electroless plating over electroplating are:

- Internal surfaces are evenly coated. The uniformity is limited only by the ability of the solution to contact the surface and be replenished at the surface.
- Deposits are less porous.
- Almost any metallic, non-metallic, non-conducting surface, including polymers, ceramics and glasses can be plated. Those materials which are not catalytic can be made catalytic by suitable sensitising and nucleation treatments.
- Electrical contacts are not required.
- The deposits may have unique chemical, mechanical, physical and magnetic properties.

However, there are some disadvantages too as compared to electroplating. They are:

- Solution instability
- Higher cost
- Slower deposition rates
- Frequent replacement of tanks or liners
- Greater and more frequent control for reproducible deposits.

4.5.3 Coating by hydro-metallurgical process

The details of the hydro-metallurgical process for production of metal powders have been described in Chapter 2. A great number of metals, metalloids, non-metals, metal alloys, oxides, natural minerals, hard metal compounds and plastic powders can be coated with one or more metals selected from Ni, Co, Mo, Cu or Ag to form composite powders.

In developing composite powders, the cores must conform to certain requirements:

- They must be non-reactive in the system selected for metal deposition.
- They must be catalytically active with respect to the deposited metal.
- They should be in a physical form that is able to remain suspended in solution in the agitated autoclave.

Many metals, alloys, carbides, nitrides and some non-metals such as graphite and phosphorus are catalytically active and are readily coated without special treatment of the core; but some cores have to be activated. The most effective way of activating is to wash with a stannous chloride solution, followed by treatment with palladium chloride. During this treatment, palladium chloride is adsorbed on the surface and is reduced in the autoclave by hydrogen to the metallic form, thus forming catalytically active sites on which the metal precipitates preferentially. Nickel is the most easy metal to coat with. Coating with cobalt is much more difficult and only the very active cores, such as carbides can be coated completely. The deposition of Cu or Ag can be achieved, but without a catalyst, a spotty coating is obtained.

There are no restrictions on the composition of composite powders, and powders containing 1% to 99% of the coating metal have been prepared. However, in order to ensure complete coating of the core, a 2–3 μm layer of the coating metal is considered to be a minimum requirement.

Composite powders produced by such methods have found commercial applications in flame and plasma spraying. Such powders are useful in preparation of alloys, dispersion-strengthened materials, porous materials, low-friction materials and hardfacing coatings.

4.6 POWDER DEGASSING

This treatment is very common for aluminium powder, in which steam is removed from the powder after hydration. Degassing is carried out in a suitable atmosphere at an ambient temperature for different periods. Powder should be degassed at a temperature equal to or greater than the temperature reached during processing, in order to obtain pore-free and surface blister–free parts. Degassing of aluminium powders in an open tray in argon at an elevated temperature is reported to prevent rehydration for up to 120 hours. The best economy in time and temperature is achieved using vacuum degassing. A temperature of 350°C and a vacuum level of 3×10^{-4} torr have been found suitable for degassing aluminium powder. In general, after degassing, the powder is not exposed to air, but is sealed in a can. The can is further hot formed directly and the can material is stripped off later. When the powder is directly compacted by preheating at 480°C for 1 h in an argon atmosphere, it gives rise to hydrogen-free compacts.

During degassing of aluminium powder, water is liberated from the hydrates by the reaction:

$$Al_2O_3 \cdot 3H_2O \rightarrow Al_2O_3 \cdot H_2O \rightarrow \gamma\text{-}Al_2O_3 \qquad (4.4)$$

The liberated water reacts with aluminium to form amorphous aluminium oxide. In case magnesium is present in the powder as an alloying additive, it gets converted to MgO crystallites. The reactions are:

$$3H_2O + 2Al \rightarrow Al_2O_3 + 3H_2 \uparrow \qquad (4.5)$$

$$H_2O + Mg \rightarrow MgO + H_2 \uparrow \qquad (4.6)$$

The above reactions take place when the degassing temperature exceeds 350°C. A higher degassing temperature is preferred as brittle crystallite of $\gamma\text{-}Al_2O_3$ is formed instead of the amorphous oxide film. Figure 4.11 illustrates the complete degassing scheme for aluminium powder.

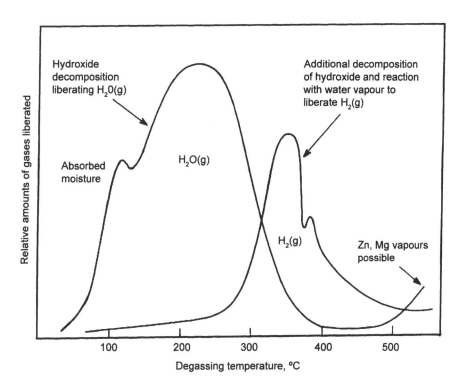

Fig. 4.11 Generalised illustration of principal aluminium powder outgassing events

SUMMARY

- The annealing operation of metal powder may be done at an elevated temperature in an atmosphere or a vacuum furnace.
- During diffusion alloying or bonding, the additive element is diffused at the surface in parent metal particles.
- Partial pre-alloying of iron powder by copper, nickel or molybdenum is common.
- During mixing segregation of particles may occur, if they have different sizes and densities.
- The mechanism of powder mixing depends on the physical properties of the powder, the flow rate of the powder and the type of handling equipment.
- The main purpose of binders is to provide strength to the moulded green body. Binders are polyvinyl alcohol, polyethylene glycol, paraffin wax or water soluble waxes.
- The electroless plating process differs from electroplating in that no external current source is required.
- Degassing treatment is very common in aluminium powder.

Further Reading

Lyman T, (ed), *Metals Handbook*: Metallography, *Structure and Phase Diagrams*, 8th edn, Vol. 8, American Society for Materials International, Materials Park, Ohio, 1973.

Masters K, *Spray Drying*, Wiley-Interscience, New York, 1976.

Onada, Jr, GY and Hench LL, (eds), *Ceramic Processing before Firing*. Wiley-Interscience, New York, 1978.

Sherrington PJ and Oliver R, *Granulation*, Hayden, Philadelphia, 1981.

EXERCISES

4.1   Describe the compressibility behaviour of the following ferrous powders in increasing order.

(a) Fe-Ni-Mo premix

(b) Fe-Ni-Mo pre-alloyed

(c) Mo diffusion bonding over Fe-Ni alloy particles.

4.2   What is the optimum speed in RPM of a cylindrical rotary mixer of 0.5 m diameter?

4.3   A water-atomised iron powder (2.95 g cm^{-3} apparent density and 7.9 g cm^{-3} theoretical density) is to be cold pressed to a density of 80% theoretical. What is the maximum weight per cent of zinc stearate lubricant which could be added during powder mixing? The density of zinc stearate is 1.09 g cm^{-3}. Is this quantity of lubricant used in practice?

4.4   Mullite has the chemical formula $3Al_2O_3.2SiO_2$. What mass fractions of Al_2O_3 and SiO_2 should be mixed for producing a mullite product?

4.5   The composition of composites is often expressed in volume per cent. A Al_2O_3 (80 vol%) and SiC whisker (20 vol%) composite is to be prepared. Express these components in mass per cent. Given: Density of Al_2O_3 is 3.98 g cm^{-3} and density of SiC is 3.19 g cm^{-3}.

4.6   Why are common mixers used for dry powder mixing not useful for preparing the PIM feedstock?

4.7   Explain why the initial density of powder used for powder injection moulding is low and non-uniform within a poured mass.

4.8   In which of the following cases are the agglomerates stronger?

(a) Capillary liquid binding

(b) Flocculation bonding.

Justify your answer.

4.9   Will lubricant addition in an inert gas–atomised metal powder enhance the flow of the powder?

4.10 What single operation for ceramic processing should be carried out to replace the serial operations of slurry filtering and conventional drying?

4.11 Write down the effect of the deflocculant on the following properties: (a) Bulk density of cast or dried out slip, (b) Viscosity of slip, (c) Casting rate of slip, (d) Dry-strength of cast, (e) Drying-shrinkage of cast, (f) Critical moisture content and (g) Rate of sedimentation of particles from suspension.

4.12 An injection moulding mix contains 20 wt% paraffin and 80 wt% Al_2O_3. What is the volume per cent of paraffin? Given: Density of paraffin is 0.91 g cm^{-3} and density of Al_2O_3 is 3.38 g cm^{-3}.

4.13 What stearate should be selected to minimise chemical contamination of the following ceramic powders: Mn-Zn ferrite, Al_2O_3 and MgO?

4.14 Explain why degassing of aluminium powder is continued till a γ-Al_2O_3 and not amorphous Al_2O_3 film is obtained over the surface.

4.15 It is observed that strontium ferrite powder after Ni-P deposition has better magnetic energy as compared to the uncoated one. What method should be adopted for coating? In case the ferrite powder is of nano size, will the coating parameter be changed? If so, how?

4.16 Reduction annealing and degassing of metal powders are both elevated temperature powder treatments. What are their distinctive features in these processes?

4.17 Powder flow is an essential criterion for better productivity in powder metallurgy. There are two aspects: the intrinsic powder characteristics and the equipment used for material transfer. List some factors pertaining to the second category.

4.18 For production of soft magnetic insulated iron powder, polymer coating is performed. Describe the coating technique.

4.19 What is the advantage of the rotary blender over the stationary horizontal blender?

4.20 A double cone blender is not recommended for underloading, while a rotary blender is. Give reasons.

4.21 Nickel-coated powders are used in electromagnetic interference shielding. Suggest a relatively cheap method for tonnage powder, other than electrodeposition.

4.22 What are the factors that affect the strength of powder agglomerates?

5

Powder Compaction

LEARNING OBJECTIVES

- Different arrangements of particles in uniform packing
- Conventional methods of powder compaction
- Different types of compaction presses
- Why withdrawal tooling has gained significance
- What factors affect tooling design
- Different PM tooling materials
- Design considerations for PM part geometry
- Cold isostatic pressing
- When to select powder roll compaction/powder extrusion
- Injection moulding and its limitations

After powder treatment, the next operation in powder metallurgy is to fill the die cavity of pre-designed shape with the powder. The die fill is an important factor and here one must be conversant with the basics of powder packing.

5.1 BASIC ASPECTS

5.1.1 Powder packing

Uniform spheres pack in five different ordered arrangements. The fraction and per cent of the bulk volume occupied by the spheres are referred to as the packing fraction (PF) and the packing density, respectively. The packing density ranges from 52% for simple cubic packing to 74% for tetrahedral and pyramidal close packing, and is independent of the size of the spheres.

The number of particles per unit volume (N_p) in a system of packed spheres of uniform size is:

$$N_p = \frac{6 \times PF}{\pi a^3} \tag{5.1}$$

The number of particle contacts per unit bulk volume (N_c) is the product of N_p and half the coordination number (CN).

$$N_c = \frac{3 \times PF \times CN}{\pi a^3} \tag{5.2}$$

For regular packing arrangements, N_c is dependent on the packing geometry.

The actual packing density depends on the powder characteristics. For monosized spherical powder, the fractional packed density is around 0.60–0.64. However, in real practice this situation is not common—except in the case of powder filter production—and the particle size will have a distribution. In a bimodal powder mixture, the large-size particles are in relatively large fraction and the remaining part pertains to small-size particles, which fill up the interstices between larger spheres. For maximum packing in such a mixture, the mixture should contain 26.6 wt% of smaller particles, and the fraction packing density would be around 0.86, i.e., greater than in the case of a monosized powder. Figure 5.1 shows the fractional packing density of the bimodal powder mixture with respect to powder composition. It is obvious that at a critical composition X, the packing density is maximum. It can be seen from the figure that the packed density of large-size powder is greater than that of small-size powder.

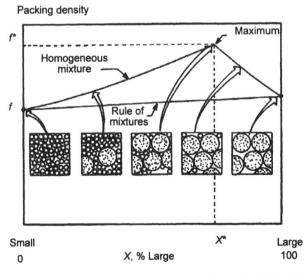

Fig. 5.1 A plot of fractional packing density versus composition for bimodal mixtures of large and small spheres (Reproduced with permission from German RM, *Powder Metallurgy Science*, 2nd edn, Metal Powder Industries Federation, Princeton, USA, 1994)

At the maximum of the curve, the larger particles are in contact with each other and all the interstitial voids are filled with small particles.

It is worth noting that the above description pertains for spherical powders only. When the shape is irregular, the packing is less efficient; however, qualitatively, the nature of variation is similar.

The above picture of a bimodal mixture can be extended to multimodal systems. As the number of size classes becomes larger, the problem in practice becomes more tedious and the gain in packing density decreases.

5.1.2 Powder bed under load

As a simple model study, it is natural to consider a spherical powder of equal size. When load is applied over a powder mass confined in a die through a punch, a number of stages may be noticed. In the early stage of compaction, sliding of particles occurs concurrently with restacking, and particles originally in contact undergo deformation. Surface deformation is at first elastic. Progressive flattening of the contacts brings the particle centres closer to each other, producing overall densification. In a bed of spheres at tap density, each particle initially has seven neighbhours on the average. As the distance between the particle centres diminishes, additional contacts are generated. The rate at which such new contacts appear will depend on the stacking geometry. The contact flats are initially circular. At a certain stage, neighbouring contacts at a given sphere will start to impinge. As compaction progresses, more and more contacts will impinge, and an increasing fraction of the volume gets work hardened.

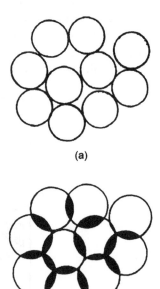

(a)

(b)

Fig. 5.2 (a) Random sphere packing; (b) Impinging of contacts in the early stage of compaction

According to the Mises Criterion, the metal will begin to deform plastically when the mean pressure reaches a value of about 1.1 times the yield stress (Y). At this stage, the region of plasticity is very small. As the contact force is further increased, the plastic domain soon extends over the whole contact area. In packing of simultaneously deforming particles, a good degree of constraint will be encountered. This leads to 'geometrical strain hardening', which plays a predominant role—a role even bigger than the strain hardening of the material itself. In other words, the strain hardening of the material alone is not enough to account for the increased densification resistance of the powder in the final stage of compaction.

Another factor to be noted is that the powder is not regularly packed within the die, but has random packing with continuous distribution of interparticle distances. In other words, during compaction of such a random packed powder, the spatial distribution of deforming particle pairs will have a substantial role in densification behaviour. Figure 5.2 illustrates how with increasing pressure many spheres will overlap. The material common to two spheres has to be redistributed in a way which approximates the actual deformation geometry. It is assumed

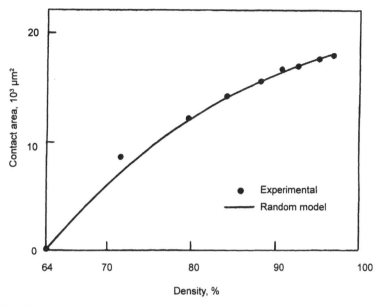

Fig. 5.3 Average area of contact flats between spheres with respect to density

that the material (in case of a ductile one) squeezed out from the contact zone is evenly deposited across the remaining surface of each sphere. Figure 5.3 shows that the linear relationship of contact area versus density variation is valid only at the beginning of the compaction. The full line curve corresponds to the calculated data based on random model. A generalised compaction formula can be given as:

$$P = \frac{CYazD}{4\pi R^2} \tag{5.3}$$

where, P is the compaction pressure, R is the average particle radius, C is the constant, Y is the yield stress, a is the contact area, z is the average coordination number and D is the relative density (volume fraction of the solid powder).

During compaction, where the powder particles are squeezed together to such an extent that the initially interconnected pores between them got transferred to small isolated pores, the stress distribution around each of them can be approximated by the stress distribution in a hollow sphere under hydrostatic outside pressure. The hydrostatic pressure required to inject plastic deformation of the hollow sphere increases as the volume of the hole decreases. In other words, an infinitely high pressure would be required to eliminate the hole inside the powder particle.

In powder premixes, the use of a lubricant is indispensable for the reduction of die wall friction, which lowers the attainable density of powder compacts. During compaction, part of the added lubricant is squeezed towards the die wall. The remaining part of the lubricant remains inside the closed pores. When the powder is being compacted in a rigid cylindrical die, the axial pressure executed on the powder by the compacting punch is partly transformed to radial pressure on the die wall. Unlike liquids, radial pressure in a powder cannot reach the level of axial pressure.

During the densification of powder mass in a die, the elastic to plastic deformation stage is not to be treated as a sudden one; the transition occurs gradually in individual particles. Apart from this, work hardening occurs in the powder particles during densification.

Frictional forces at the wall of the compacting die restrain the densification of the powder because they act against the pressure exerted by the compacting punch. With increasing distance from the face of the compacting punch, the axial stress available for the local densification of the powder decreases. The axial compressive stress in the powder mass decreases exponentially with increasing distance from the face of the moving upper punch. This effect is more pronounced—larger the frictional coefficient, smaller the inner diameter of the die. When a powder is compacted between symmetrically moving punches, the axial stresses at both ends of the compact are larger than anywhere in between. Powder compacts in such cases will have a zone of lower density, approximately midway between their end faces. This zone of lower density is referred to as the *neutral zone*.

After compaction, the next step is ejection. The ejection pressure is higher for larger *L/D* (length/diameter) ratio of the compact. This is also directly proportional to the frictional coefficient. In the beginning of the ejection process, the frictional coefficient and, consequently, the ejection pressure, assume a peak value (*adhesive friction*) considerably above the normal level (*sliding friction*). Figure 5.4 shows how the peak pressure can, in certain cases—for example, with long thin-walled bushings—exceed the maximum pressure that occurred in the compaction process. This may result in the following features:

- A certain re-densification effect at the lower end of the compact;
- A long and slender bottom punch just strong enough to withstand the compacting load may yield or break under the ejection load.

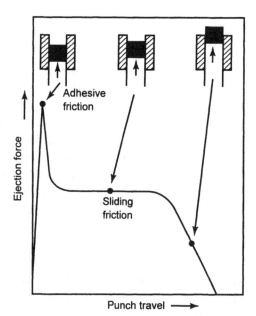

Fig. 5.4 Schematics of the ejection force as a function of the movement of the ejecting bottom punch

In case of wear or insufficient lubrication of the die, the powder compact and the die wall may tend to cold weld, which causes excessive increase in the ejection pressure. This gives rise to a typical stick–slip behaviour, which is manifested by a creaking noise.

There is another effect during powder compaction, known as *spring back*. This is the elastic expansion of the compact after ejection. This is expressed as:

$$S\% = \frac{100\,(l_c - l_d)}{l_d} \tag{5.4}$$

where, l_c is the transverse dimension of ejected compact, and l_d is the corresponding dimension of the compacting die:

The spring back depends on:

- Compaction pressure
- Powder characteristics
- Lubricant and alloying additives
- Shape and elastic properties of the compacting die.

5.1.3 Type of material

The material to be pressed needs attention from the viewpoint of the chemical nature and bonding of the powder material. All crystalline materials can have one or more types of bonds—metallic, ionic or covalent. The metallic bond is attributed to free valence electrons. So it does not have directionality; however, the properties do vary somewhat with crystallographic direction. Ionic bonds are formed by a transfer of electrons from one atom to another. On the other hand, covalent bonds arise when two atoms share electrons. Often, the powder material is compacted under an external field, e.g., magnetic field. The intermetallic rare earth-cobalt (RCo_5), which has permanent magnetic behaviour, is invariably compacted under a magnetic field. The powders of metals like lead, tin, aluminium and silver can be green compacted to fairly high density, while the hard and brittle transition metals like tungsten have poor compressibility.

From basic theoretical considerations, the atomic size of an element throws light on the interatomic forces which hold the atoms together. Other things being equal, the stronger the attraction or bonding force between two atoms, the smaller will be the interatomic distance. Moving from left to right along the Periodic table, the binding forces in the crystals of the alkali metals are comparatively weak, and increase rapidly on passing from Group I to Group IV in both short and long periods. *Compressibility* is the measure of the ease with which a metal can be compressed, and since the process is reversible, a low compressibility at zero pressure indicates strong bonding. Figure 5.5 shows the compressibilities of the metals; they follow the same general relation as atomic diameters. Alkali metals have the highest compressibilities, and hence the weakest interatomic bonding. For zinc and cadmium, the values shown in Fig. 5.5 (large open circles) are three times the linear compressibility in the direction of the close-packed hexagonal planes, in order to give an indication of the strength of the bonding in these planes. Owing to the high value of axial ratio (c/a), the volume compressibility of the crystal as a whole is much greater.

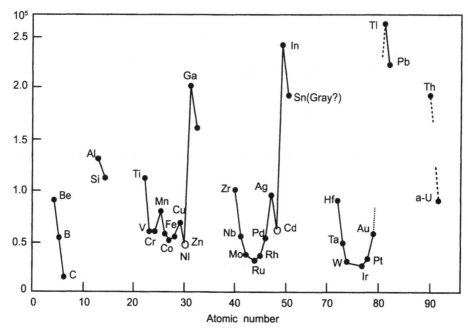

Fig. 5.5 Compressibilities of elements. To a first approximation the effect of pressure on the volume of a solid may be expressed in the form: $-\Delta V/V_o = ap \pm bp^2$. If p is in kg cm^{-2}, the values in this figure are those of a multiplied by 10^6. ΔV = change in volume, V_o = original volume, p = applied pressure, a and b are constants.

Note: The values of $a \times 10^6$ for elements of higher compressibility are as follows: Na 15, K 29, Rb 31, Ca 40, Mg 3, Ca 6, Sr 8 and Ba 10 (Source: Hume-Rothary W, Smallman RE and Haworth C, *The Stuctures of Metals and Alloys*, 5th edn, Institute of Metals, London, 1969.)

5.2 DIE COMPACTION

Die compaction is the most widely used method and is considered as the conventional technique. This involves rigid dies and special mechanical or hydraulic presses. Densities of up to 90% of full density can be achieved following the compaction cycle, the duration of which may be of the order of just a few seconds for very small parts.

Powders do not respond to pressing in the same way as fluids and do not assume the same density throughout the compact. The friction between the powder and die wall and between individual powder particles hinders the transmission of pressure. A high uniformity in green parts can be achieved depending on:

- the kind of compacting technique
- the type of tools
- the materials to be pressed
- the lubricant.

The compaction techniques used may be characterised by reference to the movement of the individual tool elements—upper punch, lower punch and die relative to one another.

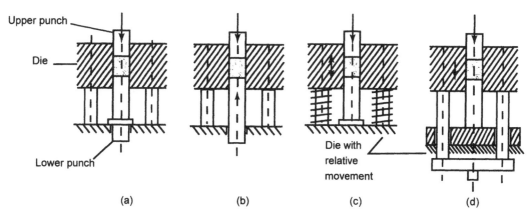

Upper punch

Die

Lower punch

Die with
relative
movement

(a)   (b)   (c)   (d)

Fig. 5.6 Various types of toolings in a rigid die compaction: (a) Single action pressing, (b) Double action pressing, (c) Pressing with spring supported die and (d) Withdrawal-type pressing

Pressing within fixed dies can be divided into single action pressing and double action pressing (Fig. 5.6). In the former, the lower punch and the die are both stationary. The pressing operation is carried out solely by the upper punch as it moves into the fixed die. The die wall friction prevents uniform pressure distribution. The compact has a higher density on top than at the bottom. In the latter type, only the die is stationary in the press. Upper and lower punches advance simultaneously from above and below into the die. The consequence is high density at the top and undersides of the compact. In the centre there will be a neutral zone which is relatively weak. In case the part has a through or a partial hole, a core rod positioned parallel to the pressing direction is used.

Regarding the tools, they may be rigid, as in the case of die compaction, or flexible, as in the case of isostatic compaction.

5.2.1 Pressing operation

The pressing operations can be sequenced as follows:

- *Filling* of the die cavities with the required quantity of powder.
- *Pressing* in order to achieve required green density and part thickness.
- *Withdrawal of the upper punch from the compact:* Here the risk of cracking of green parts is felt. As the upper punch withdraws, the balance of forces in the interior of the die ends. In the case of parts with two different thicknesses, e.g., flange with a hub, the elastic spring back of the lower punch is the greatest danger. Other problems are protrusions required on the upper face of the part. In the case of thin parts with a large projected area, cracking is common, due to elastic spring back of the lower punch and the part itself. The former pushes the part still lying in the die cavity upwards, while the latter tends to expand the part.
- *Ejection:* The tooling must be done in such a manner that the ejection of the part is feasible. Ejection of a part of complex form is problematic, as it involves friction

between the green part and tool walls. The green strength must be high to resist the bending stresses introduced by the ejection force.

There is another type of compaction involving upper punch pressing with floating die. This is characterised by a stationary lower punch; the upper punch moves into a die supported by spring. As soon as the friction between the powder and the die wall exceeds the spring power, the die wall is carried down. The friction will vary slightly from stroke to stroke. It also depends on the degree of wear in the tools so that a constant density distribution is difficult to maintain over a period.

During the Second World War, another tooling method, known as 'withdrawal tooling', was developed in Germany. In this case, the lower punch does not move during the compacting cycle. After the upper punch has entered the die cavity, both the upper punch and the die plate move downwards. After the compact has been pressed, the upper punch moves up, but the die plate and lower coupler move further down until the top of the die plate is flush with the lower punch (Fig. 5.7). The compact is ejected and can be moved out of the way by the loading shoe. Die plate and lower coupler then move back into the filling position and the cycle repeats.

Figure 5.8 shows the synchronised punch and die motions during a pressing cycle for a withdrawal tooling system. The major advantage of the withdrawal system of tooling is that the lower punches are relatively short and well supported during compaction and ejection. When there are multiple lower punches, as many of them as possible rest directly on the base plate. Withdrawal tooling can be built for very complex parts. On the other hand, in the tooling system with ejection by the lower punches, the motions of the punches are built into the multiple action presses. In many cases, no tool holders are required.

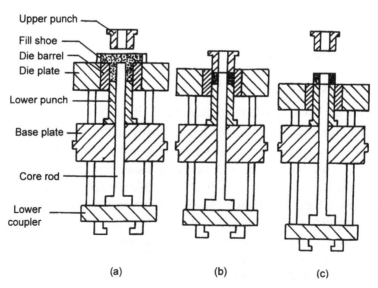

Fig. 5.7 Sequence of operations for pressing with the withdrawal system: (a) Fill position, (b) Press position and (c) Ejection position

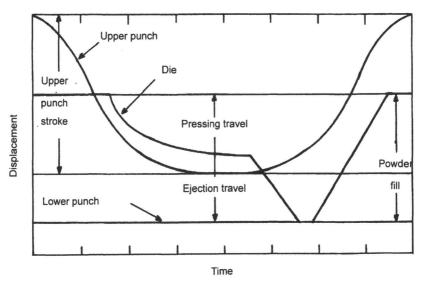

Fig. 5.8 Synchronised punch and die motions during a pressing cycle for withdrawal tool systems

The *split die system* is another rigid tooling system. It enables the compaction of parts with completely asymmetric upper and lower sections in the pressing directions. This system requires two die holding plates to carry the upper and lower dies. Each plate is controlled and moved independently.

5.2.2 Compaction presses

Compaction presses for powders are of two types: mechanical or hydraulic. There are others which are partly hydraulic and partly mechanical. Presses from 3 to 1000 tons capacity are available. The optimum operating speed is dictated by the size and complexity of the PM parts to be made.

Hydraulic presses
They produce working force through the application of fluid pressure on a piston by means of pumps, valves, intensifiers and accumulators. Inherent in the hydraulic method of drive transmission is the capability to provide infinite adjustment of stroke speed, length and pressure within the limits of press capacity. Also, full tonnage can be extended throughout the complete length of the stroke.

Mechanical presses
In mechanical presses, a flywheel stores energy, which is then released and transferred by one of many mechanisms (eccentric, crank, knuckle joint, toggle, etc.) to the main slide. In most mechanical presses, the movement or stroke of the slide is adjustable within the limits of daylight of the press. Mechanical presses are classified by one or a combination of characteristics, viz., sources of power, method of actuating the ram, type of frame, clutch, brake and control system.

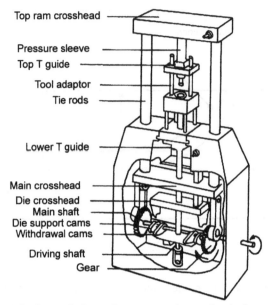

Top ram crosshead

Pressure sleeve
Top T guide

Tool adaptor
Tie rods

Lower T guide

Main crosshead
Die crosshead
Main shaft
Die support cams
Withdrawal cams

Driving shaft
Gear

Fig. 5.9 General view of an automatic mechanical press

The energy stored in the flywheel must be sufficient to ensure that the work per stroke required of the press will not reduce the flywheel's speed by more than 10%–15%. The most common type of mechanical press is the eccentric or crank type which converts rotary motion to linear motion. In the toggle type the eccentric or crank straightens a jointed arm or lever, the upper end of which is fixed at the top, while the lower end is guided for controlled accurate punch guidance into the die. In the cam type of mechanical press, pressing speed, timing and motion are controlled by changing the contours of the cams or cam inserts. Figure 5.9 illustrates a general view of the automatic mechanical press.

Rotary presses

A rotary press is a mechanically operated machine which uses a number of identical sets of tools to produce parts at high production rates. In this machine, the tool sets normally called *tool stations,* are held in a head or a turret which rotates continuously. The rotation of the head pulls the upper and lower punches past fixed surfaces called cams, and a set of pressure rolls impart compression force from above and below. All rotary presses are double action. The design of the cam surface moves the punches up or down to provide the cycle of die filling, weight adjustment, compression and ejection. Another feature of the rotary press is that all the adjustments can be made while the press is in operation. These presses can be furnished in tonnage ranges up to 100 tonnes.

5.2.3 Press selection

In hydraulic presses fluid pressure rather than a rotated crankshaft is used to actuate the slide. They are slower in operation and generally less economical to operate than mechanical presses that can efficiently perform a specific identical job. One reason is that hydraulic presses have no mechanism comparable to the mechanical press's flywheel for storing energy. In a hydraulic system, oil pressure in the cylinder drops after each stroke and has to build up in a comparatively short time. This requires the use of pumps served by motors, and these pumps draw a large amount of electric power. Therefore, the motor of a hydraulic press has several times the capacity as the motor of a mechanical press of comparable tonnage. Another disadvantage of a hydraulic drive is that the sudden release of pressure with each completed stroke is accompanied by a contraction of the cylinder and its hydraulic conduits, which in turn places great stress on pipe joints, valves, seals, etc. However, hydraulic presses offer the following advantages:

- Tonnage is adjustable
- Constant pressure can be maintained throughout the entire stroke and applied at any predetermined position
- Drawing speeds are adjustable.

In the mechanical press, the rated tonnage of a press is the maximum force that should be exerted by the ram against a tooling at a given distance above the bottom of the stroke. The higher a press is rated to its stroke, the greater the torque capacity of its drive members and the more flywheel energy it is capable of delivering. Because of the mechanical advantage of the linkage, the force actually transmitted through the clutches to rotating members (cranks or toggles) and reciprocating slides varies from a minimum at the beginning of the downward stroke to a maximum at the bottom of the stroke. A chart will show the change in tonnage that the drive is capable of delivering at various distances above stroke bottom. Provided its speed is not too high, almost any mechanical press—with sufficiently long stroke and large tool mounting space—can be suitably altered for compacting powder, either in spring-floated dies or in dies of the withdrawal type.

5.2.4 Factors affecting tooling design

The powder's response to compaction and sintering has a decisive effect on tooling design. Fill, flow, apparent density, fill ratio, compacting pressure and dimensional changes are all contributing factors.

Fill

The terms 'fill' signifies the amount of powder taken into the tool cavity prior to compaction. Powder fill is affected by many variables such as flow, apparent density, part configuration and tool design. The ideal fill should be uniform in density, free of bridging and should be fast enough to allow reasonable speed in the press cycle. Part configuration has a direct influence on the powder filling into the tool cavity. Thin walled parts have narrow cavity sections, which results in poor powder filling. Powder filling in large areas with thin walls may trap air pockets. Filling can be improved by using the three position air core. Raising and lowering the core during the filling cycle helps in releasing the trapped air. As a minimum period of time is required to obtain adequate powder filling, the speed of the press can appreciably affect the fill.

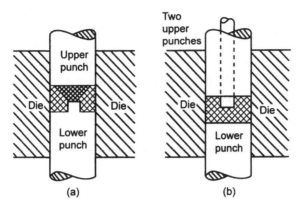

Fig. 5.10 Tool arrangement for imparting uniform compaction in single acting press. (a) One upper and one lower punch; (b) Two upper punches and one lower punch. Note: In (a) the protruding part is in the lower punch

The amount of powder fill required to produce a part is determined by multiplying the finished part thickness by compression ratio of the powder to be compacted to the required green density.

In general,

$$\frac{\text{Height of filling capacity}}{\text{Compact height}} = \frac{\text{Green density}}{\text{Apparent density}}$$

Components with shoulders or projections will therefore need to take into account the provision for different filling heights for various sections of the part.

Figure 5.10 illustrates how a double height part necessitates multiple punches in order to have uniform green density of the part after pressing.

Flow

Adequate flow of powder is essential for an ideal fill. Often, the metal powder contains a lubricant to reduce frictional drag. This prevents galling and cold welding of powder particles to the die and core walls.

Apparent density and fill ratio

The density of the finished part is usually specified. Tooling must be designed to provide enough fill to produce a part to the required compacted density. The ratio of apparent density of powder to green density is used to find the fill depth. The fill ratio accuracy can be improved further if compacts are made at the desired rate of speed for producing green parts on the press. The apparent density is affected by the time allowed for the tool cavity to fill. The true fill ratio should be evaluated after including all the variables. If a fill depth is incorrect because of incorrect tool design, it may be compensated for by a change in powder apparent density. Since apparent density adjustment in powders is limited, it should not be relied upon as a tool design solution. Normal apparent density tolerance limit for powder is 0.1 g cm^{-3}. In special cases, it can be reduced to ± 0.05 g cm^{-3}.

Compaction pressure

Although very high forces can be developed in a press, there are limits to the load a tool bears, which partially determines the density that can be obtained. Other limits to tool load are:

- Part configuration which may introduce thin wall sections
- Punch face protrusions which form areas of stress concentration

The tensile strength of tool steels is the limiting factor for the load a punch can withstand. 480 MPa (35 tons psi) is the normal tool pressure allowed for powder compaction or coining. It is advisable to add coining operation, rather than to overload the tool to obtain higher green densities and risk tool breakage. Powder manufacturers have developed powders with improved compressibility in order to achieve high green densities. Table 5.1 illustrates compaction pressure requirements for various powders.

Dimensional changes

After sintering green parts usually undergo dimensional changes; this factor must also be included in tool design. Green density variation, material variation, temperature variation, furnace load conditions and furnace atmospheres, all combine to affect dimensional change.

Table 5.1 Compaction pressure requirements and compression ratios for various materials

Type of material	Compaction pressure,		Compression ratio
	MPa	tsi	
Aluminium	70–280	5–20	1.5–1.9:1
Brass	415–690	30–50	2.4–2.6:1
Bronze	205–230	15–20	2.5–2.7:1
Copper–graphite brushes	345–415	25–30	2.0–3.0:1
Carbides	140–415	10–30	2.0–3.0:1
Ferrites	110–165	8–12	3.0:1
Iron bearings	205–345	15–25	2.2:1
Iron parts			
Low density	345–415	25–30	2.0–2.4:1
Medium density	415–550	30–40	2.1–2.5:1
High density	430–825	35–60	2.4–2.8:1
Iron powder cores	140–690	10–50	1.5–3.5:1
Tungsten	70–140	5–10	2.5:1
Tantalum	70–140	5–10	2.5:1

Note: The above pressing force requirements and compression ratios are approximations and will vary with changes in chemical, metallurgical and sieve characteristics of the powder, with the amount of the binder or die lubricants used and with mixing procedures.

Dimensional change is often affected by alloying. For example, nickel blended in iron imparts shrinkage after sintering, while copper does the reverse. In cases where a tool has been incorrectly factored for dimensional change, the powder blend can be changed to obtain a suitable growth or shrinkage. However, it must be borne in mind that the ultimate physical properties of the part are not adversely affected. Sintering temperature can also be adjusted to control growth.

5.2.5 Tooling materials

Tooling materials can be classified under three categories: steels, carbides and coated steels. Powdered metals are generally abrasive and cause tool wear. Apart from abrasion resistance, tools must have the properties of high compression strength and toughness. The steels generally used for tool making are A2-, D2-, M2- and SAE 6150-type, the details of which can be had from any Metals Handbook. 12% cobalt-containing tungsten carbide is used for solid dies and die sections. Modern techniques such as spark erosion machining make this material relatively inexpensive to work with. Due to the low pressing forces required for compacting ceramic and carbide powders, punches are tipped with carbide. Carbide material is also used for making core rods where maximum wear resistance is required. Table 5.2 illustrates various grades of cemented carbides and their typical PM tooling applications. Among the third category of tools, carburised and nitrided coatings are very common. The principle limitations of using coated steel tooling is the method of applying the coating to the tool member. Most coatings are difficult to apply uniformly on irregular surfaces. Usually, only round core rods are subjected to these coatings, although at present the technology has developed too fast.

Heat treatment of tools is to be controlled to minimise distortion. Heat treatment stresses must be relieved by proper tempering, since untempered tools will warp as they are machined. All tools should be normalised prior to machining operations to relieve heat treatment stresses.

Punches must be tough, and able to take repeated deflections. Although punch wear is important, it is sometimes sacrificed to obtain other properties in the tool. For example, large chamfers on punch faces are subjected to higher loads than the main tool body. Sharp corners at the chamfer root cause stress risers. Punches with these design features should be made of relatively ductile tool steel. On the other hand, when punches have fixed steps or depressions for protrusions or pockets in the sintered parts, a wear resistant steel is used. Auxiliary mechanisms, e.g., floating core arrangement, improves the performance of tool materials. In brief, the tool designer must consider a variety of physical properties of tool materials, selecting the ones which best suit the requirements of the particular design.

The die barrel components of compacting tools usually consist of wear resistant inserts which are held in a retaining ring made of tough steel. Usually, the inserts are shrunk into the retaining ring, which has the advantage that the compression stresses induced in the die barrel during shrinkage counteract the radial stress during compaction. Although quite complex internal contours can be produced in single piece inserts by electric discharge machining, it is more common to fix several carbide inserts together (Fig. 5.11). The entry edges to the die cavity in the die barrel are levelled or radiussed. In order to minimise spring-back strain during ejection, the die barrels are sometimes tapered, but the taper should be less than the spring-back of the part during ejection.

Table 5.2 Cemented carbides for powder metallurgy tooling: Properties and typical applications

Grade	% of Co-binder	Hardness, HRA	Transverse rupture	Compressive strength, MPa	Tool application		
					Cores	Punches	Dies
C-4	3%–6%	92.3	1220	5516	Cores	Punches	Dies
C-9		91.5	1586	4895			Bearing dies
C-10	6%–9%	90.6	1931	4482	Simple shapes short lengths		Straight through dies—simple cavity contour
C-11	12%–13%	89.7	1931	4137		Ceramics, ferrites, high polish and no face projections	
C-12	14%–15%	88.5	2344	4000	Step cores and complex contours		Complex shapes, gear forms, sectioned dies
C-13	15%–20%	87.4	2586	3792	All cores within physical limits of carbides	All within physical limits of carbides	Multi-level dies, vulnerable projections
C-14	20%–30%	82–86	2517	3241			

(Arrows in the Tool application area: "Wear resistant" pointing upward for grades C-4 through C-11; "Shock resistant" pointing downward for grades C-12 through C-14.)

Note: All property data represent average for grade.

Punch and core rod dimensions are generally relieved behind the forming face so that the close fitting portion is as short as possible to permit escape of powder.

5.2.6 Part classification

PM parts are usually classified by evaluating the complexity of part design on a range of I through IV.

Class I parts: Single level components with the compacting force applied from one direction only. In this case the part thickness is generally limited to a maximum of 6–7 mm.

Class II parts: Single level components with the force applied from two directions.

Class III parts: Two level components pressed with forces from two directions.

Class IV parts: Multilevel parts pressed with forces from two directions.

Figures 5.12–5.15 illustrate the above classes of PM parts.

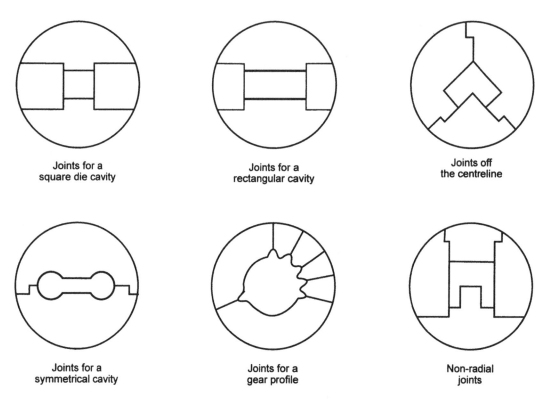

Fig. 5.11 Design of carbide inserts for fitting into retaining ring of die barrel

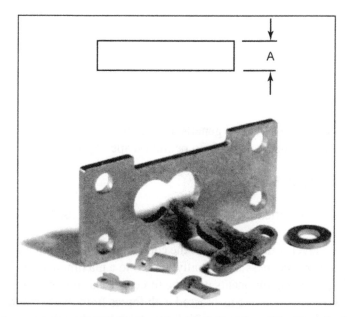

Fig. 5.12 Class 1 components (Reproduced with permission from *The Powder Metallurgy Design Manual*, Metal Powder Industries Federation, Princeton, USA, 1995)

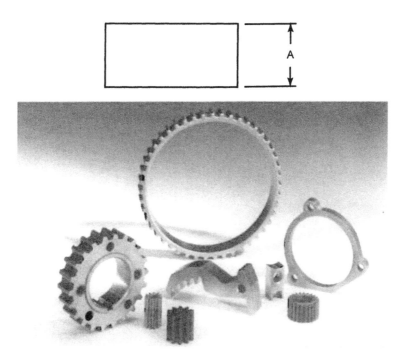

Fig. 5.13 Class 2 components (Reproduced with permission from *The Powder Metallurgy Design Manual*, Metal Powder Industries Federation, Princeton, USA, 1995)

Fig. 5.14 Class 3 components (Reproduced with permission from *The Powder Metallurgy Design Manual*, Metal Powder Industries Federation, Princeton, USA, 1995)

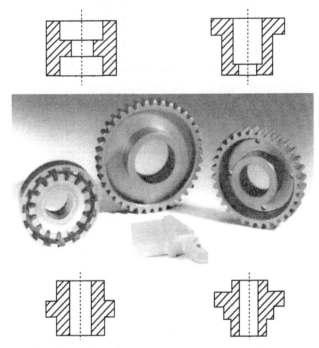

Fig. 5.15 Class 4 components (Reproduced with permission from *The Powder Metallurgy Design Manual,* Metal Powder Industries Federation, Princeton, USA, 1995)

5.2.7 Guidelines for part geometry

The following guidelines related to conventional die compaction should be highlighted.

Wall thickness: The minimum wall thickness is governed by overall component size and shape. Where the ratio of length to wall thickness is 8:1 or more, special precautions must be taken to achieve a uniform fill. The tooling required for long thin walls is quite fragile and has low life.

Holes: Holes in the direction of pressing, produced with core rods extending up through the tools, can be readily incorporated in the tooling. While round holes are least expensive, other shapes of holes, including splines, keys, keyways, D shapes, squares and hexagonals, can also be produced. Blind holes or blind steps in holes, and tapered holes which could be difficult to machine, are also readily produced. Side holes or holes not parallel to the direction of pressing cannot be made in the pressing operation and are produced by secondary machining. In big heavy parts, holes are intentionally incorporated in order to make them light.

Tapers and drafts: Draft is generally not required on straight-through PM parts. Tapered sections usually require a short, straight level to prevent the upper punch from running into the taper in the die wall or on the core rod (Fig. 5.16). When a flange-type section is made on a step in the die, a draft is desirable to assure proper part ejection. Similarly, drafts on the sides of bosses' or counter bores made by the punch face, aid tool withdrawal and minimise possible chipping of the part.

Fillets and radii: Generous radii fillets are desirable for economical design of tools and production (Fig. 5.16). A true radius is not possible at the junction of a punch face and a die wall. Therefore, a full radius is approximated by hand finishing or some other process such as tumbling.

Chamfers and bevels: Chamfers, rather than radii, are preferred on part edges to prevent burring. A 45° angle and 0.125 mm minimum flat is common practice to eliminate feather edges (Fig. 5.17). The

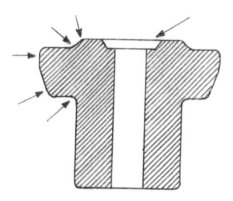

Fig. 5.16 Tapers and drafts in a PM part

preferred chamfer, and the most economical to produce, is 30° maximum from radial, in order to minimise the chance of breaking the punch protrusion. When chamfers with a large angle from the horizontal are required for punching a part, or when a step would create a problem (Fig. 5.17), the chamfer can be produced by a bevel in the core rod or die. Chamfers on irregular shapes are more costly than on a plain round part or on two sides of a square part.

Countersinks: A countersink is a chamfer around a hole for a screw or bolt head. A flat of about 0.25 mm is essential to avoid fragile, sharp edges on the punch (Fig. 5.18).

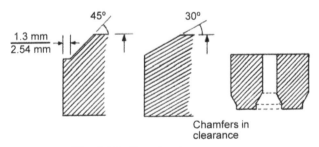

Fig. 5.17 Chamfers in PM parts

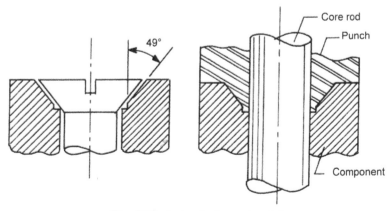

Fig. 5.18 A standard counter sink

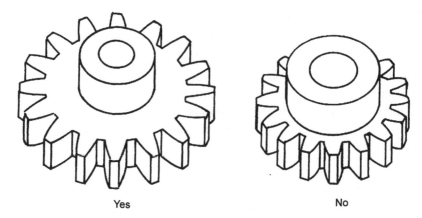

Yes   No

Fig. 5.19 Hub's location for gears root

Bosses: A boss can be located on top or at the bottom of a part. To produce bosses with perpendicular sides, special punches are required to give positive part ejection.

Hubs: Hubs provide drive or alignment rigidity in gears, sprockets and cams. They can be readily produced by the PM process. A generous radius between the hub and flange is preferred, as well as maximum permissible material between the outside diameter of the hub and the root diameter of components (Fig. 5.19).

Studs: Shallow studs with drafted sides are made in the regular tools.

Knurls: Vertical—but not diamond or angled—knurls can be made on the inside and outside diameters because they interfere with ejection.

Undercuts: Undercuts on the horizontal plane cannot be produced, as they prevent the part from being ejected from the die. A part with a reverse taper (larger at the bottom than the top) cannot be ejected from the die.

Slots and grooves: Grooves can be pressed from projections on the punch into either end of a component, within the following limits.

• Curved or semicircular grooves to a maximum depth of 30% of overall component length,
• Rectangular grooves to a maximum depth of 20% of the overall component length.

Threads: These cannot be incorporated in holes or outside diameters, as they prevent the part being ejected from the die. They are produced by secondary machining operations.

5.3 WARM COMPACTION

The warm compaction process was patented by Höganäs, Sweden, under the name 'Ancordense' in the mid-1990s. The system utilises heated tooling and powder to achieve improved green density, high green strength and improved ejection characteristics. The powder can be heated by a variety of methods, including screw heating, microwave heating or a slotted heat exchanger. The warm compacted parts can be machined and then subsequently subjected to conventional

Example 5.1: A sponge iron powder has an apparent density of 3 g cm⁻³ and a tap density of 4 g cm⁻³. The powder is die compacted to a green density of 6 g cm⁻³ at 300 MPa. If the powder is compacted to 10 mm, what is the initial powder fill depth?

Solution: Assuming the cross section of the die opening is constant,

$$\frac{\text{Apparent density}}{\text{Green density}} = \frac{\text{Height of green compact}}{\text{Height of loose powder}}$$

Height of loose powder = $\dfrac{6}{3}$ × 10

= 20 mm

The tap density value is not required for solving this problem.

Example 5.2 The designs of PM parts shown in Fig. Ex. 5.2a are improper. Give the correct versions, justifying your answer.

Solution: The correct designs are presented in Fig. Ex. 5.2b. The justification for each is as follows:

(a) The part section is too thin. This will cause tooling failure.

(b) Sharp corners on tooling can lead to tool failure. Hence radii are preferred.

(c) There is a large change in section, which will lead to density variations and hence part weakness.

(d) The re-entrant features have to be redesigned, otherwise the part cannot be ejected out from the die during the pressing cycle.

(e) Chamfers are preferable to radii in the edges on the part. Chamfer angles of 45° or less cause powders to chip at the edges.

(f) It is better to avoid the undercut, which cannot be mounted. The modified design takes care of this problem.

(g) In this case, not only is the component thin, but also has large surface area. To solve this problem the projection should be made thick and sharp edges should be given radii.

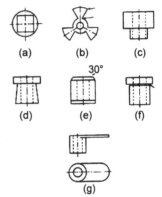

Fig. Ex. 5.2a

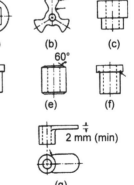

Fig. Ex. 5.2b

sintering process. Both apparent density and flow of powder are very stable and not sensitive to temperature variations. To use the warm compaction process successfully in production it is recommended that the powder be heated to 130°C and 150°C for the die and upper punch, respectively, and allow a scatter of ± 2.5°C maximum for the temperature within the whole system. High-performance turbine hubs have been produced by warm compaction of a diffusion-alloyed iron powder containing 4% Ni, 1.5% Cu and 0.5% Mo to a density of 7.25 g cm^{-3}. The use of such an alloyed powder allowed heat treatment to be avoided.

5.4 WET COMPACTION

This method is common in clay-based ceramics, which uses a soft plastic mass. The starting material is de-aired and extruded blank or as shredded granules into a steel or plastic die. As the pressure is applied, the plastic mass flows into the various contours of the die. Typical products formed in this way are electrical insulations and special refractory shapes. In one version of the process, the steel die is heated. However, this should not be confused with hot pressing. Heating causes a steam cushion to develop between the wet plastic mass and the die surface, acting as a lubricant to allow easy removal of the formed part. In case the die is not heated, some sort of lubrication is to be improvised. In some machines, the die is rotated. Presses range from small size part to big sizes. Two points to be taken into account are: moisture content of the body and the proper de-airing of the starting material.

5.5 COLD ISOSTATIC COMPACTION

In cold isostatic compaction a flexible mould is filled with the powder and pressurised isostatically using a fluid such as oil or water. Compaction pressures up to 1400 MPa have been achieved in this manner; however, cold isostatic compaction is usually performed at pressures below 350 MPa. Complex shapes can be created by using a rubber mould.

Advantages of cold isostatic compaction

- Uniform density of compacted bodies
- High green density—about 5%–15% higher than that achieved with die compaction at the same pressure
- High green strength and good handling properties of the powder body
- Reduced internal stress
- Powder can be compacted without binding or lubricant additives
- Bodies with complex shapes or with a large length to cross section ratio can be compacted with high, uniform density
- Composite structures can easily be obtained
- Low tool costs through the use of rubber or plastic moulds
- Low material and finish machining costs.

Disadvantages of isostatic compaction

- Dimensional control of green compacts is less precise than in rigid die pressing.
- Surface of isostatically pressed compacts is less smooth.
- In general the rate of production in isostatic pressing is considerably lower.
- The flexible moulds used in isostatic pressing have shorter lives than rigid steel or carbide dies.

5.5.1 Isostatic press equipment and pressing operation

The main subgroups of isostatic presses are described below (Fig. 5.20).

Pressure vessel or cavity: The pressure vessel is the most important element of an isostatic press. Of all the various high-pressure structures known, the monolith forged-type is preferred. While forging, care is taken that the grain structure is similar in all directions so that with closely controlled heat treatment the high mechanical properties obtained in tangential, radial and longitudinal directions of the forging will be as desired and within close tolerance of each other. The pressure vessel must be designed to withstand the severe cyclic loading imposed by rapid production rates and must take into account fatigue failure. Pressure vessels designed and constructed as per Section VIII, Division 2, of the ASME Code are available for pressures up to approximately 276 MPa (40,000 psi). Devices, including

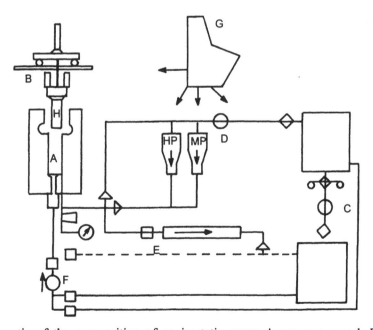

Fig. 5.20 Schematic of the composition of an isostatic press: A pressure vessel, B closure with mechanism, C reservoir(s) with filtering system, D high-pressure generator group, E depressurisation system, F fluid transfer system, G controls, H tooling

an absolute closure control, absolute pressure restrictor and an energy absorption protection shield, should be installed in direct combination with the pressure vessel.

Closure system: This seals the pressure vessel cavity. Threaded covers are extensively used. However, a proper design must be adopted, otherwise any stress concentration at the root of the first thread could restrict the life of the vessel assembly. Quicker and more economical automated closing mechanisms can be effected by closure with interrupted threads.

Reservoir with filtering system: In case of dry bags, the contamination of the fluid is avoided by the use of normal hydraulic systems reservoirs and good conventional filters. For wet bags, where mostly water is used, the contamination is acute and adversely affects the pumps and seal life.

High pressure generator: From being small high-pressure hand pumps, pressure generators have developed by the use of small air hydraulic intensifiers. In these, a large air driven piston moves a small high-pressure hydraulic piston which pumps the liquid to the desired pressure. The surface ratio of the low-pressure air piston versus the hydraulic piston can reach 600:1 or more.

Depressurisation system: There are various depressurisation systems capable of almost any decompression profile to help eliminate compact breakage that may occur by too rapid depressurisation.

Fluid transfer system: After the compaction step, the superfluous quantity of liquid in the pressure vessel must be evacuated as rapidly as possible at the end of the depressurisation stage. All such transfers can be accomplished by an appropriate standard transfer pump and the correct valving between the reservoirs and the pressure vessel.

Controls: Most simple units have manual control of the pumps, valves and other mechanisms. However, automatic controls by servo-operated valves with identical closing and opening forces are used. This enhances the service life of high-pressure valves.

Tooling
The following factors have to be considered when designing proper tooling.

1. Type of press
   - wet or dry bag
   - automatic, semi-automatic or manual
   - top or bottom ejection
2. Properties and type of powder
   - metal, ceramics, etc.
   - flowability
   - compression ratio
   - green strength
   - adhesion to bag or punches
   - density

3. Bag material
  - dipped latex, natural rubber, neoprene
  - PVC
  - moulded natural rubber, neoprene, nitrile
  - polyurethane
4. Production rate
5. Life of tooling
6. Accuracy of compact
7. Shape of compact
8. Operator's skills.

   Table 5.3 summarises the main properties of flexible die materials.

Wet bag tooling: The filled and sealed mould is immersed into a fluid chamber which is pressurised by an external hydraulic system (Fig. 5.21). After pressing, the wet mould is removed from the chamber and the compact removed from the mould.

Dry bag tooling: Both powder filling and ejection are performed without removing the bag assembly. Sealing is achieved by an upper punch which enters the bag before pressurisation. The compaction stresses are generated by isostatic compression of the bag through a fluid without loading the punch. The dry bag method is much more rapid because the bag is built directly into the pressure cavity.

Table 5.3 Type of flexible mould materials and their properties

	Natural rubber latex	Natural rubber hot moulded	Neoprene latex	Neoprene hot moulded	Silicone	PVC plastisol	Polyurethanes
Water resistance	G	G	G	G	G	F	F → Ex
Oil resistance	P	P	G	Ex	G	F	G → Ex
Aging resistance	P	F	G	G	G	P	P → Ex
Toughness	VG	VG	VG	VG	P	P	G → Ex
Tear resistance	VG	VG	VG	VG	VP	P	G → Ex
Cut resistance	VG	VG	VG	VG	VP	P	G → Ex
Resilience	Ex	Ex	G	G	VG	P	P → Ex
Set resistance	Ex	Ex	VG	VG	G	P	VP → Ex
Availability	R	S	R	S	S	R	S
Cost	L	M	L	M	H	L	L → M → H

R = Readily available; S = Subject to mould; L = Low; M = Medium; H = High; VP = Very poor; P = Poor; F = Fair; G = Good; VG = Very good; Ex = Excellent.

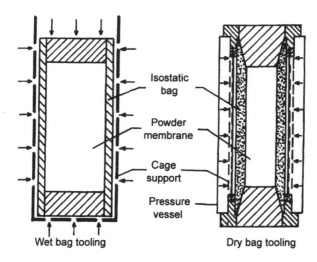

Fig. 5.21 Wet and dry bag tooling in cold isostatic pressing

In general, the more versatile wet bag process is used for the batch production of a small number of shapes, whereas the dry bag process is used for the semiautomatic or automatic production of a large number of shapes, often of smaller dimensions.

A typical isostatic pressing cycle comprises of the following.

Insertion of compact: The powder material to be pressed is sealed and placed in the partially filled vessel. This causes the liquid level in the vessel to rise.

Filling and venting: The upper closure is installed and locked. Any air remaining above the fluid level is removed. Otherwise, greater amount of energy would be used before the vessel reaches operating pressure, because, air is highly compressible. The displaced air is vented through the valve in the top closure.

Pressurising: The high-pressure pumping system pressurises the water to the operating pressure. Water is added during pressurisation in order to compensate for the volume reduction of the powder being pressed and the compressibility of water at these pressures. Table 5.4 gives some typical values of pressures required for isostatic pressing.

Table 5.4 Typical pressures required for isostatic pressing powders

	MPa	ksi
Aluminium	8–20	55–138
Iron	45–60	311–414
Stainless steel	45–60	311–414
Copper	20–40	138–276
Lead	20–30	138–207
Tungsten carbide	20–30	138–207

Note: These pressures are approximate, since every application presents its own requirements of density, configuration and size.

Depressurising: During this stage, the extra amount of water is expelled from the vessel.

Compact removal and draining system: The green part and mould are now removed from the pressure chamber. The water level drops, but subsequently overflows if another uncompacted mould is inserted.

The major defects associated with improper tooling are as follows.

Elephant footing: If the modulus of the bag is less than that of the closure, then the bag will distort or compress more than the closure. The result is elephant footing. Figure 5.22 shows the problem and a suggestion for its remedy.

Poor consolidation: If the modulus of the bag is too high, or the wall thickness is too great, then insufficient pressure will be exerted on the powder. Altering the hardness of the bag/gauge normally eliminates the problem.

Radial cracking: When a bag with a very thick wall is used, the pressing pressure causes the material to 'ruckle'. This gives differing radial pressures along the length of the preform and the resultant differing densities can cause cracking on decompression. Thinner gauge sections are usually required to reduce or eliminate this problem.

Preform cracking: If the bag properties are poor, then powder sticking can be a major problem. On decompression, a surface layer can be turned out of the preform, resulting in cracking which is noticed on sintering. A change in the grade of bag material is called for.

Some examples of cold isostatic applications are:

- *Graphite:* Electrographite, refractory graphite and graphite used in nuclear reactors.
- *Ceramics:* Cold isostatic pressing of tubes, tiles, nozzles and linings made of special ceramics

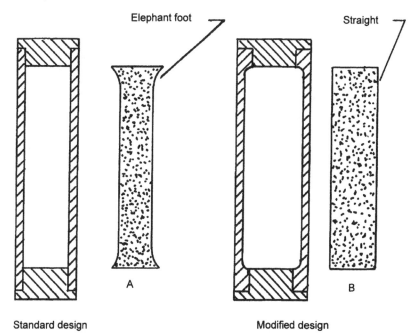

Fig. 5.22 Modified tooling design for cold isostatic pressing

- *Ferrites:* Parts used for permanent magnets, computer memories and electronics
- *Cemented carbides:* To preform powder products
- *Metal powder:* High-speed tool steels, superalloys, stainless steel, titanium alloys, beryllium, etc.

5.6 POWDER ROLL COMPACTION

Powder rolling, also called *roll compacting*, is an important process for producing metal strips. In powder rolling, metal powder is fed from a hopper into the gap of a rolling mill and emerges from the gap as a continuous compacted green strip. The rolls of the mill may be arranged vertically or horizontally. The latter type of arrangement is more common, with either saturated feed or starved feed.

The powder characteristics have the following effect on powder rolling.

Particle shape: The generation of maximum 'green' strip strength to withstand the rigors of handling through the process line requires particle shape to be irregular.

Compressibility: Good compressibility is required to ensure that sufficient particle interlocking takes place to give adequate 'green' strip strength. A good green strip has a density of at least 80%–85% theoretical. Compressibility is also of importance in determining the limiting dimensions of the roll compaction mill.

Particle size: The thickness of the finished strip and particle segregation severely restrict maximum particle size which can be tolerated in the powder feed to the compaction mill.

Flowability: The powder must flow smoothly and quickly through the hopper systems with minimum tendency to stick, slip or bridging.

Surface oxidation: This plays a significant part in determining subsequent powder behaviour.

The roll compaction operation can be divided into three distinct zones (Fig. 5.23).

1. The *free zone* where blended powder in the hopper is transported freely down under gravity. Here all the usual criteria of hopper flow apply.

2. The *feed zone* where the powder is dragged by the roll surface into the mill bite, but has not yet attained coherence.

3. The *compaction zone* close to the roll nip, where the powder becomes coherent, the density changes rapidly and air has to be expelled.

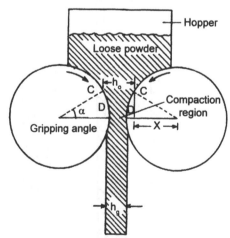

Fig. 5.23 Schematic of powder rolling with saturated feed and rolling mill rolls arranged horizontally

In contrast to conventional rolling, the thickness of the strip which can be rolled in powder rolling is closely limited by the diameter of the rolls of the rolling mill. The change in density is accomplished as the powder is transported through the feed zone and the compacting zone. The length of these zones is determined by the diameter of the rolls (D), the internal friction between the powder particles and the friction between powder and rolls. With the geometry shown in Fig. 5.23, the nip angle α may be defined as:

$$\cos\alpha = \frac{x}{D/2} \tag{5.5}$$

The dimension, h_o, will be equal to $D(1 - \cos\alpha) + h_g$. The strip thickness, h_g, is equal to h_o/C, where C is the compression ratio. The strip thickness would then be:

$$h_g = \frac{D(1-\cos\alpha)}{C-1} \tag{5.6}$$

It has been shown that, due to slipping between powder particles, the actual angle at which the powder is gripped is much lower than the calculated friction angle, α. As a result, in order to roll compact to a certain thickness of strip it is necessary to use rolls of much larger diameter than are required for producing similar strips from solid material. Roll diameters between 50 and 150 times the strip thickness are often required. The maximum strip thickness can be increased by increasing the value of μ, i.e., by roughening the roll surface, although, during rolling the conditions change as the surface ultimately gets polished. Figure 5.24 shows the strip thickness that can be obtained with different roll diameters with two types of nickel powders.

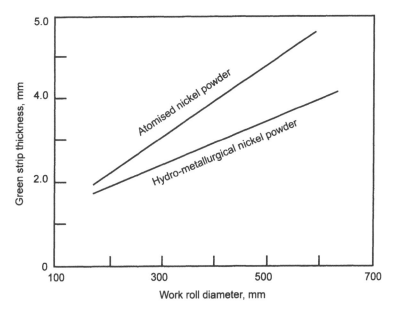

Fig. 5.24 Effect of work roll diameter on green strip thickness for two types of nickel powder

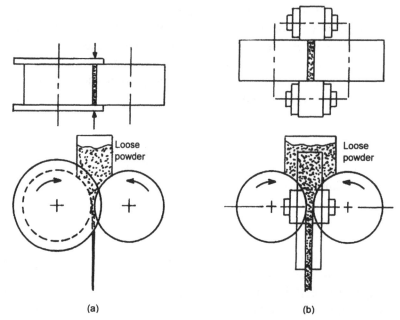

(a)   (b)

Fig. 5.25 Methods of controlling powder feed to compacting rolls: (a) Flange-edge control and (b) Bell-edge control

The green powder strip should have uniform thickness and density across the width of the strip, and its edges should be well formed and dense at the centre of the strip. Therefore, edge controlling is essential to process control. Figure 5.25 illustrates typical methods for such a control. In Fig. 5.25a, floating flanges that are attached to one roll and that overlap the other roll are used. Pressure is applied to the flanges as they approach the roll gap, thus preventing powder loss from the gap. A continuous belt that covers the gap at the edge of the rolls is also effective in preventing powder loss (Fig. 5.25b).

If the air trapped in the powder is not properly released, the disturbance of the powder in the hopper can be sufficient to interface with the smooth flow of powder to the roll nip, and the strip produced will not be of uniform density. Up to a given speed, called the 'flow transition speed', the flow rate of powder increases linearly with roll speed. Above this critical speed, relatively less powder flows into the roll gap, until at speeds considerably above the transition speed, a continuous strip can no larger be rolled.

5.7 POWDER EXTRUSION

Cold powder extrusion is not very common for metals, except for metals with relatively low melting point, e.g., lead. The product obtained through the process has a high aspect ratio. In cemented carbide drill making and various ceramic products, extrusion is very much applied. There are two types of extruders: piston type and auger type (Fig. 5.26). In the former, the feed material is compressed and forced to flow down the barrel by the

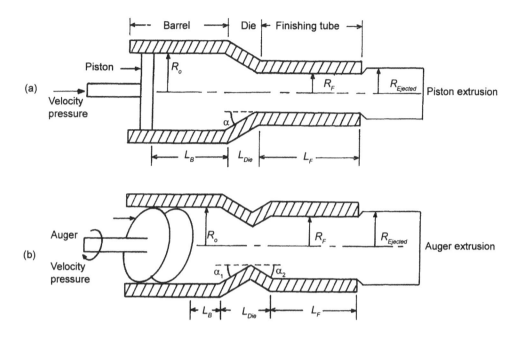

Fig. 5.26 Schematic diagram of (a) piston extruder and (b) auger extruder

moving piston. For the displacement of material in auger feeding, the material must not slip on the wall of the barrel and the yield strength of the body must be less than the adhesive strength of the body on the surface. Axial rib may increase the circumferential friction. A helix angle of approximately 20°–25° is commonly used. The required ratio of the auger diameter to the product diameter increases as the yield strength of the material increases. Geometrical parameters of the die are the entrance angle α and the reduction ratio R_o/R_f (Fig. 5.26). Complex dies may contain small channels for injecting a die wall lubricant.

During extrusion, flow occurs by the mechanisms of slippage at the wall and differential laminar flow in the extrudate. Figure 5.27 shows the pressure relationship during extrusion of an electrical porcelain body in a piston extruder. It is evident that the pressure is highest in the barrel and decreases along the axis of the extruder.

Some of the common extrusion defects are:

- Insufficient strength and stiffness
- Cracks and laminates
- Surface craters and blisters
- Non-uniform flow through the die
- Gradients in extrudate stiffness

Table 5.5 gives some examples of plastic bodies of ceramics obtained through extrusion.

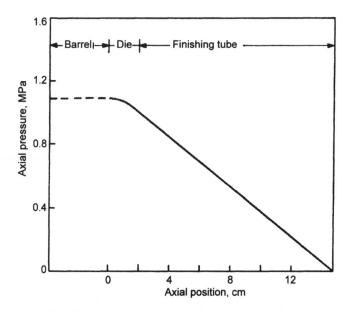

Fig. 5.27 Axial pressure gradient during the extrusion of an electrical porcelain body in a piston extruder $(R_o/R_f = 3, R_f = 0.6$ cm$)$ (Adapted from Reed JS, *Introduction to the Principles of Ceramic Processing,* John Wiley & Sons, Singapore, 1989)

Table 5.5 Some ceramic compositions of extrusion bodies

Composition, vol%					
Refractory alumina		**High alumina**		**Electrical porcelain**	
Alumina (< 20 µm)	50	Alumina (< 20 µm)	46	Quartz (< 44 µm)	16
Hydroxyethyl cellulose	6	Ball clay	4	Feldspar (< 44 µm)	16
Water	44	Methylcellulose	2	Kaolin	16
$AlCl_3$ (pH > 8.5)	<1	Water	48	Ball clay	16
		$MgCl_2$	<1	Water	36
				$CaCl_2$	<1

Note: Chlorides are coagulants, while cellulose is a binder.

5.8 INJECTION MOULDING

The introduction of powder injection moulding (PIM) has already been made in Chapter 1. Other terms like metal injection moulding (MIM) and ceramic injection moulding (CIM) are also frequently used.

In injection moulding, the feedstock pellets are heated to a sufficiently high temperature such that they melt, and the melt is forced into a cavity where it cools and assumes a compact shape. The purpose is to obtain the desired shape free of voids or other defects and with a homogeneous distribution of powder.

There are three types of moulding machines: reciprocating screw, hydraulic plunger and pneumatic. The most commonly used is horizontal reciprocating screw inside a heated barrel, which has adequate control. Figure 5.28 illustrates the reciprocating screw type injection moulding machine. After the mould is filled, heat is extracted from the feedstock through the die. Finally, the cavity is opened to eject the hardened compact. The presence of high content of solids raises the viscosity, requiring high pressures during moulding. High packing pressure can result in the compact sticking to the die wall with severe ejection problems.

In brief, moulding parameters are highly dependent on particle characteristics, binder formulation, feedstock viscosity, tool design and machine operating conditions. The probability of defects and the presence of distortion in sintering is highly dependent on the moulding step.

The capacity of injection moulding machines are specified by the following parameters:

- Quality of feedstock delivered to the mould
- Clamping force on the die and clamp opening force
- Injection pressure
- Speed of moulding in terms of shots or volume per unit time
- Space of moulding in terms of depth of moulding
- Mould opening size or travel
- Parting line versus perpendicular injection
- Open versus closed loop control.

Table 5.6 provides examples of typical ranges for PIM moulding conditions.

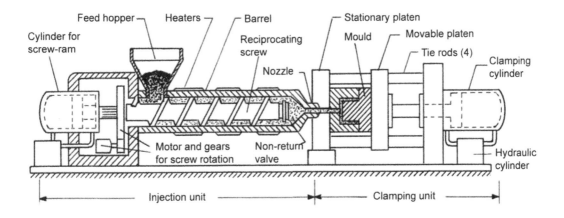

Fig. 5.28 An injection machine, reciprocating screw type (Reproduced with permission from Groover MP, *Fundamentals of Modern Manufacturing*, 2nd edn, John Wiley & Sons, Hoboken, NJ)

Table 5.6 Typical PIM machine parameters

Barrel temperature	100–180°C
Nozzle temperature	145–175°C
Mould temperature	35–40°C
Screw rotation rate	35–150 rpm
Injection pressure	14–20 MPa
Packing pressure	7 MPa
Injection time	0.4–0.6 s
Packing time	2–12 s
Cooling time	18–45 s
Total cycle time	37–65 s

5.8.1 PIM part design

The following features would suffice for the practical ranges for various design characteristics for PIM parts.

Size range: PIM processing is best suited for relatively small parts. The lower limit of size is determined only by the limitations of the injection moulding process itself. Parts as small as $6.4 \times 2.5 \times 1.3$ mm^3 are economically feasible. The upper limit of the part size is established by processing economics. As the part size increases, the powder cost becomes a significant percentage of the overall cost.

Section thickness: The part thickness must be kept relatively low in order to remove binders efficiently. Minimum wall thickness should be about 0.5 mm which makes the process very competitive for producing thin-walled parts.

Example 5.3 In some metal injection moulded parts, different wall thicknesses cannot be avoided as shown in Fig. Ex. 5.3a. Suggest some changes in the design configuration in order to ensure proper packing of the feedstock.

Solution: A graded transition from one thickness to another reduces stress concentration and poor surface appearance. It is recommended that the ratio of the transition be changed as shown in Fig. Ex. 5.3b. In addition, the mould should be gated at the heavier section to ensure proper packing of the feedstock. See Fig. Ex. 5.3c.

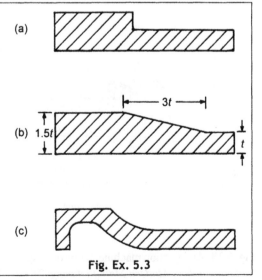

Fig. Ex. 5.3

Part complexity: Because the metal/polymer feedstock can nearly fill any volume, complex geometries can be created. Hence, the economical advantage of the PIM process is felt for more complex parts. For example, external threads can be moulded directly on the parts, although an allowance must usually be made for parting lines. While internal threads can also be directly formed, it is often more economical to thread in a subsequent tapping operation. Complex internal cavities are difficult to form, where undercuts prevent the tool member from being withdrawn.

Gates, parting lines, knockout or ejection pin marks are present in PIM parts and the design of parts should take them into consideration. For example, flats can be built into parts in the areas of parting lines, particularly on contours or external threads. The areas around the pins may be recessed if *bursts* caused by knockout pins are the problems.

5.9 GREEN PART MATERIAL HANDLING

For better productivity, the use of automation is essential for green part handling. The green parts are relatively delicate and fragile compared with sintered parts; hence, greater care is called for. The first step is unloading the compacting press. Frequently, a feeder shoe is used to push the part out of the press. Such a method is feasible for robust parts that have a smooth underside and strong outside features. Parts that have external features on the underside or outside require additional equipment. The feeder shoe can be equipped with a gripper, or unloading devices can be used. Press take out devices based on pneumatic or electrical activators are also in use. In addition, take out robots are also available and are typically employed for larger and more complex parts.

The subsequent material handling operation after compaction is dictated by whether the product flow is in batch or continuous flow mode. Batch operation is defined as moving a quantity of parts together between processes. On the other hand, the continuous or one-piece flow operation is defined as moving one piece at a time through the processes in a continuous fashion. The batch set up is typically seen in a conventional factory layout where the compacting department feeds the sintering department, and so on. In continuous process the equipment must be able to handle a steady rate of product arriving at the station.

The decision regarding the selection of batch or continuous flow models is dependent on the following factors:

- Frequency of tool change
- Time required to change press tools
- Shipment frequency and quantity
- Plant layout and physical space constraints
- Annual production requirements
- Mix of products that have to pass through the production centres
- In-house technical capability or capacity to maintain complex automation.

Typically, large volume orders (>900,000 line annual volume) are candidates for continuous flow set ups. It is possible to group similar parts into a family and use continuous flow. In batch operation, the green parts are usually gathered in the same type of tray or dunnage at the compacting press. This can be accomplished by the operator, or through the use of automation. The trays are usually racked or palletised and then transported to the sintering furnace, where they are transferred to the belt. In case of continuous operations, several types of conveyors are used for transportation to the sintering furnace. It is always advisable to avoid part-to-part contact in the green state, especially for PM parts with fragile outside features or thin walls. It is common to integrate other equipment in-line with green part conveyer. Typical items include weighing, vision inspection, a sorting station, deflashing/deburring and cleaning. For better productivity, it is advisable to use the sintering plate as a pallet to move parts from the compacting press to the sintering furnace. In case the parts are tall or difficult to handle, e.g., sprocket with small hub, they can be suitably conveyed from the compacting press to the sintering furnace by a sintering plate with a hole in the centre.

SUMMARY

- The actual packing density depends on powder characteristics.
- During the densification of powder mass in a die, a transition from elastic to plastic deformation stage occurs.
- Powders do not respond to pressing in the same way as fluids and do not assume the same density throughout the compact.
- Pressure operation of powder consists of filling, pressing and ejection.
- Mechanical and hydraulic presses are the major types of compaction presses.
- Hydraulic presses are slower in operation than mechanical presses and are less economical to operate.
- Fill, flow, apparent density, fill ratio, compacting pressure and dimensional changes of parts are responsible for tooling design.
- Flow during powder extrusion occurs by the mechanisms of slippage at the wall and differential laminar flow in the extrudate.
- The purpose of powder injection moulding is to attain a shape free of voids and defects with a homogeneous distribution of powder.

Further Reading

Arunachalam VS and Roman OV, (eds), *Powder Metallurgy: Recent Advances*, Oxford & IBH Publishing Co., New Delhi, 1989.

Bradbury S, (ed.), *Powder Metallurgy Equipment Manual*, 3rd edn, Metal Powder Industries Federation, Princeton, 1980.

German RM and Bose A, *Injection Molding of Metals and Ceramics*, Metal Powder Industries Federation, Princeton, NJ, 1997.

German RM, *Particle Packing Characteristics*, Metal Powder Industries Federation, Princeton, 1989.

James PJ, (ed.), *Isostatic Pressing Technology*, Applied Science, London, 1983.

Kunkel RN, *Tooling Design for Powder Metallurgy Parts*, American Society of Tool and Manufacturing Engineers, Dearborn, 1968.

Mutsuddy BC and Ford RG, *Ceramics Injection Moulding*, Chapman and Hall, London, 1995.

Onada GY, Jr and Hench LL, (eds), *Ceramic Processing before Firing*, Wiley-Interscience, New York, 1978.

Smith LN, *A Knowledge-based System for Powder Metallurgy Technology*, Professional Engineering Publishing, London, 2003.

EXERCISES

5.1 Calculate the following for cubic and tetrahedral configuration of sphere packings:

(a) Void fraction

(b) Volume ratio of void/sphere

(c) Radius ratio of primary sphere and interstitial sphere

5.2 Calculate the packing density of mixed spheres of diameters 1.28 mm and 0.155 mm, respectively, in the weight fractions 0.73 and 0.27.

5.3 In packing platelet and fibrous particles, on what factors does the density depend?

5.4 Polymeric binder over the surface of ceramic particles reduces the packing density. Explain why?

5.5 It is noticed that continuously rotating the pressing punch during powder compaction results in an increase in the density of the green compact and an improvement in density homogeneity. Explain why?

5.6 Why is it difficult to die compact amorphous metglass powder? What other method would you select?

5.7 Alumina powder containing 4 wt% polyvinyl alcohol binder after pressing at 90 MPa was found to have laminations in the part. The compacts had lower green density and green strength. What is your solution for this?

5.8 A 10 mm diameter final dimension powder compact is needed with a final density of 85% which is to be formed from an 82% dense pressing. What die diameter is needed for this process?

5.9 Three powders are die pressed using a compaction pressure of 450 MPa. One of the three is –325 mesh Cu, the second is –325 mesh superalloy and the third is –100 mesh copper. Which powder gives the highest green strength? Why?

5.10 A PM part (drawing shown in Fig. P. 5.10) is to be made out of iron-10% copper powder premix at a compaction pressure of 520 MPa. Determine the following:

(a) The required press tonnage to compact the part;

(b) The final weight of the part if the porosity is 10%.

Assume shrinkage during sintering to be negligable (All dimensions are in mm).

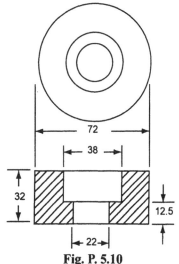

Fig. P. 5.10

5.11 You are asked to prepare Cu-10 vol% W and Cu-20 vol% W of composites from copper powder and tungsten fibres (10 μm dia) by hot pressing. The strength of the composite in tension depends strongly on the angle Ø between the tensile and fibre axis, and is maximum for Ø = 0. Describe various technological propositions during processing, so that the angle remains close to zero.

5.12 Graphite powder in PM steels acts both as a lubricant as well as an alloying element. In case you are to prepare a eutectoid plain carbon steel compact, will this premix method suffice?

5.13 A tungsten bearing alloy part is to be produced by conventional press and sinter technology. In order to avoid contamination during milling of the powder, it was proposed to prepare balls of 1 cm diameter of the same composition as the milling medium by conventional press and sinter route. Sketch the tooling design to make the green tungsten–based spheres. What special precautions would you suggest?

5.14 In powder rolling, to achieve a certain strip thickness, why are larger diameter rolls required in comparison to the strip rolled from a solid stock?

5.15 During powder milling the maximum strip thickness can be increased by roughening the roll surface. Why?

5.16 What types of extrusion presses should be selected for (a) plastic ceramic and (b) non-plastic ceramic mixes?

5.17 Describe how the long single piece high voltage electrical insulators in green form are produced.

5.18 More complex-shaped parts can be produced by PIM process than by conventional PM. Justify this statement giving examples.

5.19 The green density for stainless steel powder is 6.5 g cm^{-3}. The apparent density is 2.7 g cm^{-3}. What is the compression ratio and the required powder fill height for a final height of 4 cm?

5.20 Why is galling a major problem in powder die compaction?

5.21 When do close tolerances need not be specified in MIM processing?

5.22 For MIM parts designed with thin walls, reinforcing ribs are an effective way to improve rigidity and strength. However, there are certain disadvantages. Name them.

5.23 External and internal threads can be moulded into MIM parts. Which ones are more problematic? What is the least expensive moulding method for external threads? How can one ensure that tight tolerances are held on the thread diameter?

5.24 Designing MIM parts with internal undercuts or recesses is not recommended. Why?

5.25 A two-step part is to be produced using MIM process. The design engineer recommends modifications in the design. Suggest the changes.

5.26 What are the advantages of green machining over machining of sintered component?

5.27 Compare conventional and metal injection moulding for the following:

(a) Powder particle size

(b) Porosity before sintering

(c) Lubricant/ binder content

(d) Sintered density (%).

5.28 Between PM strips made of copper and stainless steel powders, which one will be more economical compared to the corresponding ingot metallurgy processed strip? Why?

5.29 Which of the following is not important in achieving a straight, uniform density extruded part?

(a) Tool design

(b) Rheology of the mix

(c) Wet strength after extrusion

5.31 Generally, nanocrystalline powders are less forgiving to green compact defects than conventional powders. Comment on this statement.

5.32 Cordierite ceramic honeycomb is used for catalytic converters in automobiles. Describe the method adopted for its shaping from powder.

6

Pressureless Powder Shaping

LEARNING OBJECTIVES

- Pressureless powder shaping methods
  - ◦ Slip casting/slurry moulding
  - ◦ Electrophoretic deposition shaping and its significance
  - ◦ Spray deposition
  - ◦ Solid freeform fabrication

The previous chapter described how powder mass is shaped into green shape by applying pressure within a die. This chapter deals with shaping without application of pressure. The methods include slip casting, tape casting, spray deposition and solid freeform fabrication. Slip casting is a very old method and was used by potters since ancient times, while freeform fabrication is the latest addition, where extensive use of computer aided design (CAD) is incorporated.

6.1 SLIP CASTING/SLURRY MOULDING

A slip is a slurry of very fine metal or ceramic material suspended in water. If such a slip is poured into a porous plaster mould, the mould will draw water from the slurry and will build up a deposit of particles on the mould wall. In this manner an article can be formed with an outer configuration that reproduces the inner configuration of the plaster mould. Wall build up during slip casting continues as long as any slip remains in the mould. However, the rate of build-up drops off with time, because water cannot be removed from the slip by passing through the thickening cast wall. If a hollow cast is required, the excess slip is poured or drained out of the mould after the required wall thickness has been achieved. This is sometimes

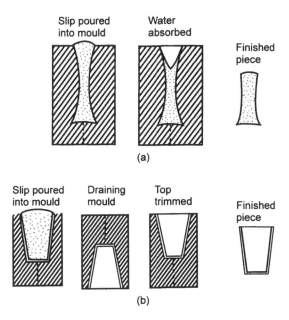

Fig. 6.1 Schematics of (a) solid slip casting and (b) drain slip casting

called *drain casting*. Figure 6.1 shows the scheme of operations carried out to produce solid or hollow casts.

Sometimes, instead of water, another liquid can be used for slip casting. For example, MgO and CaO react with water. Hence, a suspension of absolute alcohol or another inert liquid is prepared and cast. Slurry moulding is similar to slip casting, except that the mould is solid and the powder becomes rigid due to polymer freezing.

The casting behaviour of the slip depends on the specific gravity and viscosity of the slip, and more importantly, on how the viscosity changes with time. The specific gravity of the slip should be kept as high as possible, but consistent with the appropriate viscosity parameters. The flow behaviour of slips is generally controlled by means of small amounts of dispersion additives or deflocculants. In clay-based ceramics, the common deflocculants used are sodium silicate, sodium carbonate, sodium phosphate, and a number of organic substances such as polyacrylates. For non-clay-based slips, strong acids on bases often give dispersions, but certain organic substances are also suitable for this purpose. In case the particle size distribution of a slip is wide, the slip will be difficult to disperse and will settle rapidly, giving uneven casts. On the other hand, if the particle size is too fine, the slip will cast very slowly.

The mould may consist of more than two parts, depending on the complexity of the part. The mould parts must be tightly held together during casting, as a hydrostatic pressure builds up in the mould due to the weight of the slip. The mould is subsequently disassembled after casting to remove the cast.

Casting is followed by drying and sintering. Section 6.1.2 discusses the drying operation, while sintering technology is discussed in Chapter 8.

Slip castings are most effective for longer shapes, since the equipment and tooling costs are not high.

6.1.1 Viscosity of a slurry

The viscosity of the fluid is important in slip casting slurries of metal or ceramic powders. Viscosity is also important in the injection moulding process. It is also of significance in pumping, filtering, pouring, mixing, spray drying and settling of slips, and in drying operations. The viscosity of a fluid represents the relation between the shearing stress and the rate of shear (Fig. 6.2). When the resulting consistency curve is linear, the fluid is known as a *Newtonian fluid* and the reciprocal of the slope is the viscosity. Most slips exhibit nonlinear consistency curves; the two types are shown in Fig. 6.2. Curve 'a' represents a *dilatant fluid*, that is, one which becomes stiffer at high shear rates than at low shear

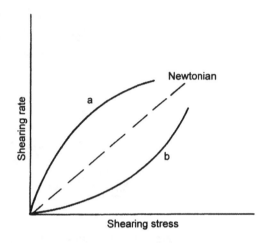

Fig. 6.2 Consistency curve for non-Newtonian fluids: (a) Dilatant behaviour and (b) Pseudoplastic behaviour

rates. Curve 'b' represents a *pseudoplastic fluid*. In this case the slip becomes less stiff at high shear rates than at low rates. Non-Newtonian fluids cannot be characterised by a single viscosity, and therefore, single point measurement offers only limited information. Viscosities of slips depend significantly on temperature, specific gravity, the presence of deflocculants, and the particle size distribution of the powder in the slurry.

In the case of a non-Newtonian fluid the apparent viscosity is expressed as:

$$\eta_a = K\gamma^{n-1} \qquad (6.1)$$

where, K is the consistency index, n is the shear thinning constant which indicates the departure from Newtonian behaviour and γ is the shear rate. For pseudoplastic fluid, $n < 1$.

The ranges of shear rate observed in metal/ceramic powder operations are shown in Fig. 6.3.

One of the simplest methods of determining viscosity is the capillary tube or Poiseuille viscometer (Fig. 6.4). Here the liquid is introduced through H until it reaches level G. Suction is applied at A to draw the liquid above mark B. The liquid is then allowed to flow under its own head through capillary DE, and the time required for the liquid level to drop from B to C is a measure of the viscosity.

6.1.2 Drying

Drying is the removal of mechanically combined water, and sometimes includes vaporisation of organic additives such as plasticisers and binders.

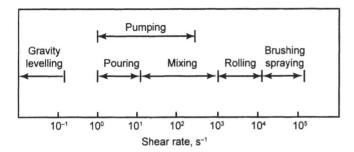

Fig. 6.3 Ranges of shear rate for different processing operations

Figure 6.5 shows a typical drying curve for a cast slip. The drying process starts at point A and proceeds towards the left. In the AB zone, the rate of evaporation is independent of moisture content. This is known as the constant rate period. BC zone reflects a linear dependence between drying rate and moisture content. After C, the relationship is not linear and the evaporation rate depends on the diffusion rate of water from the interior of the cast body.

In case of thin cross sections of the cast part, even good air circulation would suffice for drying. However, most parts are dried in a drier, where the temperature, humidity and air flow can be controlled. Many sources of heat like gas/oil combustion or electricity may be used. Electrical resistance heating is expensive, but becomes necessary if no surface contamination of the part is envisaged. Microwave drying and vacuum assisted drying are also not uncommon. There are two types of driers: batch and continuous (tunnel). In the latter type, various zones of constant temperature and humidity are maintained along the length of the tunnel.

An overall decrease in the dimension of the cast body occurs during drying—greater the amount of water added originally to form the slip, greater will be the amount of drying shrinkage when water is removed. The shrinkage must be taken into account in the mould design. As the surface of the casting will get dry first, the case often gets hardened, while the interior remains unchanged. The solution for this problem is to impede the rapid evaporation of water from the surface of the part. This can be achieved by heating the part in an enclosure,

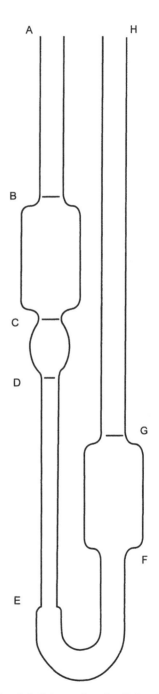

Fig. 6.4 Schematic of a Poiseuille viscometer

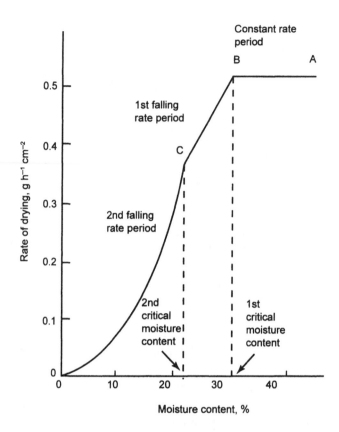

Fig. 6.5 A typical drying curve for a cast slip

where the relative humidity is initially kept high. With progress in drying, the temperature can be slowly raised and the humidity gradually lowered. The final stage of drying is done at very low humidity and temperature greater than 100°C. Such careful scheduling will result in a reasonably hard part that is free from shrinkage cracks. In case the cross section of the part does not have uniform thickness, greater care should be taken during drying. Sometimes, fans are used to circulate air over a drying part to sweep away the saturated layer near the surface. Control of the dew point of air introduced in the drier is critical for successful drying.

6.2 TAPE CASTING

Tape casting is also known as *doctor blade forming*, in which the product is in sheet form. Generally, parts of intricate shapes can be machine stamped from pre-dried sheets of plasticised powdery materials. The process is referred to as *continuous tape casting* when the blade is stationary and the supporting surface moves (Fig. 6.6), and as *batch doctor blade casting* when the blade moves across the stationary supporting surface covered with slurry. The cast film is dried to an elastic, leathery state and slit for proper handling width.

The first step in the tape casting process is the preparation of the casting slurry. The binders for tape casting are similar to those described earlier in Chapter 4. The composition of a typical tape casting slurry for barium titanate ceramic is given in Table 6.1. Mixing and dispersion of the slurry are done in two stages. The idea is to maximise dispersion but minimise degradation of the binder. After milling, the slurry is heated, filtered and de-aired to remove agglomerates and bubbles. The slurry is cast on a clean, smooth, impervious and insoluble surface such as cellulose acetate. The thickness of the tape is a function of the height of the blade, viscosity of the slurry, speed of carrier film and drying shrinkage. The casting speed is dependent on

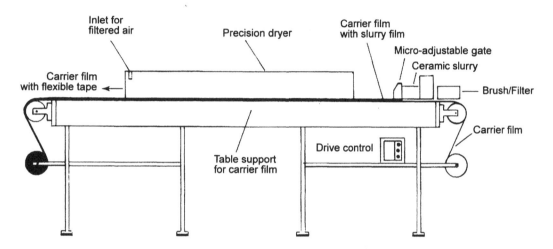

Fig. 6.6 Schematic of continuous tape casting unit

Table 6.1 Composition of a typical tape casting slurry for barium titanate

Component	Material	Composition, vol%
Ceramic powder	Titanate powder*	28
Liquid system	Methylethyl ketone	33
	Ethanol	116
Deflocculant dispersant	Menhaden fish oil	1.7
Binder	Acrylic	6.7
Plasticiser	Polyethylene glycol	6.7
	Butylbenzyl phthalate	6.7
Surfactant	Cyclohexanone	1.2

*< 5 μm, includes sintering aids and grain growth inhibitor

the thickness of the tape, evaporation rate and length of the machine, all of which control the drying time. The drying shrinkage occurs through the thickness, and the dried thickness is commonly about one half the blade height. Dried tape may be used directly or stored on a take up reel.

Tape defects include cracks, camber, local regions of low density, and surface defects consisting of unacceptable roughness and large surface pores. Tape cast products find use as battery electrodes, brazing layers, microelectronic substrates, coatings and thin foils.

6.3 ELECTROPHORETIC DEPOSITION

In Chapter 4, electrophoretic coating on powders was discussed briefly. In electrophoretic deposition, small particles dispersed in an electrolyte become charged, and can be packed into a shape in the presence of an electric field. The particle suspension is rather dilute, being only about 10 vol% powder. Since in such deposition, the green density of the deposit is independent of the particle size, it is possible to deal with very small size particles, which is otherwise very difficult in conventional compaction. Generally, non-aqueous electrolytes are required. The voltage ranges over 100 V and is kept higher for rapid deposition. Since the deposition rate is low, the composition of the deposit can be changed with time, thus resulting in a composition gradient. The main achievement of electrophoretic deposition has been in forming shapes from nanoparticles, coatings and functional gradient parts. For further consolidation, the bulk parts so generated are subjected to sintering operation.

6.4 SPRAY DEPOSITION/FORMING

Spray forming is a hybrid of atomisation powder spraying and conventional ingot metallurgy processing. Hence, it is often not categorised as a conventional powder metallurgy consolidation method.

In all spray deposition/forming processes from the melt, the first objective is to atomise the liquid metal into small liquid particles with mean diameter between 100 and 200 μm. The liquid particles are propelled rapidly towards the surface of the substrate, either by a high-speed stream of gas or by centrifugal force. As the particles move towards the substrate they become spherical in shape and also cool in flight by radiation and convection. It is important that they reach the substrate whilst still fully liquid or at least mostly liquid. The liquid particles flatten on the substrate to form adherent splats which deposit on one another, thus building up a substantial layer. Because of the thinness of the splat and the excellent contact with the cool substrate, the splats cool rapidly at speeds up to 10^6 °C s^{-1}. Typical splats are 10–20 μm thick and 500–1000 μm long.

Surface preparation of the substrate is an important feature of spray forming processes. A very rough surface causes mechanical interlocking, making the subsequent removal difficult. On the other hand, deposition on very smooth surfaces leads to serious difficulties owing to porous non-adherent coating. The success of spray forming lies in the fact that each deposited particle should splat at its point of impact and stay exactly in its original position.

The integrity of spray deposits is greatly influenced by the entrained or adsorbed oxygen in the form of air or oxygen-containing compounds. Although all commercially used metals can form undesirable oxides, it is particularly difficult when the molten metal contains aluminium, chromium or titanium. The metal oxide so formed, even in very small quantity, can significantly prevent the effective welding of one splat to the next during deposition.

The spray process is used to produce a wide variety of preformed shapes and sizes. Typical preform shapes are tubes, rings, cylinders, discs or simple billet. For a particular product line, the size of the part is mainly dictated by the melt facility, atomiser type, including the chamber size, and above all by the economic considerations. The largest preform size produced weighs about 540 kg. Typical deposition rates range from 10 to 50 kg min^{-1}.

Spray deposition (~98% theoretical density) is more homogeneous in comparison to casting or conventional ingot metallurgy forging. In many cases the product can be used directly, while in others some additional mechanical working like hot forging, hot extrusion, hot pressing or machining may be necessary. The added advantage of the spray forming process is that ceramic particulates can be reinforced in the spray plume, thus permitting the production of particulate composites.

Figure 6.7 illustrates the spray process for production of parts from spray-formed preform. The left-hand side of the scheme resembles the atomisation process, while the right-hand side shows the scheme for hot forging, which will be described in Chapter 9.

In this process the first step is to make a preform by atomising molten metal with an inert gas and spray depositing into an open mould. The deposit fills the mould, exactly replicating the mould surface. The operation is continued until the mould is sufficiently filled, when the atomisation is either stopped or transferred to another mould. The mould cavity is shaped in such a way that the preform can readily be separated from it. After separation, the preform is reheated to the forging temperature, transferred to a press and forged hot, usually in a single operation.

Hot rolling or forging has the advantage of closing any residual porosity and modifying the internal structure as well as obtaining a more acceptable external surface. Hot rolling has the additional advantage that it can be operated continuously or semi-continuously. A wide range of metals and alloys has been made by spray rolling, including aluminium and its alloys, copper and copper base alloys, some super plastic zinc base alloys and

---

Example 6.1: Describe the positive and negative aspects of spray formed metal matrix composites (MMC) containing ceramic particle reinforcements.

Solution: MMC tends to exhibit inhomogeneous distribution of the ceramic particles. Ceramic-rich layers, in spray formed MMC, approximately normal to the overall growth direction are often seen. This may be the result of hydrodynamic instabilities in powder injection and flight patterns, or due to the repeated pushing of particles by advancing solidification front in the liquid or semisolid layer, until the ceramic content is too high for this to continue.

Microstructure, in general, reveals that the interfacial bond is strong, with little or no interfacial reaction layer and very low oxide content.

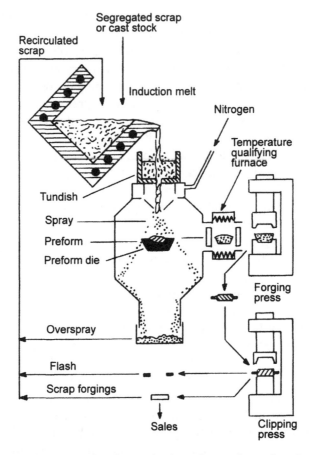

Fig. 6.7 Osprey process for production of spray-formed preform

various steels. The product after hot rolling is of fine grain size and generally shows a much finer structure in terms of unabsorbed or insoluble constituents. The mechanical properties of the products are at least equal and often superior to conventional materials.

6.5 SOLID FREEFORM FABRICATION

This method of shaping falls within the ambit of rapid prototyping, which is based on computer-aided design (CAD). *Rapid prototyping* allows the designer to complete a design iteration loop significantly faster than what was traditionally possible by reducing the time and cost required to produce a prototype. Manufacturing processes fall into four categories:

- Substractive processes
- Forming processes
- Additive processes
- Hybrid processes.

Most traditional manufacturing processes fall into one of the first two categories, while the most rapid prototyping processes fall into one of the last two categories. Figure 6.8 illustrates these categories. For additive processes, parts are produced by adding material to create a part. Some techniques are considered hybrid, as they require an initial substractive process to cut each layer from a sheet and then additively bond the layers together to create the part. These processes are often referred to as three-dimensional printing or desktop manufacturing. The original rapid prototyping process—known as *stereolithography*—used a liquid vat of photopolymer resin. When the liquid polymer is exposed to laser light, it solidifies. Modifications of this method use a bed of metal or ceramic powder coated with polymer binder. Other rapid prototyping processes form the solid by extruding the polymer from a small orifice, or build up the solid by depositing droplets of polymer binder, as from an inkjet printer head, into a bed of powder.

Selective laser sintering is another process of free forming, which is based on the sintering of powders selectively into an individual object. A thin layer of powder is first deposited into the part build cylinder. A laser beam guided by a process control computer, using instructions generated by the 3D CAD program of the desired part is then focussed on that layer, tracing and sintering a particular cross section into a solid mass. The cycle

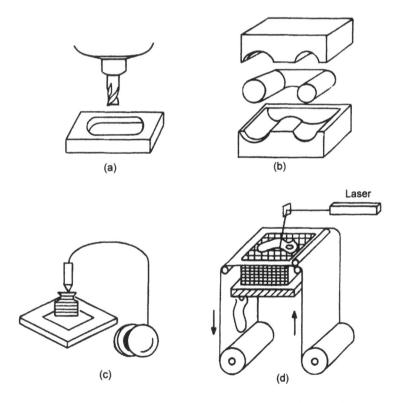

(a)   (b)

Laser

(c)   (d)

Fig. 6.8 Four manufacturing paradigms: (a) Substractive process, (b) Forming process, (c) Additive process and (d) Hybrid process (Reproduced with permission from *ASM Metals Handbook*, Vol. 20, ASM International, Ohio, 1997)

of deposition is repeated until the entire three-dimensional part is produced. The loose particles are then shaken off and the part is removed. It is more common to use polymer, because of the smaller, less expensive and less complicated lasers required for sintering. With ceramics and metals, it is common to laser sinter only a polymer binder, followed by conventional sintering of the green shape in a furnace. This is done after the debinding step. Part infiltration, for example, porous steel by molten copper, can be performed in a manner similar to any powder metallurgy product. For example, the dies used in electro-discharge machining are made by such a process.

The laser lamination process is relatively fast, but the debinding, sintering and infiltration step can take many days. More recently, high-power lasers capable of directly sintering or melting the powders are being used, thereby forming a three-dimensional object without polymers. The patented LENS® (Laser Engineered Net Shaping) technique is an extension of rapid prototyping technologies into direct fabrication of parts. It is operated in a sealed chamber filled with argon gas. In this process, Nd-YAG laser is applied to create a molten pool on a metal surface. Powder is delivered through nozzles in an argon gas stream, into the molten pool. The laser and nozzles remain stationary, while the substrate is moved, providing controlled deposition of a thin layer of finite width. The part shape is controlled through a pre-programmed CAD model. The most significant features of this method are its high cooling/solidification rate (up to $1000°C$ s^{-1}), rapid prototyping capability and proper shape control. Thus, it has distinct advantages to control grain growth, increase process efficiency and save machining costs. The LENS® technique has been used to fabricate a broad range of materials, including stainless steels, nickel-based superalloys, copper alloys, titanium alloys, and composites with improved physical and mechanical properties. Typical process parameters are 200 W power, 2 mm s^{-1} speed and layer thickness of 0.1 mm. The major advantage of the high-power laser process is the provision of multiple powder feeders. Different particles can be mixed to generate composites. The resultant fine grain size, caused by very rapid cooling rates, gives rise to higher strengths in the parts as compared to the conventional sintered ones.

SUMMARY

- Slurry moulding is similar to slip casting, except that the mould is solid and the powder becomes rigid due to polymer freezing.
- The flow behaviour of slips are generally controlled by means of a small amount of dispersion addition or deflocculant.
- Slip casting is most effective for large shapes, since the equipment and tooling costs are not high.
- Drying is the removal of mechanically combined water and sometimes includes vaporisation of organic additions such as plasticisers and binders.
- In tape casting (doctor blade forming), the product is in sheet form.
- Spray forming is a hybrid of atomisation powder spraying and conventional ingot metallurgy processing.

Further Reading

Jones JT and Bernard MF, *Ceramics*, 2nd edn, Iowa State University Press, Iowa, USA, 1993.

Keey RB, *Drying of Loose and Particulate Materials*, Hemisphere Publishing, New York, 1992.

Keey RB, *Introduction to Industrial Drying Operations*, Pergamon Press, Oxford, 1978.

Keicher D, Sears JW and Smugersky JE, (eds), *Metal Powder Deposition for Rapid* Manufacturing, Metal Powder Industries Federation, Princeton, NJ, 2002.

Onada Jr. GY and Hench LL, (eds), *Ceramic Processing before Firing*, Wiley-Interscience, New York, 1978.

EXERCISES

6.1 Why are clay-based ceramic materials generally more difficult to slip cast than non-clay-based ceramic materials ?

6.2 What modification would you suggest to prepare a ceramic honeycomb structure for heat exchangers by tape casting method?

6.3 A 'pseudoplastic' slip of a powder cannot be characterised by a single viscosity value. Why?

6.4 How many grams of binder polyvinyl butyral (PVB) are required to produce 1 kg of tape consisting of 50 vol% Al_2O_3 and 50 vol% PVB? Density of Al_2O_3 is 3.98 g cm^{-3} and that of PVB is 1.08 g cm^{-3}.

6.5 How will you partially dewater a slip?

6.6 Enumerate three major variables to control the drying of a ceramic shape.

6.7 What are the criteria to select a tunnel dryer and a batch dryer for a ceramic?

6.8 During conventional drying of monolithic refractories, explosive spalling is sometimes witnessed. Could you suggest some additive to take care of this problem? Give reasons.

6.9 Alumina tape after drying, containing 12 wt% organics has a mean bulk density of 2.55 g cm^{-3} and an apparent mean density of 2.95 g cm^{-3}. Estimate the open porosity.

6.10 During electrophoretic deposition, if particles have a tendency to settle, what deposition parameter needs to be controlled?

6.11 What operating parameter should be controlled for getting a density gradient (i.e., controlled porosity) during electrophoretic deposition?

6.12 Wood and Honeycomb after extensive research on rapid solidification, found that the following solute elements extended the maximum solid solubility limit in iron

Solute	Max. equil. solubility, at %	Extended solubility, at %
Cu	7.2	15.0
Ga	47.0	50.0
Ge	20.0	25.0
Mo	26.0	40.6
Rh	50.0	100.0
Sn	9.8	20.0
Ti	9.8	16.0
W	13.0	20.8

Select some convenient alloying additions available in the PM production plant, and describe how you will produce a rapidly solidified PM ferrous part commercially. How do the mechanical properties of these alloys differ from that of normally solidified alloys?

6.13 What is the difference in conventional melt atomisation and the thermal spraying during spray forming?

6.14 For many electronic components layered, single- or multi-laminates made of ceramic or metallic powders are used. These layers may be punched and metallised by thick film techniques or stacked. Describe a cost-effective process to prepare these laminates.

7

Sintering Theory

LEARNING OBJECTIVES

- How and why pore closure happens during sintering
- The analytical approach to sintering
- Solid sintering without any external pressure
- Liquid phase sintering
- Mechanisms of activated sintering and pressure-assisted sintering
- The role of electronic structure in sintering

The sintering process has been defined by various scientists. It is the thermal process for consolidating powder particles into a coherent structure via mass transport on the atomic scale. The bonding leads to improved properties, like strength, electrical and magnetic properties. Sintering theory is most accurate for single phase powder sintering by solid-state diffusion. However, many sintering systems consist of multiple phases, possibly forming a liquid, and may even be subjected to an external pressure to enhance densification.

Chemical kinetics deals with the course of processes, expressed in terms of structure, in space and time. The formal treatment of kinetics is basically phenomenological, but it often needs detailed atomistic modelling for constructive and appropriate formal frame. Kinetic parameters, such as, rate constants and transport coefficients, are directly related to the number and kind of point defects. Except for the case of homogenous reactions in infinite systems, the reaction path is determined by the state of the boundaries, like surfaces, solid–solid interfaces and other phase boundaries. Often, during sintering we take recourse to an exothermic reaction in a mixture of dissimilar powders. The reaction not only gives rise to a

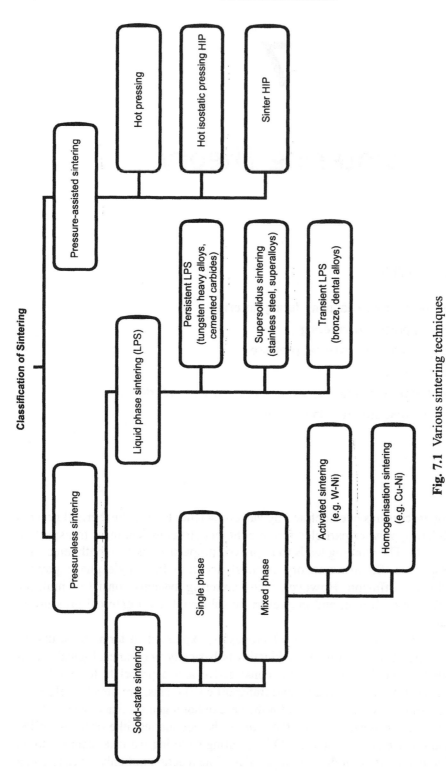

Fig. 7.1 Various sintering techniques

new compound, but at the same time the heat of reaction is used to sinter the resultant phase into a coherent solid.

Figure 7.1 summarises the various sintering techniques: the major classification is based on pressureless and pressure-assisted sintering. As a consequence, the mechanisms may be more involved, depending on the specifics of the processing.

7.1 SOLID-STATE SINTERING: PRESSURELESS

Solid-state sintering of single component materials is the best understood form of sintering. Table 7.1 summarises various approaches that have been used in the theoretical analysis of sintering; the analytical models have received the greatest attention. It is worth mentioning that these models suffer from the limitation that they do not fully describe the real powder systems, where the particles are of varying shapes and sizes. Most theories assume a monosized spherical powder of good purity. The green compacts of powder mass do not have uniform density. Apart from this, they may have organic binders, lubricants, additives and other impurities from the sintering atmosphere. Powders are never flat; they have curvatures. It is therefore logical to begin our treatment from this origin. A curved surface always has some stress. The Laplace equation gives the stress σ associated with a curved surface as:

$$\sigma = \gamma \left[\frac{1}{r_1} + \frac{1}{r_2} \right] \tag{7.1}$$

where, γ is the surface energy and r_1 and r_2 are the principal radii of curvature for the surface. When the radius is located in the vapour phase, i.e., when the surface is concave, the convention is to assign a negative sign and the surface is under compression. As a corollary, convex surfaces are under tension. The flat surface is invariably stress free. During sintering, the atoms from the convex region are expected to move to the concave regions.

Table 7.1 Main approaches for the theoretical analysis of sintering

Approach	Comments
Scaling laws	Not dependent on specific geometry. Effects of change of scale on the rate of single mechanism derived.
Analytical models	Greatly oversimplified geometry. Analytical equations for dependence of sintering rate on primary variable derived for single mechanism.
Numerical simulations	Equations for matter transport solved numerically. Complex geometry and concurrent mechanisms analyzed. Results not easily visualised.
Topological models	Analysis of morphological changes. Predictions of kinetics limited. More appropriate to microstructural evolution.
Statistical models	Statistical methods applied to the analysis of sintering. Simplified geometry. Semi-empirical analysis.
Phenomenological equations	Empirical or phenomenological derivation of equations to describe sintering data. No reasonable physical basis.

(Reproduced from Rahaman MN, *Ceramic Processing and Sintering*, 2nd edn, Taylor and Francis, London, 2003.)

The atoms and vacancies within a curved surface will have their chemical potential altered by the curvature of the surface. The difference in the chemical potential from one region of the surface to another leads to a diffusional flux of atoms to reduce the free energy of the system. The flux of atoms is equal and opposite to that of the vacancies. Figure 7.2 illustrates the above facts schematically

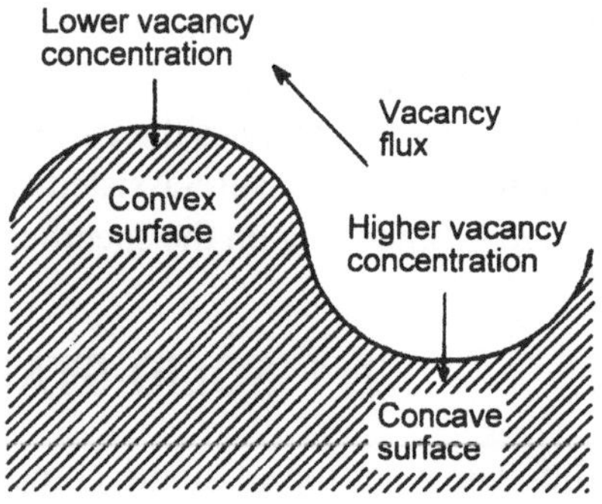

The diffusional flux of atoms determined by gradients in chemical potentials of atom and vacancy can be expressed as:

Fig. 7.2 Schematics of the direction of flux for vacancies in a curved surface

$$J_a = -\frac{DC_a(\Delta\mu_a - \Delta\mu_v)}{\Omega kT} \tag{7.2}$$

where, D is the self-diffusion coefficient for atoms, k is the Boltzmann constant and Ω is the atomic volume. The chemical potential for vacancy μ_v, which incorporates the pressure and surface curvature effect can be expressed as:

$$\mu_v = \mu_{vo} + (p + \gamma_{sv}K)\Omega + kT\ln C_v \tag{7.3}$$

where, K is the curvature, Ω is the volume of vacancy and C_v is the concentration of vacancy.

Similarly, the chemical potential of atoms can be expressed as:

$$\mu_a = \mu_{a_o} + (p + \gamma_{sv}K)\Omega + RT\ln C_a \tag{7.4}$$

Equations 7.3 and 7.4 show that the chemical potential of the atoms or vacancies depend primarily on the hydrostatic pressure in the solid and the curvature of the surface. Since the curvature term $\gamma_{sv}K$ has the units of pressure or stress, it will produce the same effect as an equivalent external applied pressure. Therefore, the pressure and curvature effect can be treated by the same formulation.

From the above treatment it is obvious that the driving force for sintering is the excess surface free energy. Having seen 'why' sintering is necessary, the next section will discuss 'how' sintering is performed.

7.1.1 Analytical approach to sintering: Stages in sintering

Let us consider a mass of loosely packed spheres of a polycrystalline material. The three stages of sintering, viz., initial, intermediate and final stages can be visualised such that in the initial stage, there is a neck formation between two touching spheres, which in the final stage culminates into small closed pores.

In polycrystalline materials, atomic (ionic) transport takes place along definite paths that define the sintering mechanisms. From the previous section, it is evident that the atoms are transported from regions of higher chemical potential to regions of lower chemical potential.

Some of the major mechanisms for sintering a polycrystalline material are:

- Evaporation and condensation
- Diffusion: Surface, grain boundary and lattice
- Plastic flow
- Viscous flow.

Evaporation and condensation mechanism operates for materials with high vapour pressure at elevated temperatures, e.g., chromium. For most of the materials, this mechanism is not of much significance.

Initial stage

Kuczynski studied the sintering of a polycrystalline sphere on a plate of similar material. As sintering proceeds, a neck will grow between a sphere and a plane and there will be large negative curvature of the neck. The curvature is a reciprocal of the neck radius, ρ, with a comparatively small positive curvature of the sphere. Kuczynski analysed the rate of neck growth for various material transport mechanisms from the particle surface. Herring extended the results of Kuczynksi to include the effect of particle size.

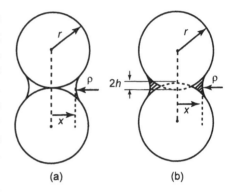

(a)   (b)

Fig. 7.3 Schematic two-sphere model showing (a) non-densifying mechanism and (b) densifying mechanism

Figure 7.3 represents a two-sphere model, where two equal-sized spheres are in contact. Figure 7.3a shows the non-densifying mechanism, such that there is no change in the centre-to-centre distance. On the other hand, Fig. 7.3b represents the scheme, when the spheres penetrate each other, in addition to neck growth. This is the densifying mechanism. The main geometrical parameters of the model are the principal radii of the curvature of the neck ρ, area of the neck surface A, and volume of the material transported into the neck, V. It is assumed that ρ is uniform over the section of the neck. For all sintering mechanisms operating in the initial stage, Kuczynski's results can be written in the form:

$$\left(\frac{x}{r}\right)^n = \frac{Ct}{r^m} \tag{7.5}$$

where, x is the radius of the neck, r is the sphere radius of the material, t is the isothermal sintering time and C is the constant. x/r is know as the neck size ratio. C can be expressed in the exponential form as:

$$C = C_0 \exp\frac{-Q}{RT} \tag{7.6}$$

where, C_0 is the cumulative constant for a material at a particular temperature and given particle geometry, R is the gas constant, T is the absolute temperature and Q is

Table 7.2 Initial stage sintering $(x/r)^n = Ct/r^m$

Mechanism	n	m
Viscous flow	2	1
Plastic flow	2	1
Evaporation–condensation	3	2
Lattice (volume) diffusion	5	3
Grain boundary diffusion	6	4
Surface diffusion	7	4

the activation energy associated with the atomic transport process. The values of n, m and C in equation 7.5 depend on the mechanism of material transport and are given in Table 7.2. Equation 7.5 is valid for a neck size ratio below 0.3. It is worth mentioning that the values of m and n are plausible ones.

Although all the mechanisms may contribute to grain growth, only plastic or viscous flow and grain boundary/lattice diffusion can result in densification, because, these are the only mechanisms that can cause interparticle distances to decrease. A schematic presentation of these mechanisms is shown in Fig. 7.4. The relative change in vacancy concentration due to the change in the surface curvature can be expressed as:

$$\frac{\Delta c}{c} = \frac{2\gamma\Omega}{\rho kT} \tag{7.7}$$

where, γ is the surface energy, Ω is the volume of the vacancy, ρ is the radius of curvature, k is Boltzmann's constant and T is the absolute temperature. The equation for neck growth may be obtained by considering the vacancy diffusion. This depends on the assumptions made about the source, sink and form of diffusional flux. In general, it can be expressed as:

$$\left(\frac{x}{r}\right)^n = \frac{BD\gamma\Omega t}{r^m kT} \tag{7.8}$$

where, n, m and B are constants, D is the concerned diffusion coefficient, γ is the surface energy, Ω is the volume of the vacancy, t is the time, k is Boltzmann's constant and T is the absolute temperature.

In real systems, with progression in sintering, additional new necks form. It is easier to measure the compacts' dimensional change rather than the neck size. Thus, shrinkage is approximately related to the neck size by the expression:

$$\left(\frac{\Delta L}{L_0}\right) = \left(\frac{x}{2r}\right)^2 \tag{7.9}$$

where, ΔL is the compact length change and L_0 is the initial length of the compact undergoing sintering. Shrinkage is a negative quantity but the sign is usually ignored.

Plastic flow mechanism: As we know, the green powder compacts are under stress, since the powders were subjected to plastic deformation during compaction. In other words, the

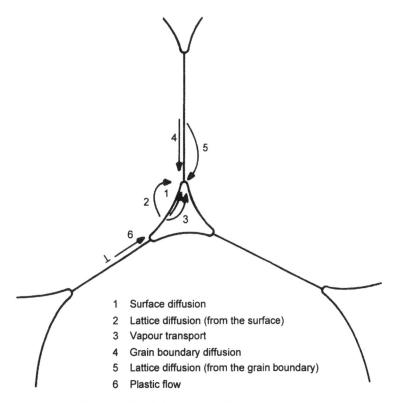

1 Surface diffusion
2 Lattice diffusion (from the surface)
3 Vapour transport
4 Grain boundary diffusion
5 Lattice diffusion (from the grain boundary)
6 Plastic flow

Fig. 7.4 Six distinct mechanisms for sintering

dislocations involved would play their role, particularly during early stages of sintering. Two events are possible:

- Dislocation climb due to vacancy absorption
- Dislocation glide due to surface stresses exceeding the flow stress at the sintering temperature.

Schatt and co-workers showed shrinkage rate improvements because of dislocation climb with the rate of pore elimination given as:

$$\frac{dV_P}{dt} = \frac{\sigma_e \Omega D_v}{kTL^2}$$
(7.10)

where, V_p is the fractional volume of porosity, t is the time, σ_e is the effective surface stress, Ω is the atomic volume, D_v is the volume diffusion coefficient, k is the Boltzmann's constant, T is the absolute temperature and L is the mean distance between mobile dislocations.

There is another type of plastic deformation mechanism caused by diffusional flow creep. This will be described in the section related to pressure-assisted sintering.

Viscous flow mechanism: Frenkel proposed the viscous flow mechanism, mainly for glassy spherical powders. He presented the expression for neck growth for a two-sphere model as:

$$\frac{x^2}{R} = \frac{3}{2} \times \frac{\gamma}{\eta} \times t \qquad (7.11)$$

where, x is the neck radius, R is the particle radius and η is the viscosity. According to him, the surface tension is the driving force for sintering and the solids could behave like Newtonian liquids at elevated temperatures. The viscosity η is expressed as:

$$\eta = \frac{kT}{D\Omega} \qquad (7.12)$$

where, k is the Boltzmann's constant, T is the absolute temperature, D is the self-diffusion coefficient and Ω is the atomic volume of the solid material.

Intermediate stage

Kingery and Berg carried out an extensive study of the intermediate stage, which was subsequently modified by Coble. It should be borne in mind that more than one mechanism may be operative at the same time. In such cases, the rate of one mechanism influences the other. The interaction between the mechanisms is very important. Coble idealised the powder system as a space-filling array of equal-sized tetrakaidecadra, each of which represents one particle. The pores are cylindrical and occur along the edges of the tetrakaidecadra (Fig. 7.5). A tetrakaidecahedron is constituted from an octahedron by trisecting each edge and joining the point to remove the six vertices (Fig. 7.6). The resulting structure has 36 edges, 24 corners, and 14 faces (8 hexagonal and 6 squares). Since the model assumes that the pore geometry is uniform, the chemical potential is the same everywhere on the pore surface. Here, the densifying mechanism will operate taking into account grain boundary and lattice diffusion, the former being more important. As long as the grain boundary remains attached to the pores, the grain size will increase as the porosity decreases.

By introducing a solution for the diffusion equation 7.7 suitable for the geometry, and using equation 7.7 to determine the vacancy concentration near the pores, Coble obtained an equation for the rate of change of porosity (P) as:

$$\frac{dP}{dt} = -\frac{CD\gamma\Omega}{G^3 kT} \qquad (7.13)$$

where, C is a constant and G is the edge length of a grain. From equation 7.13, it is evident that the densification rate is the inverse of the cube of the grain edge, or in other words, inversely proportional to the volume of the average grain. Hence, equation 7.13 can be written as:

$$\frac{dP}{dt} = -\frac{C'D\gamma\Omega}{kTt} \qquad (7.14)$$

where, C' is a constant. Equation 7.14 integrates to give the relation that the porosity decreases linearly with time.

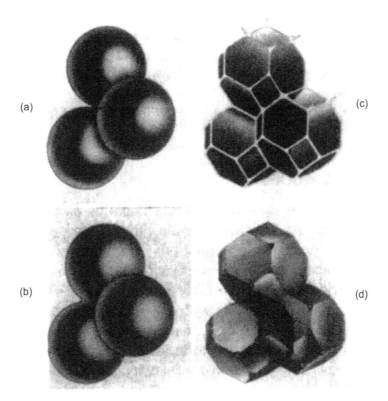

Fig. 7.5 Model structures for various stages during sintering: (a) Spheres are in contact, (b) Initial stages of sintering involving neck growth, (c) Intermediate stage: dark grains have adopted the shape of tetrakaidecahedron, enclosing white pore channels at the grain edges and (d) Final stage: pores are tetrahedral inclusions at the corners where four tetrakaidecahedra meet (Adapted from Coble RL, *Journal of Applied Physics*, 32:793, 1961)

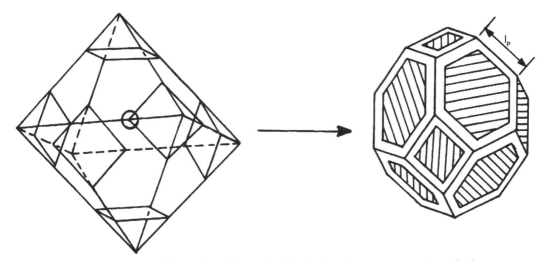

Fig. 7.6 Schematic of formation of a tetrakaidecahedron from a truncated octahedron

Final stage

After prolonged sintering in any polycrystalline system, the open pores become closed or isolated. They become lens-shaped or spherical and shrink by lattice or volume transport processes if they are connected to the grain boundaries. From Fig. 7.6 it can be visualised that each corner is occupied by a spherical pore. Lens-shaped pores are evident when they are situated at the grain boundary, and where the equilibrium between the grain boundary energy and solid–vapour surface energy will lead to a dihedral angle groove. Spherical pores result when the pores are separate from the grain boundary. Differences in the pore curvature and vacancy concentration lead to pore coarsening, i.e., growth of larger pores at the expense of smaller and less stable ones. This is akin to Ostwald ripening of solid grains.

The pore removal rate during the final stages of sintering depends on the balance between the surface energy and the pore gas pressure. The densification rate equation can be expressed as:

$$\frac{d\rho}{dt} = \frac{12D_v\Omega}{kTG^2}(\frac{\gamma}{r_p} - p_g) \tag{7.15}$$

where, ρ is the fractional density, t is the sintering time, Ω is the atomic volume, D_v is volume or lattice diffusivity, k is the Boltzmann's constant, T is the absolute temperature, G is the grain size, γ is the solid–vapour surface energy, r_p is the pore radius and p_g is the gas pressure within the pore.

In real polycrystalline materials, it should be borne in mind that there is no sharp transition between the intermediate and final stages of sintering, because of the pore size/particle size distribution. In some areas the final stage occurs; the other areas may still be subjected to an intermediate stage of sintering.

Table 7.3 summarises various features associated with different stages of solid-stage sintering for polycrystalline materials. As an illustration from a real system, Fig. 7.7 shows

Table 7.3 Various features associated with the stages of solid-state sintering for polycrystalline materials

Stage	Typical microstructural feature	Relative fractional density range	Idealised model
Initial	Rapid interparticle neck growth	Up to ~ 0.65	Two monosize spheres in contact
Intermediate	Equilibrium pore shape with continuous porosity	~ 0.65–0.90	Tertrakaidecahedron with cylindrical pores of the same radius along the edges
Final	Equilibrium pore shape with isolated porosity	~ ≥ 0.90	Tetrakaidecahedron with spherical monosize pores at the corners

(Reproduced with permission from Rahaman MN, *Ceramic Processing and Sintering*, 2nd edn, Taylor and Francis, London, 2003.)

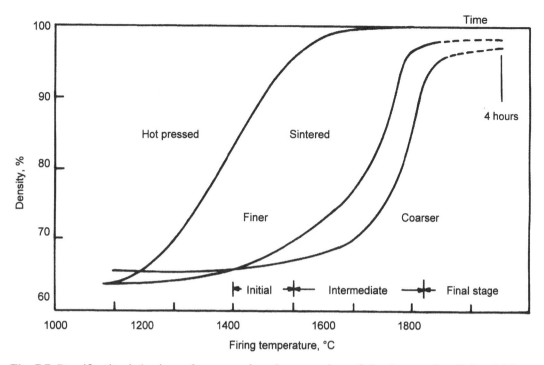

Fig. 7.7 Densification behaviour of compacts based on two sizes of alumina powders (1.3 and 0.8 μm average sizes) showing initial, intermediate and final stages of sintering. The hot pressing data are imposed.

the densification behaviour of a magnesia-doped alumina powder compact of two different sizes during the constant heating rate. The three stages of solid-state sintering are lineated in the plot.

To determine the activation energy for the sintering process, it is important to perform the sintering experiment for time dependence as well as temperature dependence. Such a combination of studies are required to isolate the mechanism and kinetics of sintering.

We have earlier described the significance of diffusional processes in sintering. Figure 7.8 gives a schematic plot of the diffusion coefficient with respect to $1/T$, corresponding to different types of diffusion. The slope of each segment of the plot gives the activation energy value. It tends to be in the region of a few electronvolts in magnitude. Surfaces tend to be the regions of relatively high disorder and therefore the activation energy for diffusion tends to be low. The activation energy at grain boundaries is generally higher, and that in the lattice diffusion is still higher. Figure 7.8 confirms this, as the plot corresponding to lattice diffusion has the highest slope and the one for surface diffusion is the lowest.

7.1.2 Non-isothermal sintering

Sintering experiments perfomed for identifying the mechanism are carried out under isothermal conditions. However, during such heating, the green compacts are subjected to thermal shock, internal strain and uneven thermal expansions. These effects can be corrected

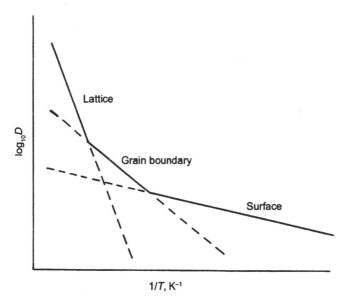

Fig 7.8 Schematic diagram showing the dependence on temperature of surface, grain boundary and lattice diffusion

in constant rate of heating (CRH) experiments by using small constant rates of heating. This is a typical example of non-isothermal sintering. The CRH method amply indicates that particle size distribution in powder mass affects its sintering kinetics. Non-linearity in sintering plots arises mainly from size distributions. The limitation of the CRH method is that it cannot separate parallel mechanisms contributing to sintering kinetics. However, such experiments are useful in describing effective activation energy for sintering, and to define effective densification parameters of real powder compacts.

It is worth mentioning that the sintering rate depends not only on diffusivity but also on the density ρ, and the grain size d. The general expression for the instantaneous sintering rate ρ, can be expressed as:

$$\dot{\rho} = \frac{d\rho}{dt} = \frac{A \exp(-Q/RT)}{T} \times \frac{f(\rho)}{d^n} \qquad (7.16)$$

where, $A = C\gamma V^{2/3}/R$, d is the grain size, $f(\rho)$ is the function of only density, Q is the activation energy, γ is the surface energy, V is the molar volume, R is the gas constant, T is the absolute temperature, C is a constant and A is the material parameter that is insensitive to d. The grain size dependence of the sintering rate is given by d^n, where $n = 3$ if sintering is controlled by lattice diffusion, and $n = 4$ if it is controlled by grain boundary diffusion.

The complication in the determination of Q lies in the fact that the rate of sintered density changes not only with temperature but also with density and grain size as given by $f(\rho)$ and d^n in equation 7.16. Unambiguous estimates of activation energy Q, are possible only if ρ and d remain constant.

7.1.3 Microstructural evolution

The binding energy between pore and grain boundary decreases during sintering. Grain growth causes the boundary to be curved, leading to an increase in the grain boundary area. A critical condition occurs when it is favourable for the boundary to breakaway from the pore. In brief, we meet two types of mobilities: grain boundary and pore mobility. The dihedral angle is formed between the grain boundary and a pore. This is a measure of the relative interfacial energies. The angle is defined as

$$\cos\phi/2 = \gamma_{gb}/2\gamma_{sv}$$

where, γ_{sv} and γ_{gb} are the interfacial tensions at the pore surface and grain boundary interfaces, respectively (Fig. 7.9). It is worth mentioning that the dihedral angle may not be the same everywhere, since the interfacial energies of solids are anisotropic. Grain boundary energy is a function of grain misorientation while solid–vapour interfacial energies vary with crystal orientation. In addition, even a minor impurity level in the system may change the angle.

As the pore dihedral angle increases, grain boundary migration is inhibited. Large pores are unable to remain attached to moving grain boundaries and become occluded in the grain interior. Rapid grain growth is noticed once the grain boundary breaks away from the pore (Fig. 7.10). Successful sintering occurs when the pore size/grain size ratio is small, giving faster pore removal and higher pore mobility. This way the pore remains attached to the moving grain boundary. It is important to minimise the breakaway by careful temperature control during sintering.

The widely different rate of migration of grain boundary and pores in a porous polycrystalline compact during sintering gives rise to abnormal grain growth. This feature will not result in dense material. For example, sintered alumina does show porosity and abnormal growth. In case the alumina is doped with MgO, one gets almost full density with normal grain growth. Figure 7.11a shows the normal growth, which is characterised by an increase in the average size while retaining similar size distribution. On the other hand, discontinuous growth (abnormal, exaggerated) results in broad bimodal distribution with some grains growing much faster than the other average (Fig. 7.11b).

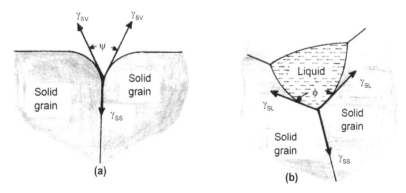

Fig 7.9 Dihedral angle at equilibrium when the grain boundary meets a surface: (a) Pore and (b) Liquid

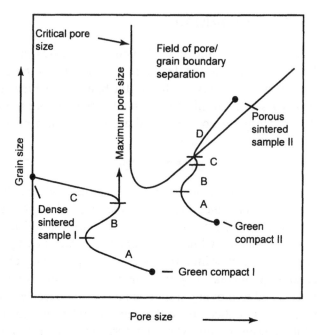

Fig 7.10 Variation of grain size with pore size on sintering, showing the condition for breakaway from the pores. Sample I: favourable path for high final density. Sample II: path for porous product (Adapted from Spears MA and Evans AG, *Acta Metallurgica*, 30:1281, 1982)

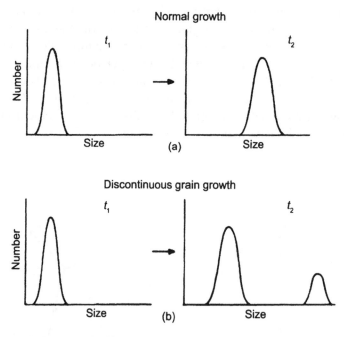

Fig 7.11 Single model distribution of grain sizes during sintering: (a) Normal grain growth and (b) Discontinuous grain growth

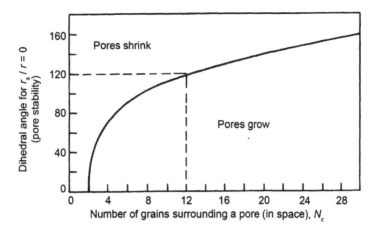

Fig 7.12 Condition for pore stability. Here, r_s = radius of the circumscribed sphere around a polyhedral pore surrounded by grains; r = radius of curvature of the pore

Pore coordination number is defined as the number of grains surrounding the pore. The critical coordination number is one, where the pore neither has convex nor concave surface. Figure 7.12 illustrates the values of N_c as a function of the dihedral angle. In summary, knowing the dihedral angle and the pore coordination number, the curvature of the pore surfaces and the tendency for the pore to shrink or grow can be obtained.

The pore size observed during sintering of a real system depends on complex interaction of various phenomena, such as, grain coarsening, coalescence and shrinkage. The disappearance

---

Example 7.1 Chromium metal powder has a surface energy of 2.2 J m^{-2}. Assuming a particle diameter of 1 µm, calculate the relative change in vapour pressure with respect to equilibrium at 1000°C. Atomic weight of Cr is 52 g per mole and density of Cr is 7.2 g cm^{-3}.

Solution: Based on LaPlace equation, the expression for the vapour pressure is

$$RT (\ln P - \ln P_0) = \frac{2\gamma\Omega}{d}$$

where, d is the diameter of the particle, Ω is the atomic volume, γ is the surface energy, P is the vapour pressure, P_0 is the equilibrium vapour pressure, R is the gas constant and T is the absolute temperature.

Rearranging, $\ln P/P_0 = 2\gamma\Omega /dRT$ (i)

where, d = 1 µm, R = 8.314 J mol^{-1} k^{-1}, T = 1000 + 273 = 1273 K

The atomic volume of Cr = 52 g mol^{-1}/7.2 g cm^{-3} = 7.22 × 10^{-6} m^3 mol^{-1}

Substituting these values in equation (i) above,

$\ln P/P_0$ = 2 × 2.2 × 7.22 × 10^{-6} × 8.314 × 1273 = 3.0 × 10^{-3}

Thus P/P_0 = 1.003

Example 7.2 In solid iron the cohesive energy per atom in the bulk is 4 eV. Estimate the energy of the atom in the surface and the surface energy in J m^{-2}. Given: the lattice parameter of iron in FCC crystal is 35.4 nm.

Solution: Surface energy is about 1/3 the cohesive energy, viz., 4/3, i.e., 1.3 eV.

Surface area of the unit cell of FCC Fe = $(3.54 \times 10^{-10})^2$ m^2

Each corner atom in on FCC atom is shared by 4 faces.

Therefore, the effective number of atoms/cell = $4 \times 1/4 + 1$, i.e., 2 atoms for every $(3.54 \times 10^{-10})^2$ m^2 of surface.

Therefore, number of atoms per square metre is ~10^{19}

Surface energy = energy of one atom × number of atoms per square metre

= 1.3×10^{19} eV m^{-2} = $1.3 \times 10^{19} \times 1.6 \times 10^{-19}$ = 2.08 J m^{-2}

Example 7.3 Represent schematically the configuration of the grain boundary moving along with a pore during sintering. Why is the leading surface of the pore less strongly curved than the trailing surface?

Solution: Figure Ex. 7.3 shows the schematic of a moving pore indicating the atom flux from the leading to the trailing surface. Small isolated pores can be dragged by the grain boundaries. The reason is that the grain boundary moving under the influence of its curvature, applies a force on the pore trying to drag it along. The force causes the pore to change its shape. The leading surface of the pore becomes less strongly curved than the trailing surface. The difference in curvature leads to a chemical potential difference which causes material transport from the leading surface to the trailing surface.

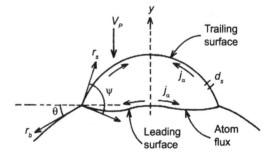

Fig. Ex. 7.3

of smaller pores during sintering is governed by the fact that there are more vacancies, due to more curvature than for larger pores.

7.1.4 Numerical simulation of sintering

The analytical models described in Section 7.1.1 generally assume that the cross section of the neck surface is a circle. This circle is also assumed to be tangential to both the grain boundary and the spherical surface of the particle. This assumption is not tenable, because the surface curvature must change discontinuously at the point of tangency between the neck surface and the spherical surface of the particle. Such a discontinuous change would give rise to a sudden change in chemical potential. Bross and Exner attempted numerical simulation for the sintering of a row of cylinders of the same radius. They considered two situations:

- Material transport by surface diffusion only (Fig. 7.13a)
- Simultaneous operation of surface and grain boundary diffusions (Fig. 7.13b).

Here, the undercutting and the continuous change in the curvature of the neck surface are predicted by the numerical analysis. The numerical simulation method is more useful when more than one type of diffusion mechanism is operative.

Numerical modelling of solid-state sintering of heterogeneous packed particles has been reported. Starting from size distribution and fitting parameters, it is possible to obtain same densities and the same configurations as those observed in the real case. Figure 7.14 shows an example of a heterogeneous complex shape obtained after numerical modelling of sintered powder packing of spherical particles. In spite of having good potential, the numerical simulation has two major limitations compared with the analytical models:

- The calculations are fairly complex.
- The results cannot be transformed into a useful form showing the dependence of kinetics on the sintering parameters.

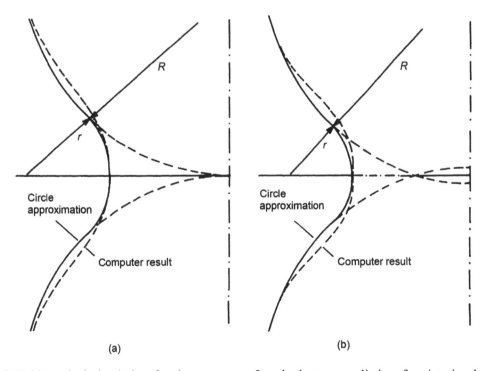

(a)   (b)

Fig 7.13 Numerical simulation for the contours of necks between cylinders for sintering by (a) surface diffusion and (b) simultaneous surface and grain boundary diffusion (Adapted from Bross P and Exner HE, *Acta Metallurgica*, 27:1013, 1979)

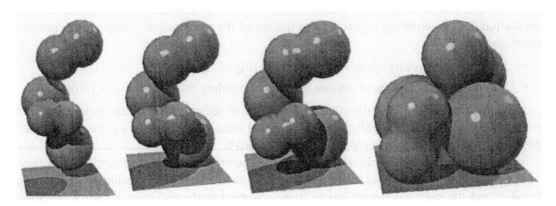

Fig 7.14 An example of a heterogeneous complex shape obtained after numerical simulation of a loose spherical powder assembly undergoing sintering [Reference: Leclerc H and Gelin JG, New developments on powder technology, In Torralba JM, *Proceedings of the International Conference on Advanced Materials Processing Techniques (AMPT'01)*, Universidad Carlos III de Madrid, Spain, 2001]

German and co-workers have reported studies on the atomic level simulation for sintering nanocrystalline tungsten powders using molecular dynamics method and modified embedded atom method. They compared two-particle simulation and full three-dimensional simulation with six coordination number. It is logical that the activation energy following surface diffusion mechanisms (158 kJ mol^{-1}) for two-particle model for nano-sized tungsten powder is far lower than that corresponding to conventional micron-range powders (293 kJ mol^{-1}). However, in contrast to conventional powder for either two-particle or three-dimensional models, the decrease in activation energy for nanocrystalline tungsten powder sintering follows the order: Surface → grain boundary → volume diffusion → evaporation and condensation → viscous or plastic flow.

According to them, the emergence of the pre-melting layer enhanced viscous flow during sintering.

7.1.5 Phenomenological approach to sintering

In the phenomenological approach, empirical equations are developed to fit sintering data, usually in the form of density versus time. These equations should not be assumed to amplify the sintering mechanism. Their significance is in numerical models, which incorporate equations for the densification of a powder compact. A simple expression fitting sintering data could be given as:

$$\rho = \rho_0 + K \ln \frac{t}{t_0}$$

$$(7.17)$$

where, ρ_0 is density at time t_0, ρ is density at time t and K is the temperature dependent constant. The above equation is also known as semi-logarithmic law.

Coble's intermediate stage sintering equation can be written in the form

$$\frac{d\rho}{dt} = \frac{AD_l\gamma_{sv}\Omega}{G^3kT} \tag{7.18}$$

where, A is a constant that depends on the stage of sintering. Assuming that the grain growth is according to a cubic law of the form

$$G^3 = G_0^{\ 3} + at \sim at \tag{7.19}$$

and assuming that $G^3 >> G_0^{\ 3}$, equation 7.17 becomes

$$\frac{d\rho}{dt} = \frac{K}{t} \tag{7.20}$$

where, $K = \dfrac{AD_l\gamma_{sv}}{\alpha kT}$

Since equation 7.17 has the same form in the intermediate and final stages of sintering, the semi-logarithmic law is expected to be valid in both these stages. Coble's semi-empirical law has been useful in fitting many sintering data.

7.1.6 Sintering maps or diagrams

In Section 7.1.4, we described the numerical simulation approach for sintering mechanisms, but the results cannot be easily formulated into a practically useful form. Ashby, based on the two-particle model, prepared a diagram to exhibit the predominant sintering mechanism for a given set of temperature and neck size. Later, a second type of diagram was presented in which the density rather than the neck size was evaluated. Figure 7.15 shows the sintering diagram for copper spheres with a radius of 57 μm. The axes are the normalised grain size X/R and the homologous temperature T/T_m, where R is the radius of the sphere, and T_m is the melting point of the material in absolute temperature. The diagram is divided into three fields corresponding to surface diffusion, grain boundary diffusion and lattice diffusion. At the boundary between two fields, shown as solid lines, two mechanisms contribute equally to sintering. Superimposed on the fields are contours of constant sintering time. The diagrams are constructed by numerical methods. It is assumed that the total neck growth rate is the sum of all neck growth rates for individual mechanisms. The field boundaries at which one mechanism contributes 50% of the neck growth rate is calculated. The process is carried out in steps for small increases in X/R and T/T_m.

The sintering diagrams are useful, but their applicability is rather limited. Firstly, they are based on a geometrical model, which is different from the real system; secondly, the materials data used may not be accurate. For solids like ceramics, these data are often not accurate.

7.1.7 Sintering of nanopowders

In Chapter 2 the relevant methods for producing nano-scale powders have been described. Mechanical milling is the convenient method for producing nano-sized powders where

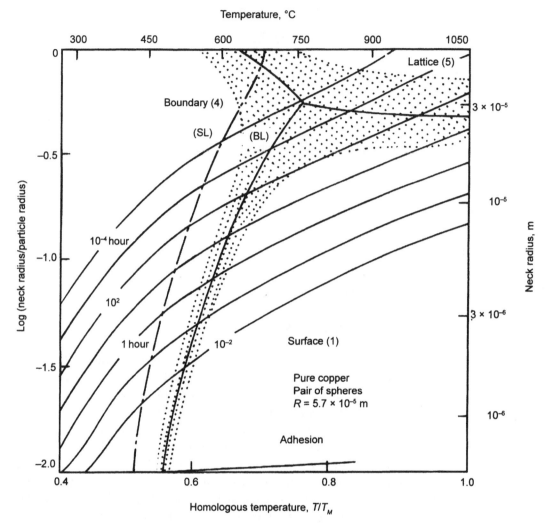

Fig 7.15 Sintering diagrams for copper spheres

a decrease in the powder particle size with concurrent increase in the specific surface area results. The steps involved are:

- Deformation of powder, creating a number of dislocations
- Extremely dense dislocation structure, resulting in grain subdivision
- Emergence of nano-grained structure with high angle boundaries.

Since the driving force for densification is dependent on the surface area of the powder, it is natural that small particle size is important in enhancing the sintered density. In case the grain sizes in two sets of powders are identical, the powder lot with finer powder particle size achieves higher densification earlier. Refractory metal nanopowders are being successfully prepared by mechanochemical method followed by hydrogen reduction treatment of their trioxide.

The main purpose of nanopowder consolidation is retention of initial nanocrystalline structure in the final stage. Often, the extremely fine grain sizes, lead to inherent metastable structure and a departure from equilibrium. This fine grain size may cause other deviations from the equilibrium, such as, alternate crystal structure, extended solubilities, or changes in physical properties. In sintering nanopowders, it is necessary to define the conditions under which metastability is lost.

There is varying opinion on whether the basic mechanism for sintering nanocrystalline powder is similar to conventional powders. The question is whether the physics of the sintering process of nano-scale particles is size-scale dependent. This has special significance in sintering of ceramic or other brittle particulates, where the densification process is most important and deformation by pressure plays a lesser role.

The specific features of sintering nanopowders are related to a high tendency of agglomerate, significant interparticle friction, high reactivity and associated contamination, high densification and coarsening kinetics. Densification of nanopowders takes place at temperatures consistently below those of large-grained powders by up to several hundreds of degrees. The onset of sintering in nanopowders may be as low as $0.2T_m$, compared to about $0.5T_m$ for their conventional counterparts.

In nano-scale powders, sintering during shrinkage displays an asymptotic characteristic as the compact nears full density. This behaviour has been represented by a sigmoidal function given by the following empirical relation:

$$\frac{\Delta l}{l_0} = f_1 + \frac{f_2}{1+\exp\left[(f_3 - Y)/f_4\right]} \quad (7.21)$$

where, $f_1, f_2, f_3,$ and f_4 are constants; l_0 is the height of the as-pressed (green) compact, and Y is the densification factor that can be calculated from an Arrhenius type equation as follows:

$$Y = \frac{1}{d^v}\exp\left[B_s t^w - \frac{Q}{RT}\right] \quad (7.22)$$

where, d is the particle size, w and v depend on the diffusion mechanism, B_s is the material parameter, t is the sintering time, Q is the activation energy for diffusion and R is the gas constant.

Johnson, while sintering compacts (compaction pressure: 320 MPa) of nano- and submicron-sized tungsten powders found that despite the smaller particle size of the 46 nm tungsten powder, the compact showed poor densification response and the sintered densities had much lower values than the model predictions. It was noticed that at a sintering temperature of 1200°C, this particular size of powder achieved a higher sintered density than the 1.65 µm tungsten powder. Both sinterings were carried out in hydrogen for one hour. It is worth mentioning that optimisation of impurity removal from the powder during heating must be done carefully.

The most important differences in the densification behaviour of nanopowders as compared to conventional powders occur in the early stages of sintering. In these stages, lower activation energies than required for conventional diffusion have been reported in sintering Y-TZP, Al_2O_3, TiO_2 and tungsten. Mere surface diffusion cannot explain fast kinetics. Other mechanisms, like dislocation motion, grain rotation, viscous flow and grain boundary slip, have been suggested. It is noticed that after neck formation the adjacent particles rotate to achieve minimum grain boundary energy.

The consolidation of nanopowders may be carried out either by the conventional pressureless method (powder compaction/sintering) or by pressure-assisted methods. In conventional methods, a large number of the initial powder contacts—small pores in a high green density compact and a uniform pore distribution—favour a high final sintered density. The major problem is the elimination of large pores, if any, that originate in the green compact. A lower green density may be achieved as compared to the conventional micrometre-sized powders. Thus, in nanopowders, a relatively high compaction pressure is required to achieve increased green density. If the pore size is small and the pore distribution is uniform, conventional sintering of nanopowders may give rise to a high density, in conjunction with retaining a final grain size in the nanometre-scale.

In pressure-assisted sintering, pressure levels vary from low (< 100 MPa) in hot pressing to moderate (100–500 MPa) in HIP and high (> 0.5 GPa) in high pressure methods, such as piston cylinder, diamond anvil or severe plastic deformation method and magnetic and shock compaction.

Table 7.4 summarises the effect of sintering methods on the final grain size of consolidated compacts made of oxide ceramics TiO_2 and an intermetallic TiAl.

Table 7.4 Effect of sintering methods on final grain size of TiO_2 and TiAl-based nano-sized powders

Material	Initial Particle size, nm	Conventional pressureless sintering		Pressure-assisted sintering	
		Method	Final grain size, nm	Method	Final grain size, nm
TiO_2	~ 6 (sol–gel process)	A	< 60	A'	46
TiAl	10–20 (partially amorphous)	B	15–20	B'	25–50

A: Sintering temperature 873 K, time 2 h in vacuum

A': Hot-pressed at 873 K, at 500 MPa for 2 h

B: Rate controlled sintered at 723–773 K for 2 h in vacuum

B': HIPed at 1000 K at 310 MPa for 1 h

(Adapted from Groza JR, Sintering of nano-crystalline powders, *International Journal of Powder Metallurgy*, 35(7):59, 1999)

7.1.8 Solid-state sintering of premixed/pre-alloyed powders

Premixed powder sintering may encounter the following situations:

1. Mixtures of powders of same composition but differing sizes
2. Mixtures of powders of different compositions which are prone to alloying
3. Mixtures of powders, where one component is inert and does not sinter, e.g., oxide dispersion-strengthened metals.

In case (1), although the packing density of the premix is improved, it does not necessarily mean that the highest sintered density will be achieved in the premixed compacts. Bimodal distribution of the particle size poses a problem during densification. A similar situation may be witnessed in the case of (3), because, an inert component, even with a similar particle size, does hamper the densification of the matrix materials.

Case (2) mentioned above can be classed as homogenisation during sintering. This is an alternative to forming sintered products from pre-alloyed powders. Let us consider two components A and B, which form an isomorphous solid solution. The diffusion rates between A and B will dictate the homogenisation process. Heckel caried out extensive work assuming spherical powder geometry. Initially, the concentration gradient is steep, but with further heating (sintering), the gradient levels out and finally reaches a constant value. Generally, fine particle sizes, higher sintering temperatures and longer sintering periods promote better homogenisation.

Interdiffusion rates between the individual components complicate the events. For example, in Cu-Ni system at elevated temperatures, the rate of diffusion of Cu into Ni is faster than the rate of Ni diffusion into Cu. These differing diffusivities can offset sintering due to swelling (compact growth). Kirkendall was the first to demonstrate it in α-brass–Cu couple—the lower melting zinc diffuses out of the alloy more rapidly than the copper atoms diffusing into it. A similar effect is manifested during the sintering of the Cu-Ni system.

In conclusion, the formation of a solid solution can lead to densification by:

- Enhancing diffusion coefficient for the controlling species in the lattice, or parallel to the grain boundaries, by affecting the point defect concentration in the boundary or lattice;
- Slowing grain boundary movement by forming a segregated layer at the boundary which must then be pulled along by the boundary;
- Altering the overall driving force by altering the ratio of grain boundary energy to the free surface energy;
- Slowing intrinsic grain boundary movement by reducing the diffusion coefficient for atom movement across the grain boundary, again by affecting the defect chemistry.

On the other hand, in case the additive remains as a distinct second phase, it can affect the densification process by:

- Providing a high diffusivity pathway, e.g., liquid phase at the boundaries
- Providing a continuous low diffusivity pathway at the boundaries for diffusion across the boundaries, which then acts to restrain grain boundary movement.

Example 7.4 A green compact made of a powder mixture of Al-5.5Cu is sintered at 575°C to achieve full densification.

(a) Describe the sequence of phase formation and densification response.

(b) What particle size of Cu powder will you select in comparison to that of Al?

(c) What is the effect of fast thermal cycle?

(d) Suggest a simple method to remove any residual porosity. (Refer Metals Handbook for the binary Al-Cu phase diagram and interdiffusivity data).

Solution

(a) The binary Al-Cu phase diagram shows a eutectic reaction at 548°C, between the α-solid solution and the Al_2Cu (θ) phase. The present scenario of sintering pertains to that of transient liquid phase sintering. The sequence of events during the heating cycle is as follows:

- In the initial stage of heating a series of intermetallics will be formed as a result of the diffusion between Al and Cu.
- The first liquid phase appears at eutectic temperature (548°C).
- Cu is drawn from the liquid into the Al matrix and is replaced by the dissolution of the intermetallics. These intermetallics are replenished by solid-state diffusion by adjacent copper particles.
- The intermetallics disappear when all the Cu is completely dissolved.

If we look into the diffusivity data, the diffusivity of Cu into Al is about 5000 times faster than that of Al in Cu. At 600°C, in the former case it is 5.01×10^{-9} cm^2 s^{-1}, whereas, for the latter it is 1.14×10^{-12} cm^2 s^{-1}. The faster diffusivity of Cu into Al promotes faster alloying, but at the same time gives rise to compact swelling.

(b) The particle size of copper must be smaller than that of aluminium. The fine spread of the powder over the Al surface will hasten sintering.

(c) In case the thermal cycle is fast, one may expect the presence of the intermetallic phase in the microstructure. This would affect the mechanical properties of the final sintered alloy.

(d) Any residual porosity in the compact can be successfully removed by repressing and resintering.

7.2 LIQUID PHASE SINTERING

An introduction to liquid phase sintering has already been given in Chapter 1. During solid-state sintering of polycrystalline powders, an important factor that determines the densification rate is the grain boundary diffusion coefficient multiplied by the grain boundary thickness. In case the grain boundary is replaced by a film of liquid, many times thicker than the grain boundary, the densification rate greatly increases because the diffusion rate in a liquid is higher than in a solid. It is this feature which is achieved in liquid phase sintering. However, only the presence of a liquid phase is not sufficient; the melt must also wet the solid surface. The wetting would cause spreading over the solid

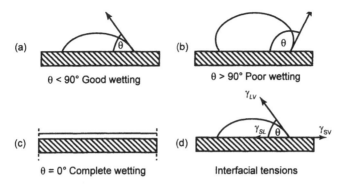

Fig 7.16 Wetting behaviour between a liquid and a solid showing (a) good wetting, (b) poor wetting, (c) complete wetting and (d) equilibrium between the interfacial tensions for a liquid with a contact angle

grains. The wetting liquid with contact angle (θ) is defined by the equilibrium of the surface energies (Fig. 7.16) as follows:

$$\gamma_{sv} = \gamma_{sl} + \gamma_{lv} \cos\theta \tag{7.23}$$

where, γ_{sv} is the solid–vapour surface energy, γ_{sl} is the solid-liquid surface energy and γ_{lv} is the liquid–vapour surface energy.

Figure 7.1 schematically summarises various forms of liquid phase sintering.

7.2.1 Stages of liquid phase sintering

The three stages of liquid phase sintering are:

- Rearrangement or liquid flow
- Accommodation or dissolution–reprecipitation
- Coalescence or solid phase bonding.

These stages follow in the approximate order of their occurance but there may be significant overlapping for a specific system (Fig. 7.17).

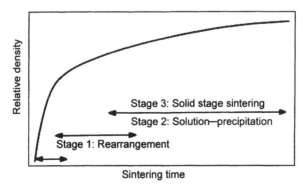

Fig 7.17 Schematics of the three stages of liquid phase sintering

Rearrangement or liquid flow

In this stage, the melt must wet the solid substructure, which is accompanied by rapid initial densification due to the capillary force exerted by the wetting liquid on the solid surface. The porosity is drastically reduced as the system minimises its surface energy. During subsequent particle rearrangement, the compacts respond as a viscous solid to the capillary action. The process depends on the amount of liquid, particle size and solubility of the solid in the melt. Usually, finer-sized particles give better rearrangement.

Figure 7.16 illustrates three conditions of wetting: poor, good and complete (ideal) wetting. At the contact angle of zero, spreading of the liquid takes place. Depending on the amount of liquid, a major amount of densification can occur in the first stage. If the amount of liquid is sufficient to fill up the pores in the green compact, full densification will be achieved by rearrangement alone.

The driving force for liquid phase sintering is the reduction of the liquid/vapour interfacial area. For a spherical pore of radius, r, in a liquid, the pressure difference (Δp) across the curved surface is expressed as:

$$\Delta p = \frac{2\gamma_{lv}}{r} \tag{7.24}$$

where, γ_{lv} is the surface energy of the liquid–vapour interface. The above relationship infers that with decrease in the pore size, the driving force for sintering is high.

The dihedral angle concept related to solid state sintering has been discussed in Section 7.2. The same concept is now to be revoked in liquid phase sintering too. Here, the dihedral angle (ϕ) is expressed as:

$$\cos\frac{\phi}{2} = \frac{\gamma_{ss}}{2\gamma_{sl}} \tag{7.25}$$

where, γ_{ss} is the solid/solid interfacial tension, i.e., the grain boundary tension and γ_{sl} is the solid/liquid interfacial tension.

When the dihedral angle is zero, γ_{ss}/γ_{sl} is equal to 2. This means that under this condition complete penetration of the grain boundary by the liquid takes place. This promotes grain–grain separation, giving rise to some swelling, for example, the Fe-Cu system during the very first instance of liquid phase sintering. These separated grains, disintegrated from the original particles, would eventually promote secondary rearrangement akin to primary rearrangement of powder particles.

Accomodation or dissolution–reprecipitation

In this stage, the solubility and diffusivity effects become dominant: a microstructural change, i.e., coarsening, occurs. This is also known as *Ostwald ripening*, where the small grains dissolve faster and deposit over the bigger grains, resulting in coarsening. The densification in this stage is not of the same magnitude as in the first stage (Fig. 7.17). The size of the particle plays a critical role in its solubility in the melt. The concentration of the dissolved solid, c, surrounding the spherical powder of radius, r, is given by:

$$\ln(\frac{c}{c_o}) = \frac{2\gamma_{sl}\Omega}{kTr} \tag{7.26}$$

where, c_o is the equilibrium concentration of solid in the liquid, Ω is the atomic volume, γ_{sl} is the solid–liquid interfacial tension, T is the absolute temperature and k is the Boltzmann's constant.

It is obvious from the above equation that the solubility increases with decreasing particle radius.

Because of better microstructural adjustment, the enhancement in properties—for example, strength—is much more noticeable in this stage than in the first stage of liquid phase sintering. The effect of solubility on densification can be demonstrated by two binary systems, Fe-Cu and W-Cu, where the densification in the former is greater than the latter system. This is attributed to the fact that iron is soluble in the copper melt, in contrast to the zero solubility of tungsten.

Coalescence or solid phase bonding

In this stage, densification is slow, because of the existence of the solid skeleton. It is evident in two cases:

- When the volume fraction of the melt is too little and the melt does not ideally wet the solid particle;
- During incipient liquid phase sintering, where the low melting point component completely dissolves in the solid particle.

For example, in Cu-Sn premixed bronzes, elemental tin dissolves in the copper matrix (temperature greater than the melting point of tin), resulting in the solid bronze skeleton at the end. On the other hand, there are systems, for example, heavy alloys (W-Ni-Fe) or cemented carbides (WC-Co), where persistent liquid phase sintering occurs. Here, the last portion of the melt solidifies. All solidification defects, such as, porosity and gas entrapment, may be present in such types of systems.

Analysis of liquid phase sintering kinetics shows that the shrinkage rate is dependent on sintering time to the half or one-third power. For solution–reprecipitation controlled shrinkage, the material transport path is through the melt and the time for the one-third power for round-shaped grains is appropriate. In the final stage, i.e., solid-state bonding, the densification is quite small.

7.2.2 Wetting aspects of liquid phase sintering

In Section 7.2.1, the mechanisms were described, taking into account a pair of arbitrary polycrystalline materials and the melt. The first stage of liquid phase sintering is highly dependent on good wettability of the melt. The wetting angle needs to be tailored, sometimes,

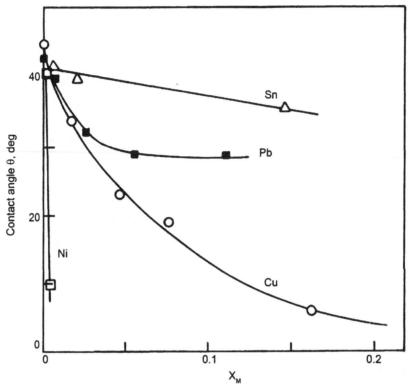

Fig 7.18 Influence of Sn, Pb, Cu and Ni dissolved in silver as the contact angle of the Ag melt/solid Fe system at 965°C in an Ar-5%H₂ atmosphere (Reproduced with permission from Eusthopoulos N, Nicholas MG and Drevet B, *Wettability at High Temperatures*, Pergamon, Oxford, 1999)

by alloying the melt with solute. Figure 7.18 illustrates how the effect of nickel in silver melt is much stronger on the contact angle over a solid iron substrate. Another example is when there is some dissolution of the solid substrate in the molten phase. A further modification may be when both dissolution and reaction take place. Figure 7.19 illustrates how the contact angle at 1500°C of nickel-based alloys on alumina gets affected. The addition of chromium to the melt promotes dissolution of alumina and a fall in the contact angle. At the same time, the addition of titanium in nickel causes the dissolution of alumina followed by precipitation of titanium oxide.

Selection of sintering temperature during liquid phase sintering is another factor which affects the wettability of the melt. It is generally found that the increase in temperature and time improves wettability. Figure 7.20 illustrates the above features for the WC-Cu system.

7.2.3 Microstructural evolution during liquid phase sintering

In conventional liquid phase sintering, the presence of two phases is essential. All the problems faced in green compaction of premixed powder, as described for solid-state sintering, is equally valid here. For ideal distribution of the two phases, it is necessary for

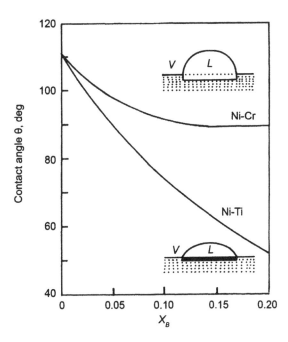

Fig 7.19 Contact angles at 1773 K of Ni-based alloys on Al$_2$O$_3$ plotted as a function of Cr(Ti) molar fraction for Ni-Cr and Ni-Ti (Reproduced with permission from Eusthopoulos N, *Wettability at High Temperatures*, Pergamon, Oxford, 1999)

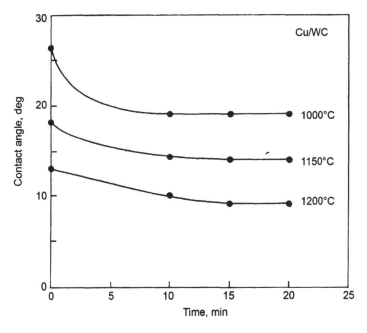

Fig 7.20 Contact angle of copper melt on WC as a function of time

the high–melting point particles to be coated by the low–melting point phase. Examples are Fe or W powders coated with copper. In the case of premixed real systems, the distribution of the two phases may not be uniform. Moreover, there is a particle size distribution in the powders of either phase.

It is evident from Table 7.5, that out of various mechanisms of liquid phase sintering, the mechanisms responsible for change in grain shape are pore elimination by liquid flow, contact flattening and Ostwald ripening with shape accommodation.

In order to illustrate the microstructure of alloys sintered by liquid phase sintering, two systems—iron-copper and alumina-borate glass—are selected here. Figure 7.21 shows the schematics of the microstructure along with the alloying mode of Fe-10Cu premix compact liquid phase sintered at 1120°C. It is worth noting that the entire Cu melt in this case is not dissolved in the iron grains, suggesting that sintering is not yet fully accomplished. In the second system, i.e. alumina–borate glass (Fig. 7.22), the melt is in a glassy state. The premix was sintered at two different temperatures, 1400°C and 1700°C for 1 hour. It is evident from the microstructure (Fig. 7.22b) that alumina spheres get disintegrated at a high sintering temperature of 1700°C, due to better attack by the glassy phase. This definitely enables secondary rearrangement of particles.

The microstructural accommodation in the system has already been mentioned while describing the second stage of liquid phase sintering. The significance of the dihedral angle has been highlighted. Figure 7.23 shows the microstructural evolution in three cases: (i) when $\phi = 0°$, (ii) when $\phi < 60°$ and (iii) when $\phi > 60°$. Some of the pictures pertain to polished section, while others are isometric views. In case (ii), the liquid is unable to penetrate

Table 7.5 Basic mechanisms of liquid phase sintering: Contribution to rearrangement or shape change

Driving force	Mechanism	Contribution to densification by	
		Rearrangement	Shape change
Decrease of interfacial energy by 1–100 J mol^{-1}	Ostwald ripening	M	N
	Ostwald ripening with grain shape accommodation	M	E
	Contact flattening	N	E
	Particle disintegration	E	N
	Coalescence	M	N
	Pore elimination by liquid flow	M	E
	Grain/liquid mixture flow	E	M
Decrease of free energy <1000 J mol^{-1}	Directional grain growth	M	M

E: Essential; M: Modest; N: Negligible

(Adapted from Petzow G, In Upadhyaya GS, (ed.), *Sintered Metal-Ceramic Composites,* Elsevier, Amsterdam, 1984)

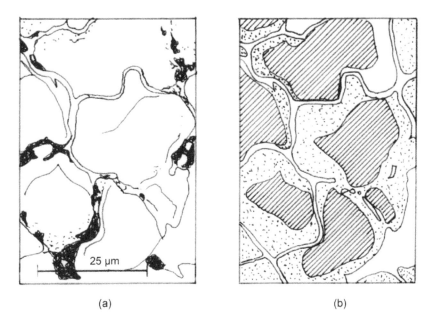

(a)   (b)

Fig 7.21 Schematics of (a) microstructure and (b) alloying mode in Fe-10Cu premix compacts liquid phase sintered at 1120°C. The white regions correspond to melt; the dotted area represents solid solution, and hatched areas indicate unalloyed iron.

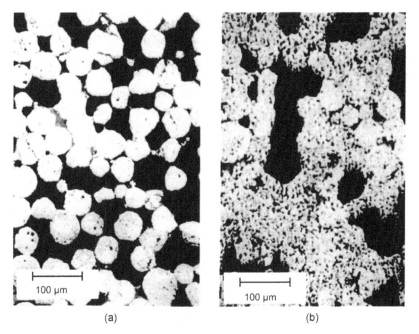

(a)   (b)

Fig 7.22 Microstructural changes during the liquid phase sintering of Al_2O_3-alkali borate glass sintered for 1 h at (a) 1400°C and (b) 1700°C (Reproduced with permission from Upadhyaya GS, (ed.), *Sintered Metal-Ceramic Composites*, Elsevier, Amsterdam, 1984)

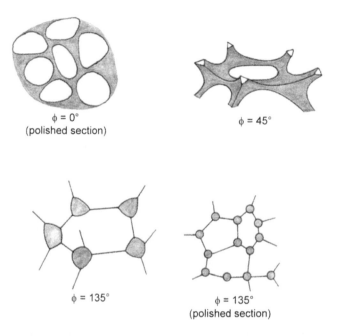

$\phi = 0°$
(polished section)

$\phi = 45°$

$\phi = 135°$

$\phi = 135°$
(polished section)

Fig 7.23 Idealised liquid phase distribution (in three dimensions or polished section) under three conditions of dihedral angles

indefinitely along the three grain edges. The microstructure should therefore consist of two continuous interpenetrating phases. When ϕ is greater than 60°, the liquid begins to form isolated pockets at the corners of three grains. It may be mentioned that these shapes are idealised ones. In real systems, some departure from these shapes may be visualised.

After careful microstructural examination of the plane section of a liquid phase sintered material, it can be concluded what possible mechanisms are the controlling factors. For quantitative assignment, a rigorous stereological study is required. Microstructural control is an excellent tool to achieve the desired properties.

7.2.4 Supersolidus sintering

Supersolidus sintering is a form of liquid phase sintering. It involves densification by melt formation in a pre-alloyed powder compact on heating it above the solidus temperature. In this type of sintering, solid and liquid phases have much closer compositions. A limited amount of densification occurs through solid-state diffusion until the temperature is within a few degrees of the optimum sintering temperature, when the fine particles melt preferentially to form a small quantity of the viscous liquid phase. Melting also occurs over the surfaces of large particles and to some extent at the grain boundaries within particles. Densification occurs through rearrangement with a substantial contribution by particle deformation. Figure 7.24 shows a typical example of supersolidus sintering of high-speed PM steels. Initially, the powder particles are sintered in the solid phase to form a skeleton before the formation of the

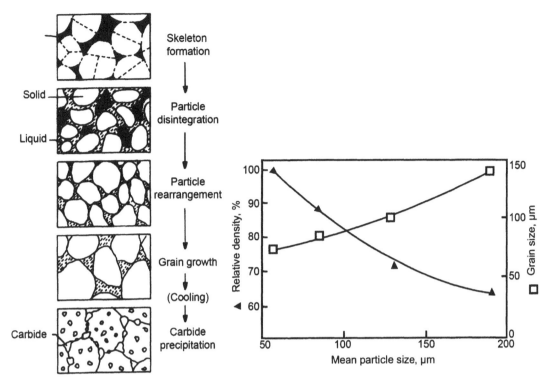

Fig 7.24 Mechanism of supersolidus liquid phase sintering of high-speed steels

Fig 7.25 Influence of particle size distribution with grain size and relative density of sintered astroloy

liquid phase. When a liquid phase appears, it disintegrates the skeleton into almost individual solid grains. The grains rearrange their configuration and rapid densification follows. The solid grains grow by a solution–reprecipitation process in which material transport occurs through the liquid phase.

Another example of a real system subjected to supersolidus sintering is nickel-based superalloys. When atomised loose-packed powders.of astroloy (15.2Cr-16.63Co-5.19Mo-3.5Ti-4.07Al-bal.Ni) were argon sintered above 1350°C, it showed that finer the particle size, better the densification (Fig. 7.25). However, the mean final grain size was approximately identical to the prior particle size, but the final grain size distribution was narrower than the prior particle size distribution. This is due to the fact that the prior particles dissolve during the second stage of liquid phase sintering, i.e., solution–reprecipitation.

7.3 ACTIVATED SINTERING

Activated sintering is the process by which sintering kinetics is increased and sintering temperature is reduced, thus, giving rise to improved properties. By this definition, even liquid phase sintering can be classified as activated sintering. But, here is a rider. The amount of

Example 7.5 Figure Ex. 7.5 shows the variation of a relative linear shrinkage of WC-10Co cemented carbide powder compact during non-isothermal (room temperature to 1400°C) and isothermal (1400°C for 30 min and 60 min) sintering as studied by Exner.

(a) On the plot, highlight the various solid and liquid phase sintering events and their effects.

(b) Why is the shrinkage appreciable during a particular non-isothermal mode of sintering?

(c) What powder characteristics would improve the sinterability?

(d) Suggest an activator to enhance sintering.

Solution

(a) See Fig. Ex. 7.5.

(b) The densification during solid-state sintering appears due to a very slow heating rate.

(c) Small particle size and purity of the powders will improve sinterability.

(d) Small amount of phosphorous addition (e.g., 0.3%) was found to decrease the sintering temperature from 1400°C to 1275°C for liquid phase sintering.

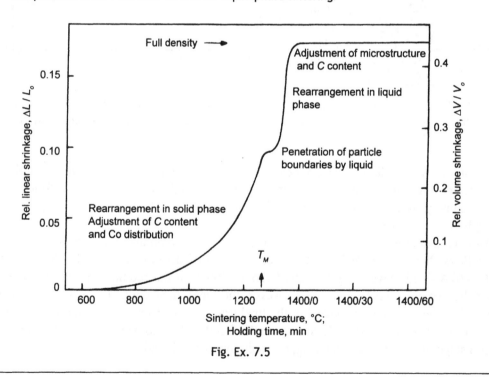

Fig. Ex. 7.5

activator is very small, sometimes, even as low as 1%. This kind of processing was introduced to consolidate refractory metals with very high melting points, and hence low diffusivities. The refractory metals pertain to early transition metals. They are invariably activated by minor addition of another transition metal, preferably, the later ones in the Periodic Table.

The electronic concept behind such a response will be discussed in a later section where a generalised approach to sintering is discussed. A few tenths of a percent of nickel lowers the sintering temperature of tungsten from 2000°C to ~1300°C. Nickel diffuses on to tungsten particles and enhances the grain boundary self-diffusion of refractory metals by a factor of up to 5000 at 1300°C. It is impossible for the activating element to remain concentrated at the grain boundary during sintering. Figure 7.26 shows shrinkage data for a 0.5 μm tungsten powder with respect to the activator amount for three transition metal additions: Ni, Pd and Pt, sintered at 1200°C and 1300°C for 1 h in hydrogen. The arrows in the lower right-hand corner of the figure indicate the shrinkage for undoped tungsten at these temperatures.

German postulated criteria for activated sintering based on the phase diagram of the activator (A) and the base metal (B) (Fig. 7.27). The criteria are:

- High solubility of base B in A
- Activator A must have lower melting point, thus, giving a large melting point difference
- As low a solubility of the activator A in base metal B as possible
- High diffusivity
- Segregation of activator at the grain boundary of the base metal.

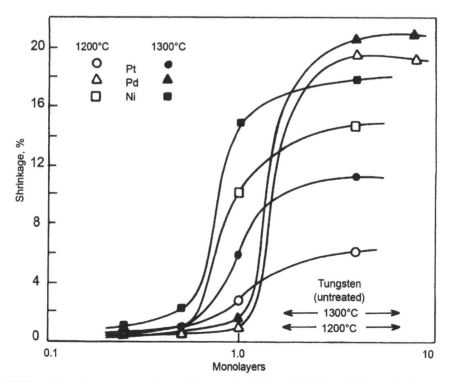

Fig 7.26 Effect of varying amounts of transition metal additives (Ni, Pd, and Pt) on the linear shrinkage behaviour of 0.5 μm tungsten powder (Adapted from German RM, *Sintering Theory and Practice*, John Wiley and Sons, New York, 1996)

A careful visualisation of the schematic binary phase diagram (Fig. 7.27) suggests that activated sintering can be classified into two categories: solid state and liquid phase activated sintering. The figure illustrates that at a temperature slightly above the activated sintering range, a liquid forms. In practice, the term *activated liquid phase sintering* is not common, since liquid phase sintering in itself signifies enhanced sintering.

Table 7.6 classifies various categories of activation criteria for sintering, in the case of iron.

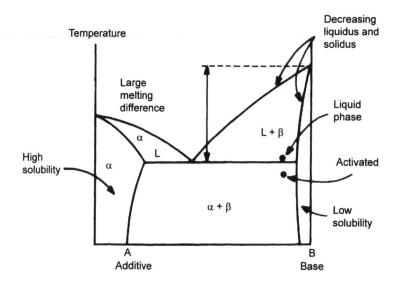

Fig 7.27 An idealised binary phase diagram schematically showing the effect of an additive A on the sintering enhancement of the base metal B (Reproduced with permission from German RM, *Powder Metallurgy and Particulate Materials Processing*, Metal Powder Industries Federation, Princeton, USA, 2005)

Table 7.6 List of activator candidates for iron

Selection criterion	Suitability as an activator		
	Good	Neutral	Poor
Electron structure difference	Mo, V, W, Ta, Ti, B, C, P, Si, Ru, Rh, Pt, Au	Cr, Ni, Cu, Zn	-
Phase diagram considerations	B, C, P, Si, Ti, Mo	-	Al, Cu, Sn, Zn
Grain boundary cohesion			
(i) smaller atom size	B, C, P	Ni, Cu, Zn, Cr,	Sn
(ii) high sublimation energy	B, C, Co, Mo, W, V, Sn, Ti	Mo, W, Ti, Si, V, Cr, Ni, Cu	P, Si, Zn

(After Madan DS and German RM, in *Modern Developments in Powder Metallurgy*, Vol. 15, Metal Powder Industries Federation, Princeton, NJ, USA, 1988)

Recently, the availability of better microstructural characterisation tools has offered inputs in the activated sintering of refractory metals and ceramic systems. It is established that in enhanced diffusion in pre-wetting, instead of wetting, a thin intergranular film (nano-scale thickness) is responsible for solid state activated sintering. ZnO ceramic doped with Bi_2O_3 is an excellent example of activated sintering where an amorphous film on the $\{11\bar{2}0\}$ surfaces of ZnO is known to activate the sintering process. Figure 7.28a shows the densification plot for nickel-added tungsten sintered isothermally at 1400°C—is 95° below the eutectic temperature for W-Ni binary system—for 2 hours in flowing $95N_2$-$5H_2$ atmosphere. The densification increases monotonically with increasing dopant percentage for specimens containing less than 0.2 at% Ni and approaches a plateau for specimens doped with 0.3 at% or more nickel. It was noticed that the optimal doping level was slightly greater than the solid solubility (~0.2 at%) of nickel at 1400°C. This is attributed to the fact that the formation of equilibrium interfacial film requires the nickel to be transported into the grain boundary region. Figure 7.28b shows that the nickel-rich crystalline secondary phase does not wet the tungsten grain boundaries in the solid state. Instead, a nanometre-thick, nickel-enriched, disordered film is formed at grain boundaries (Fig. 7.28c). These studies have shown that bulk phase diagrams are not adequate and there is a need for developing 'grain boundary phase diagrams' as a tool for studying activated sintering. (For more details refer Luo, 2007).

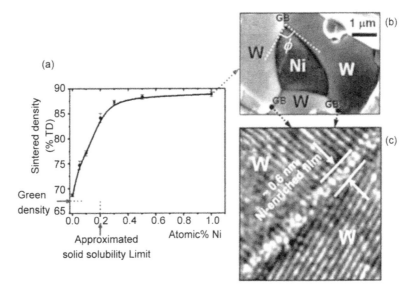

Fig 7.28 (a) Sintered density vs nickel content (at.) in tungsten sintered isothermally at 1400°C for 2 h in flowing $95N_2$-$5H_2$ atmosphere; (b) SEM picture showing that the Ni-rich secondary crystalline phase does not completely wet the grain boundaries; (c) High resolution TEM image of a subeutectic disordered intergranular glassy (amorphous) film (Reference: Luo J, Stabilization of nanoscale quasi-liquid interfacial films in inorganic materials: A review and critical assessment, *Critical Reviews in Solid State and Material Sciences*. 32:67–109, 2007. Reproduced with permission from Taylor and Francis Group LLC, London)

Example 7.6 Alumina with small additions of MgO attains full density during solid-state sintering. (a) Give the main features

(b) What is the effect of fast sintering on undoped alumina?

Solution

(a) The addition of MgO
- decreases the rate of grain coarsening in the early stage of sintering
- increases the rate of densification
- increases pore mobility such that pore-grain boundary separation does not occur.

Figure Ex. 7.6 shows the actual grain size—relative density trajectories.

It is clear that the rate of grain growth is lowered by MgO addition relative to sintered density.

(b) Fast sintering of undoped alumina increases the sintered density for the same grain size.

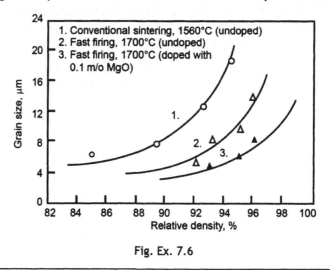

Fig. Ex. 7.6

7.4 PRESSURE-ASSISTED SINTERING

Polycrystalline materials like ceramics and dispersion-strengthened metals and alloys are difficult to densify by conventional solid state (pressureless) sintering. The former, due to the presence of strong covalent bonds, and the latter, due to the presence of hard refractory inert dispersoids in the matrix. In conventional sintering, with prolonged sintering the pores become stabilised and are difficult to remove from a compact via diffusion, especially if the pores are filled with a gas. The inability to attain full density can be offset by the application of external pressure on a polycrystalline powder compact. In pressure-assisted sintering, the enhancement of densification depends on the applied pressure, compact density, temperature and particle size. In most pressure-assisted sintering, the applied pressure is nearly constant; thus, the effective pressure at the particle contacts falls continuously with densification. It is the effective pressure which determines the mechanism of densification.

Like pressureless solid-state sintering, pressure-assisted sintering also undergoes three stages of sintering: initial, intermediate and final. Since the powder in a real system has a particle size distribution, a preliminary stage of particle rearrangement is expected when the external pressure is applied. In Fig. 7.7, the three stages of solid-state sintering for alumina powder were designated. In hot pressing, the nature of the plot is similar, but the curve shifts to the left-hand side, indicating enhanced densification. Many authors combine second and third stages of solid-state sintering into one, while describing the stages in pressure-assisted sintering. During the later stage of pressure-assisted sintering, when the relative density is greater than 90%, the compact can be modelled as a homogeneous solid containing isolated spherical pores. The effective stress responsible for densification is then identical with the applied external pressure, unless the gas trapped in the closed pores causes a back pressure which may prevent the compact from reaching full density.

Figure 7.29 schematically shows the mechanisms—plastic yielding, power law creep and lattice/grain boundary diffusion—for spherical powders of the same size during pressure-assisted sintering. The left-hand side scheme is for the initial stage, while the right-hand side scheme pertains to the later (intermediate/final) stage. These features will be described in the following sections.

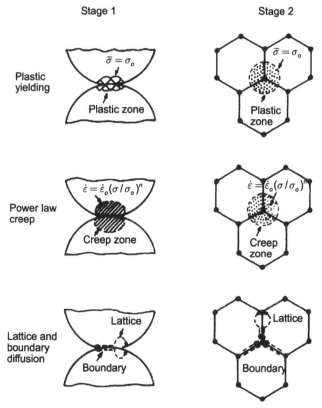

Fig 7.29 Schematics of initial and later stages of pressure-assisted sintering involving various mechanisms

7.4.1 Plastic yielding mechanism

Plastic yielding or plastic flow occurs at pressures above the yield strength, which decreases at elevated temperatures. This provides instantaneous densification by plastic flow of the powder contacts. The densification is governed by the deformation of the particle contacts by the local effective stress acting on the contact area. In isostatic compaction, this effective stress (σ_{eff}) is proportional to the applied pressure, p, such that

$$\sigma_{eff} = \frac{4\pi a^2}{AZ\rho} p \qquad (7.27)$$

where, a is the particle radius, ρ the relative density of the compact, and A and Z are the average contact area and number of contacts per particle, respectively. Under certain assumptions concerning powder packing and contact geometry, A and Z can be expressed as functions of ρ.

For yielding to occur, the following relation is considered to be applicable.

$$\sigma_{eff} \geq 3\sigma_y \qquad (7.28)$$

where, σ_{eff} is the effective stress and σ_y is the yield stress of the powder material.

Yielding increases the contact area between the particles and as a consequence of the densification, increases the number of contacts per particle. If the pressure is high enough for the compact to enter into at a later stage by plasticity alone, the picture can be visualised as the plastic collapse of a thick spherical shell.

7.4.2 Creep mechanisms

There are three submechanisms under this category: dislocation motion, grain boundary diffusion (Fig. 7.30a) and lattice diffusion creep (Fig. 7.30b). In the dislocation mechanism,

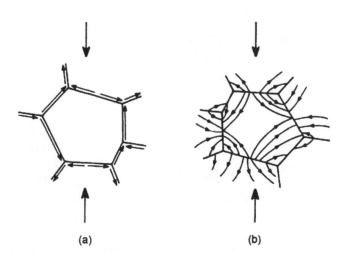

(a)   (b)

Fig 7.30 Schematics of (a) grain boundary diffusion and (b) Nabarro-Herring-type creep

Table 7.7 Hot-pressing equations obtained by modification of the creep equations

Mechanism	Intermediate stage	Final stage
Dislocation motion[a]	$(1/\rho)(d\rho/dt) = (AD\mu b/kT)(P_{eff}/\mu)^n$	$(1/\rho)(d\rho/dt) = (BD\mu b/kT)(P_{eff}/\mu)^n$
Grain boundary diffusion (Coble)	$(1/\rho)(d\rho/dt) = (47.5 D_{gb}\delta_{gb}\Omega/G^3 kT)$ $(P_{eff}+\gamma_{sv}/r_p)$	$(1/\rho)(d\rho/dt) = 7.5 D_{gb}\delta_{gb}\Omega/G^3 kT$ $(P_{eff}+2\gamma_{sv}/r_p)$
Lattice diffusion (Herring)	$(1/\rho)(d\rho/dt) = (40/3)(D_l\Omega/G^2 kT)$ $(P_{eff}+\gamma_{sv}/r_p)$	$(1/\rho)(d\rho/dt) = (40/3)(D_l\Omega/G^2 kT)$ $(P_{eff}+2\gamma_{sv}/r_p)$

[a] A and B are the numerical constants; n is an exponent that depends on the mechanism of dislocation motion and has a range of 3–10
D_l = Lattice diffusivity, g_{sv} = surface energy, D_{gb} = Grain boundary diffusivity, b = Burger's vector, μ = Shear modulus, D = diffusion coefficient, G = Grain size, r_p = pore radius, k = Boltzmann constant, r = density, T = Absolute temperature, t = time, Ω = atomic volume, P_{eff} = effective pressure, δ_{gb} = grain boundary width
(Adapted from Rahman MN, *Ceramic Processing and Sintering*, 2nd edn, Taylor and Francis, London, 2003)

climb is the main contributing factor. Table 7.7 summarises the pressure-assisted sintering equations. The exponent n depends on the mechanism of the dislocation motion.

Coble proposed that the grain boundary surfaces need to be included, such that the effective stress on the grain boundaries is related to the externally applied stress. According to him, the total driving force is a linear combination of the applied stress and the surface curvature. In real powder systems, Coble's concept does require modifications. In the creep model, the grain boundary area remains constant and is related to the grain size. However, in reality, both the grain boundary area and the path lengths for diffusion change during densification. Table 7.7 includes the equation for grain boundary diffusion creep-controlled mechanism. It is suggested that readers consult physical metallurgy textbooks to understand terms such as Burger's vector, grain boundary width, etc.

Nabarro and Herring proposed another creep mechanism applicable to pressure-assisted sintering, which operates at high temperature and low stress. In this case, it is a lattice or volume diffusion, such that atoms diffuse from grain boundaries subjected to a compressive stress towards those subjected to a tensile stress. Herring found that the creep rate is equal to linear strain rate. Table 7.7 includes the typical densification equations pertaining to this category of mechanism.

A summary of densification processes occurring during hot pressing of MgO powder at 1300°C, 2.7 MPa (400 psi) is shown in Fig. 7.31. It is interesting that in the early stage the mechanism's particle rearrangement and plastic flow are overlapping.

Like pressureless sintering maps, Ashby extensively studied the HIP maps. An example is given in Fig. 7.32 for hot isostatic pressing (HIP) of a tool steel powder (25 μm diameter) at 1200°C. The data correspond to typical industrial HIP cycle with time marked in hours. It is clear that the initial densification is due to plastic yielding of the particle contacts, while at intermediate densities, power law creep in the contact zone dominates. Diffusion may finally lead to full density. The thin lines in the map are contours of constant time.

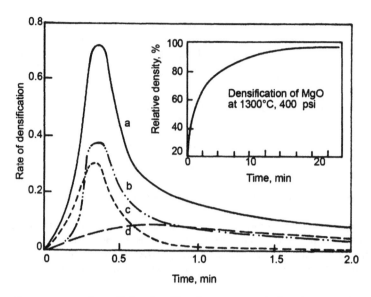

Fig 7.31 Schematic representation of the densification process during hot pressing with respect to the sintering period. (a) Total rate of densification, (b) Plastic flow, (c) Particle rearrangement and (d) Diffusion (Reproduced with permission from Notis MR and Spriggs RM, *Science of Sintering*, Vol. 10, 1978)

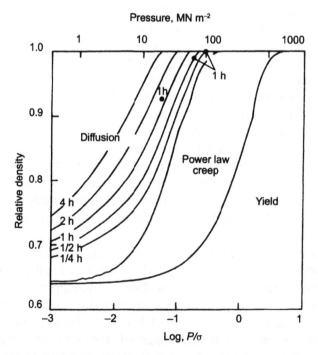

Fig 7.32 Density–pressure diagram for hot isostatic pressing of a tool steel powder (radius 25 μm) at temperature 1200°C (σ_y is the yield stress of the powder material) (Adapted from Arzt E, Ashby MF and Easterling KE,, *Metallurgical Transactions*, 14(A):211, 1983)

7.4.3 Viscous flow mechanism

Viscous flow is applicable for glass or other amorphous materials subjected to external pressure at elevated temperature. The flow is inversely proportional to viscosity (η), hence, higher temperature aids densification. The expression is

$$\frac{dV_s}{dt} = \frac{3P_E V_P}{4\eta} \tag{7.29}$$

where, V_s is the relative density, P_E is the effective pressure on the sintering body and V_p is the volume fraction of the porosity. If the effective pressure is low, the inherent sintering densification becomes predominant. In this case, the viscous flow mechanism in solids should not be confused with Nabarro–Herring microcreep mechanism.

In many ceramic systems, the ceramic phase is associated with the glassy phase. Such systems (solid grains and liquid phase) undergo viscous flow, in which the glassy phase is the deformable phase.

Example 7.7 One kg of spherical-shaped iron powder was subjected to pressure at 100 MPa.

(a) What depression in the melting point of iron do you expect under this pressure?

(b) Will the solid powder microweld under this pressure during HIPing at 1100°C?

Given: Density of Fe = 7860 kg m^{-3}, heat of fusion = 27 × 10^4 J kg^{-1}, T_m = 1808 K, volume change at melting point = 3%.

Solution

(a) We have to apply Clausius–Clapeyron equation to determine ΔT

$$\frac{\Delta T}{\Delta T_m} = \frac{\Delta p \Delta V}{\Delta H_m}$$

since, $\dfrac{\Delta V}{\Delta T_m} = 3$ vol%, $\Delta V = \dfrac{0.03}{7860}$

Substituting the values in the above equation:

$$\Delta T = \frac{T_m \Delta p \Delta V}{\Delta H_m}$$

$$= \frac{1800 \times 10^8 \times 0.03}{27 \times 10^4 \times 7860} = 2.5 \text{ K}$$

(b) No; for microwelding the pressure should be about two orders of magnitude higher. This is possible in dynamic powder compaction, where the powder is subjected to very high pressures.

7.5 ELECTRONIC THEORY OF SINTERING: A UNIFIED APPROACH

Many attempts have been made in the past to close the gap between the theory and practice of sintering. Most of the sintering mechanisms described in the earlier sections are based on simplified models like the two-sphere model. In the real system, one would prefer to have a generalised or unified approach towards sintering theory. The appeal has forced many to think in terms of the basic physicochemical bonds prevailing in crystalline materials and to judge how far it is useful in understanding the sintering process.

Depending on the structure of the outer shells of free atoms, which governs the possibility of the formation of stable configurations of localised valency electrons, all the elements can be divided into three classes:

- The *s*-elements, which have outer electrons and completely filled or completely empty inner electron levels;
- The *ds*- and *fds*-elements, which have partly filled *d*- or *fd*-levels;
- The *sp*-elements, which have *sp*-outer electrons.

Samsonov proposed the following features for the electronic configurational model of matter:

1. When free atoms are condensed to form a solid or a liquid, the valency electrons of these atoms split into localised and collective-state groups.
2. The localised fraction of the vacancy electrons in a solid form a spectrum of configurations.
3. The most stable configuration in a given spectrum can be empty, half-filled, or completely-filled. Thus, the most stable configurations are s^2, sp^3, s^2p^6, d^0, d^5, d^{10}, f^0, f^7 and f^{14}.
4. The energy stability of the electron configuration is a function of the principal quantum number of the valency electrons from which these configurations are formed.

The energy stability of the *s*- and *sp*-configurations decreases while that of the *d*- and *f*-configurations increases with increase in the principal quantum number of the valency electrons. The electronic exchanges between stable and non-localised configurations are responsible for the attractive force between the atom cores, whereas, electron–electron interactions between the non-localised electrons is responsible for the repulsive force.

With increase in temperature, which is relevant for the sintering study, thermal excitation of $s \rightarrow d$ (e.g., Sc) and $d \rightarrow s$ (e.g., Cr) transitions becomes increasingly important. The results of the fundamental calculations of the electron energy spectra of the transition metal at high temperatures and pressures are in agreement with the conclusions drawn from the configurational model. A similar feature is also noticed in *sp*-type elements, where $s \rightarrow p$ transition is favoured to form stable sp^3 configuration (e.g., carbon). The above features are manifested in the relationship governing the formation, stability and physicochemical properties of many systems with the donor–acceptor behaviour.

Table 7.8 Activation energy of diffusion in some transition metals containing chromium

System M-Cr	Cr content, at%	Activation energy, kJ mol^{-1}
Fe-Cr	0	284.5
	3.5	313.8
	7.3	376.6
Co-Cr	0	261.5
	5	317.9
	10	368.2
Ni-Cr	0	267.8
	5	257.3
	8	274.0

(Reference: Samsonov GV, Pryadko IF and Pryadko LF, *A Configurational Model of Matter*, Consultants Bureau, New York, NY, USA, 1973)

The basic atomic features of various sintering mechanisms described in the earlier sections of this chapter are directly related to the electronic structure of the materials. For example, the difficulties encountered in sintering covalent solids such as graphite, BN, SiC and diamond is explainable under the above features. At high temperature, the directed bonds get weakened and for re-establishing them the role of external pressure (e.g., hot pressing, hot isostatic pressing, etc.) becomes essential.

The correlation of the self-diffusivity data and hetero-diffusion with electronic structure confirms that the electronic concept fully describes the process, irrespective of the fact to which class of electronic analog the partners of the system belong. From Table 7.8 it is evident that the non-localised electrons of chromium donate themselves easily to stabilise the stable d^{10} configurations of nickel, thus causing the activation energy to be lowered. The value increases when one passes from Ni → Co → Fe.

The electronic approach contributes to the uniform view of the sintering process, although it still does not offer sufficient possibilities for quantitative treatment. Transition from the atomic theory to the electronic one should be examined for a uniform quantitative description of the fundamental phenomenon in diffusion and creep. One of the possibilities is the use of the activated volume, especially because it allows in a definite way, simultaneous consideration of the problem concerning both the stress and material transport.

7.5.1 Liquid phase sintering

The role of phase diagrams was highlighted by German (see Section 7.3) for understanding enhanced sintering. The basic type and formation of phase diagrams were studied by Samsonov and his co-workers on the basis of the stable electronic configuration model. The primary requirement of liquid phase sintering is the wettability of the melt over the solid substrate. As an example, we may take the wetting of the refractory carbides, which are essential constitutents in the development of hard metals. It is known that refractory carbides are practically not wetted by melts of Group IIIB–VB elements, but are generally wetted by transition metals. As the atoms of metals, which do not melt carbides, have completely filled or empty d-levels, one can conclude that the d-levels play a decisive role. The better wetting of the group V–VI metal refractory carbides by iron group melt is explained by the possible capture of the non-localised valence electrons of the latter metal atoms by the disturbed configurations of the carbon atoms in the carbides (Table 7.9). Alloying TiC in WC decreases wettability. Copper, on the other hand, partially wets WC, and thus gives rise to grain coarsening of WC. Good wettability of Ni melt over TiC substrate has given rise to alloy tailoring from non-WC-based cemented carbides. Another observation that the wetting in TiC-Ni is improved by addition of molybdenum or Mo_2C is also justified on the basis of their electronic stabilities.

Oxide ceramics (s^2p^6 electronic configurations), in contrast to refractory carbides, have good wetting with sp-type metals like aluminium. Transition metal melts, with less deficiency in d-level, are not effective in wetting such ceramics. This confirms how a similar degree of stabilities offered by the electronic configuration of the partner atoms is necessary for good wettability.

The wetting aspect of non-oxide covalent ceramics (B_4C, SiC, BN) by different melts with a wide range of melting points is illustrated in Table 7.10. From this it is evident that a better wettability of the ceramic is noticed with comparable covalency of metals: for example, Ge → Al → Si. With increase in covalent bonding in the ceramic, the wettability, in general, decreases.

Table 7.9 Wetting of some refractory carbides by liquid metals

Carbide	Wetting melt	Temperature, °C	Contact angle, deg	Atmosphere
TiC	Fe	1550	39	Hydrogen
	Co	1500	36	Hydrogen
	Ni	1450	17	Hydrogen
WC	Fe	1500	0	Vacuum
	Co	1500	0	Hydrogen
	Ni	1500	0	Vacuum
	Cu	1100	30	Argon
TiC-WC	Co	1477	26	Hydrogen
	Co	1477	4	Vacuum
	Ni	1477	16	Hydrogen
	Ni	1477	6	Vacuum

[Reference: Samsonov GV, Vitryanuk VK and Chapligin FI, *Tungsten Carbides* (in Russian), Naukova Dumka, Kiev, 1974]

Table 7.10 Contact angles of metals with non-oxide covalently bonded ceramics

Melt	Temperature, °C	Contact angle, degrees			
		Diamond	B₄C	SiC	BN
Cu	1150	146	134	-	136
Ag	1100	120	137	128	145
Sn	1150	124	133	135	135
Ge	1050	132	129	113	138
Al	1000	75	60	65	140
Si	1500	0	0	36	95

(Reference: Upadhyaya GS, (ed), *Sintered Metal-Ceramic Composites*, Elsevier, Amsterdam, 1984)

7.5.2 Activated sintering

Lenel was the first to de-emphasise the term 'activated sintering' and called it a misnomer. He viewed the process entirely in the framework of generalised sintering theory. Kuczynski agreed, in principle, with Lenel; but found it difficult to abandon the term, as its usage had come to stay in technology. Samsonov, in the stable electronic configuration model, confirmed this particular aspect in a generalised manner, which was later adopted by German (Section 7.3).

According to Samsonov, activated sintering is always due to some chemical or physical excitation of a substance which disturbs the electronic configurations and reduces their statistical weight. Small admixtures of materials which can accept or donate localised electrons to the atoms of the material being sintered can activate the process. In this way, the free energy of the whole system is reduced. As an illustration, Table 7.11 shows the activated sintered density values of tungsten after adding an optimum amount of additives at optimum sintering temperatures, such that the addition of Ni or Pd exhibits maximum effect.

7.5.3 Case study: Sintering of refractory compounds

The refractory compounds (early transition metals, carbides, borides and silicides) are excellent examples, where the degree of thermal excitation required for sintering is dictated by the energy needed to disturb the sp^3 configurations which are responsible for the strongest M–X bonds. Carbides and borides are sintered only at high temperatures, irrespective of which early transition metal they belong to. Refractory silicides are less stable and can therefore be sintered at rather lower temperatures. Samsonov and his co-workers discussed hot pressing of refractory carbides in the framework of the electronic structure of their components. The activation energy variation of hot pressing of transition metals and their carbides against the SWASC (Statistical Weight of Atoms having Stable Configuration)

Table 7.11 Sintered density of tungsten alloyed with various elements under optimal conditions

Alloying element	Optimal content, wt%	Optimal sintering temperature, °C	Sintered density, g cm^{-3}
Cr	2.0	1600	16.6
Fe	0.5	1600	17.9
Co	0.4	1600	18.1
Ni	0.4	1600	18.4
Ru	1.0	1600	17.4
Rh	0.5	1600	17.8
Pd	0.4	1800	18.4
Os	3.0	2000	16.1
Ir	1.0	2000	16.7
Pt	0.4	2000	17.4

(Adapted from Samsonov GV, Pryadko IF and Pryadko LF, *A Configurational Model of Matter*, Consultants Bureau, New York, USA, 1973)

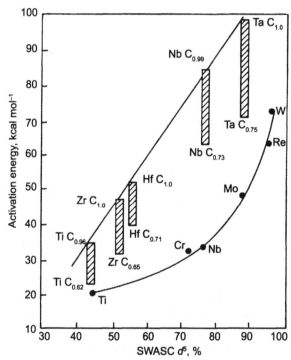

Fig 7.33 Variation of the activation energy for hot pressing of early transition metals and their monocarbides with SWASC (Statistical Weight of Atoms with Stable Configuration) d^5 (Adapted from Upadhyaya GS, *Nature and Properties of Refractory Carbides,* Nova Science Pub. Inc., Commack, NY, 1996)

d^5 is shown in Fig. 7.33. It is apparent that the activation energy increases with increase in SWASC d^5, which is related to the decrease in the fraction of non-localised electrons participating in electronic exchanges. In case of Group V transition metal carbides, the activation energy of the process increases in comparison with that for Group IV transition metal carbides.

In concluding this section on the electron theory of sintering it is worthwhile to summarise the results (Fig. 7.34) of the relative sintered densities of various crystalline materials (metal, ionic solid, non-metals, refractory compounds and covalent ceramics). All these materials could be reasonably well sintered, except highly covalently-bonded Si_3N_4, in which case one has to introduce some additives. Metals with reasonably non-localised electrons require lower homologous temperatures for sintering as compared to compounds. In compounds possessing ionic bonds, for example, CaF_2 and MgO, a relatively higher homologous temperature is required. Refractory carbides with mixed bonds require a still higher degree of homologous temperature. Covalently-bonded solids are followed by non-metallic crystalline materials, like boron and silicon. In a nutshell, the atomic approach for sintering postulated in the mid 20th century has become more inclusive by taking into account the electronic structural aspects of materials. In a way, it has amicably solved the rivalry between theories based on diffusion and plastic flow concepts.

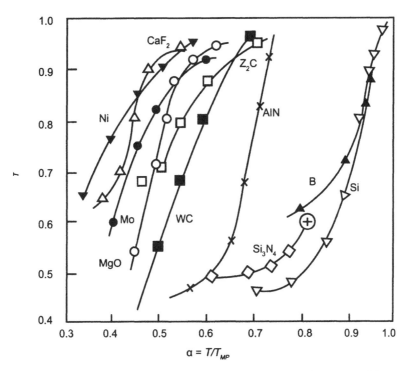

Fig 7.34 Densification behaviour (relative density) of various crystalline material powders during sintering with respect to homologous temperature (T/T_m). [Adapted from Andrievskii RA, *Introduction to Powder Metallurgy*, ILIM Publishing House, Frunze, 1988 (in Russian)]

SUMMARY

- Sintering is a thermal process for consolidating powder particles into a coherent structure via mass transport on the atomic scale.
- The difference in the chemical potential from one region of the surface to another leads to a diffusional flux of atoms to reduce the free energy of the system.
- Major mechanisms for sintering a polycrystalline material are evaporation, diffusion, plastic flow and viscous flow.
- The binding energy between pore and grain boundary decreases during sintering.
- The dihedral angle formed by the grain boundary with a pore is a measure of the relative interfacial energies.
- During sintering of nanopowders there is a high tendency of agglomeration, significant interparticle friction, high reactivity and associated contamination, high densification and coarsening kinetics.
- Differing diffusivities in mixed powder systems can offset sintering due to swelling.
- Liquid phase sintering is achieved through rearrangement, dissolution—reprecipitation and colescence steps.
- Supersolidus sintering is a form of liquid phase sintering, which involves densification by melt formation in a pre-alloyed powder compact by heating it above the solidus temperature.
- Activated sintering is classified into two categories: solid state and liquid phase activated sintering.
- In pressure assisted sintering, the densification depends on the applied pressure, compact density, temperature and particle size.
- The basic atomic features of various sintering mechanisms are directly related to the electronic structure of the metals.

Further Reading

Ashby MF, *HIP 6.0 User Manual*, Engineering Department, Cambridge University Press, Cambridge, UK, 1990.

Budworth DW, *An Introduction to Ceramic Science,* Pergamon Press, Oxford, 1970.

Chiang Y-M, BirnieIII D and Kingery WD, *Physical Ceramics,* John Wiley & Sons, New York, 1997.

Coble RL and Cannon RM, Palmour III H, (ed), Current Paradigms in Powder Processing, *Processing of Crystalline Ceramics,* American Ceramic Society, Westerville, OH,USA, 1978.

Evans JW and De Jonghe LC, *The Production and Processes of Inorganic Materials,* TMS, Warrendale, 2002.

Exner HE, In *Proceedings of Recent Advances in Hardmetal Production,* Vol. 2, Metal Powder Report, London, 1979.

German RM, *Liquid Phase Sintering,* Plenum Press, New York, 1985.

German RM, *Sintering Theory and Practice,* John Wiley and Sons, New York, 1996.

Heckel RW, Lanam RD and Tanzilli RA, In Hirschhorn JS and Roll KH, (eds), *Advanced Experimental Techniques in Powder Metallurgy,* Plenum Press, New York, 1970.

Herring C, *J.Appl. Phys,* 21:437, 1950.

Johnson JL, In Bose A, Dowding RJ and Shields Jr. JA, (eds), *Proceedings of 2008 Int. Conference on Tungsten, Refractory and Hard Materials VIII,* Metal Powder Industries Federation, Princeton, 2008.

Kang SJ-L, *Sintering: Densification, Grain Growth and Microstructure,* Elsevier, Amsterdam, 2005.

Kingery WD and Berg M, Sintering of the initial stage of sintering by viscous flow: Evaporation-condensation and self-diffusion, *J. Appl.Phys.,*26:1205–12, 1955.

Koch CC, Ovid'ko IA, Seal S and Veprek S, *Structural Nanocrystalline Materials,* Cambridge University Press, Cambridge, 2007.

Kuczynski GC, Nooten NA and Gibson GF, (eds), *Sintering and Related Phenomena,* Gordon and Breach, New York, 1967.

Lenel FV, *Powder Metallurgy Principles and Applications*, Metal Powder Industries Federation, Princeton, 1980.

Luo J, In *Critical Reviews in Solid State and Material Sciences*, Vol. 32, 67:109, 2007.

Luo J, Stabilization of nanoscale quasi-liquid interfacial films in inorganic materials: A review and critical assessment, *Critical Reviews in Solid State and Material Sciences*, 32:67–109, 2007

Nabarro FRN, *Theory of Cyrstal Dislocations,* Oxford University Press, Oxford, 1967.

Rahaman MN, *Ceramic Processing and Sintering,* 2nd edn, Taylor and Francis, London, 2003.

Samsonov GV, Prydko IF and Prydko LF, *Configurational Model of Matter*, Consultants Bureau, New York, 1973.

Sanderow H, (ed.), *High Temperature Sintering,* Metal Powder Industries Federation, Princeton, NJ, 1990.

Savitskii AP, *Sintering of System with Interacting Components,* Trans Tech Publications, Stafa-Zuerich, Switzerland, 2009.

Schmalzried H, *Chemical Kinetics of Solids,* VCH, Weinheim, 1995.

Smigelskas AD and Kirkendall EO, *Trans. AIME,* 171:130, 1947.

Somiya S, Shimada M, Yoshimura N and Watanabe R, (eds), *Sintering 87,* Vols I and II, Elsevier Applied Science, Barking, UK, 1988.

Upadhyaya GS, (ed.), *Sintering of Multiphase Metal and Ceramic Systems,* Sci-Tech. Publications, Vaduz, 1990.

Upadhyaya GS, (ed.), Sintering Fundamentals, Trans Tech Publications, Stafa-Zuerich, Switzerland, 2009.

Uskokovic DP, Samsonov GV and Ristic MM, *Activated Sintering,* International Institute for the Science of Sintering, Belgrade, 1974.

Waldron MB and Daniells RL, *Sintering,* Heyden and Sons, London, 1978.

EXERCISES

7.1 A 2024-Al/30 vol% SiC composite powder compact was degassed at different temperatures (see table on next page) prior to HIP operation to achieve the given mechanical properties. Guess the degassing temperatures A and B. Justify your answer.

Degassing temperature °C	UTS, MPa	Young's modulus, GPa
380	420	90
A	600	130
B	460	107

7.2 Figure P. 7.2 shows the dimensional change behaviour during air sintering of different alumina compacts based on different types of powder preparation routes through chemical methods. Based on the compositions of these powders (refer Table 2.5 in text), identify the preparation routes. (Note: Each compact has 2% polyvinyl alcohol binder.)

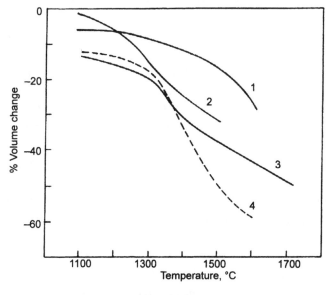

Fig. P. 7.2

7.3 Many workers gave various attributes for activated sintering of metals by an activator: electron structure difference, phase diagram criteria and grain boundary cohesion. Present a unified picture of all these attributes.

7.4 An iron powder compact is sintered for 1 h at 890°C and is found to have superior mechanical properties to the same material sintered at 930°C, with all variables remaining the same. Explain the reason.

7.5 A green compact of 68% density is to be sintered to 87% density. What is the densification parameter? Estimate the linear shrinkage.

7.6 A Cu-5Sn bronze is compacted to 90% of theoretical density. When sintered at 200°C for 2 h the compact shrinks and densifies to 95% of its theoretical density. However, when sintered at 300°C for 2 h, the compact had only 82% of the theoretical density. Explain this trend. Calculate the densification parameter for compacts sintered at 300°C.

7.7 At 1400°C, W-1Cu alloy undergoes liquid phase sintering, whereas W-1Ni alloy is solid-state sintered. The densification of the latter was found to be greater than that of the former. Explain. (Hint: Refer W-Cu, W-Ni phase diagrams)

7.8 The activation energies for surface and grain boundary diffusion of thorium in tungsten are 55% and 75%, respectively, of the activation energy, for lattice diffusion of thorium in tungsten. Comment on the statement.

7.9 AlN ceramic is currently replacing BeO advantageously for heat-sink application in electronic devices, since the latter is toxic. Y_2O_3 addition is done as a sintering aid; but the sintering temperature continues to be quite high even for liquid phase sintering. It has been observed that molten Al_2O_3-Y_2O_3 phase has good wettability with AlN, which increases with increase in Al_2O_3 content. Referring to the Y_2O_3-Al_2O_3 phase diagram, suggest a composition for the additive mixture and the possible sintering temperature. (Note: A relatively low sintering temperature contributes to process economy.)

7.10 A diffusion couple is made by joining Cu to Cu-10% Al with inert markers in the interface. Annealing gives marker motion indicating $D_{Al} > D_{Cu}$.

(a) Draw a plot of the concentration of Al versus distance for the annealed couple. Be sure to indicate the difference in penetration distance on the two sides of the marker interface resulting from $D_{Al} > D_{Cu}$.

(b) Show which way the markers at the interface will shift and where the Kirkendall porosity will grow rapidly for the $C(x)$ curve drawn in (a).

7.11 It is observed that when Cr_2O_3, doped with 0.1 wt% MgO, is sintered at 1700°C, the sintered density near theoretical is achieved and the grain growth is reduced. What is the reason for this?

7.12 Describe the microstructural defects that might be expected after sintering ceramic compacts containing (a) clusters of pores much larger than the grain size, (b) powder aggregates that densify more rapidly, and (c) powder aggregates that densify less rapidly.

7.13 Both Cu-Ni and Ge-Si binary systems form a complete range of solid solubility. During homogenisation of mixed powder compacts, why does the former system homogenise more readily?

7.14 Green compacts of MgO powder when sintered at 1380°C in air containing water vapour, were found to have far superior densification accompanied by grain coarsening than when sintered in dry air. Give reasons.

7.15 For a sintered Cu-15Sn bronze alloy made of pre-alloyed atomised powder, plot the amount of melt formed at different temperatures during heating from solidus to liquidus temperature. Refer Cu-Sn binary phase diagram. Why is this information useful to powder metallurgists?

7.16 During sintering of polycrystalline materials we often end up with pores of an assortment of sizes, located both within the grains and at grain junctions, as shown schematically below (Fig. P. 7.16).

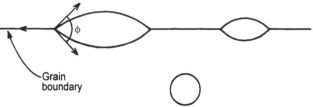

Fig. P. 7.16

(a) Which pore will disappear first and why? Which pore disappears last and why?

(b) If the three pores shown are close enough for atomic diffusion to take place readily, will there be any mass transport between the pores? (Adapted from Chiang et al, 1997)

7.17 Sketch the fractional grain size distribution curves for a sequence of times for regular grain growth and on the appearance of exaggerated grains.

7.18 The mean pore size should decrease when sintering a compact containing uniform pores smaller than the grain size, but the mean pore size may increase when a compact containing a wide range of pore sizes is sintered. Explain.

7.19 Calculate the initial rate of sintering of 1-μm spheres of Al_2O_3 at 1400°C, given the following diffusion coefficients: $D_{gb}^{Al} = 5.6 \times 10^{-17}$ cm^2 s^{-1}, $D_{gb}^{O} = 7 \times 10^{-14}$ cm^2 s^{-1}, $D_{lattice}^{Al} = 4 \times 10^{-14}$ cm^2 s^{-1}, $D_{lattice}^{O} = 1 \times 10^{-17}$ cm^2 s^{-1}. The molecular weight of Al_2O_3 is 101.96 g mol^{-1}, and the density is 3.97 g cm^{-3} (Adapted from Chiang et al, 1997).

7.20 Why are hot pressing mechanisms not very sensitive to the exact shape and curvature of the contact necks?

7.21 During HIP, which mechanism prevails at (a) low pressure and (b) high pressure?

7.22 Why does sintering alumina green compact in an atmosphere of nitrogen not improve sintering response?

7.23 Why are agglomerated powders difficult to sinter? What measures would you take to process such powders to minimise the negative effect?

7.24 What is the effect of an inert dispersoid in a metal PM component on (a) densification and (b) grain growth during sintering? Give reasons.

7.25 Decreasing porosity during sintering of a metal powder green compact does not necessarily signify that pore size will decrease. Comment.

7.26 In case the pore is associated with two adjacent solid particles A and B in a powder mass, write an expression for interfacial energy between A and B. Will the pore shape be symmetrical?

7.27 Why is the solute in a polycrystalline powder material, which lowers the dihedral angle, beneficial for densification?

7.28 While sintering a metal powder of constant particle size, is a uniform change in the grain boundary area expected during every sintering stage? Explain.

7.29 If grain growth is to be reduced during sintering, at the same time achieving high densities, what thermal history would you plan?

7.30 What result do you expect from the combination of large pore size and large grain size in a compact during sintering?

7.31 In a poorly compacted powder mass, if the pore size is large compared to the grain size, would densification be achieved, particularly if the dihedral angle is small?

7.32 For a ceramic powder, calculate the concentration change in vacancy from the following data: surface energy = 1 J m^{-2}, radius of curvature = 10^{-6} m and sintering temperature = 2000 K. Will this value cause a flux?

7.33 An oxide ceramic powder is to be sintered. Mention at least three ways in which the sintering kinetics can be affected.

7.34 A mass of NaCl spherical powder is to be sintered. Which mechanism of mass transport will be operative? Will it offer densification to the material?

7.35 Why did Kuczynski select relatively bigger-sized copper spheres in his model experiments for sintering?

7.36 In a model sintering study, why does a wire–wire configuration have some advantage over other configurations?

7.37 Is the grain boundary diffusion mechanism for sintering exactly the same for copper and alumina powder compacts? Give reasons for you answer.

7.38 Calculate the excess energy for alumina powder of 1 μm particle size, given its density is 4000 kg m^{-3}, surface energy is 1 J m^{-2} and molar mass is 0.1 kg mol^{-1}. What will be the value when the particle size is decreased ten times? For NaCl crystals do you expect the value to be lower or higher than for alumina? Give reasons for your answer.

7.39 Coble, while describing the intermediate stage sintering, considered the vacancy flow in cylindrical pores to be independent of radius r. What reasons do you suggest?

7.40 Why do pores with many sides tend to grow, in contrast to those which have few neighbouring grains?

7.41 Referring to Fig. 7.3, calculate the volume transported into the neck when the mechanism is (a) non-densifying and (b) densifying.

7.42 In a premix powder system of two components A and B, what difference would you face during homogenisation when (a) A-B is an isomorphous solid solution and (b) A-B is a partial solid solution with intermediate phases present?

7.43 It is a known fact that fine particles sinter faster than coarser particles. In a real system, why do fine powder compacts, in general, reach a lower limiting density than coarse powder compacts?

7.44 A porous solid compact contains equal-sized pores. Show that the spatial distribution of the pore does not affect the sintering stress.

7.45 (a) Differential densification at the microlevel occurs when the green compact consists of particles of two different sizes. Such a feature gives rise to a back stress. On what factor does the back stress depend?

     (b) In case the differential stress is large and opposed to the local sintering stress, what will be its effect on the pore?

7.46 Why is it practically impossible to sinter pure silicon carbide powder compacts to full density?

7.47 A 50:50 ZrO$_2$-Al$_2$O$_3$ particulate composite was sintered for different periods. Similarly, pure ZrO$_2$ and pure Al$_2$O$_3$ compacts were also sintered. Do you expect the results for the composite to be intermediate with respect to the pure consitutents?

7.48 What methods do you suggest to remove close porosity in sintered compacts?

7.49 For a pore dihedral angle of 60°, what is the critical coordination number?

7.50 In liquid phase sintering, one has to be more careful in selecting the particle size of the constituents. Explain why.

7.51 "There is no benefit from liquid phase sintering if the solid phase is not soluble in the liquid." Comment on the statement.

7.52 Determine the pressure difference between the pore and the melt during liquid phase sintering, where liquid/vapour interfacial energy is 1 J m^{-2} and the pore radius is 0.5 μm.

7.53 Binary Co–Mo system undergoes eutectic reaction at around 1340°C. Also, nickel addition has been noticed to activate sintering of refractory metals. Propose an alloy composition of Co-Mo binary which can be liquid phase sintered along with a minor Ni addition. What sintering atmosphere would you select? What special precautions are required during sintering? (Refer *Metals Handbook* for Co-Mo phase diagram).

7.54 In case of liquid phase sintered tungsten heavy alloy system W-Ni-Fe, the grain size is proportional to $t^{1/3}$ where t is the sintering time. But, in case of liquid phase sintered WC-Co alloys it is proportional to $t^{1/2}$. Why?

7.55 Tungsten is not soluble in the copper melt. In preparing a liquid phase sintered W-Cu electrical contact material, a minor additive is required. Explain the selection criterion for such additive.

7.56 A Mo-1.5Ni powder mixture was sintered at 1370°C in hydrogen for 1 h. The sintered density of Mo increased from 86% to 97.5%. What is the reason for the densification enhancement? In another compact, the Ni content in the above alloy was partially replaced by copper and sintered at 1300°C for the same time. The sintered density rose to 99.1%, despite the sintering temperature being lower. If we add more nickel in Mo would the densification be even higher?

7.57 Why does pore closure occur early in hot pressing?

7.58 An alumina powder of irregular shape is hot pressed under two conditions: (a) At normal atmospheric pressure and (b) Under vacuum. Describe the difference in the results and give reasons.

7.59 During hot pressing the stress intensification factor depends on the shape of the pores. What relative difference do you expect between the elliptical and the spherical pores?

7.60 During hot pressing, under what condition is the effective stress (a) infinite and (b) equal to the applied stress?

7.61 High external pressure during HIP gives rise to full densification but the same consolidated material, when heated without any external pressure, often gives rise to swelling. Give reasons.

7.62 In both HIP of powder mass and hot deformation during powder preform consolidation (forging, extrusion, etc.) plastic deformation takes place. What is the major difference in these two processes.

7.63 In case alumina is to be hot pressed below 1000°C with the presence of some liquid phase, which of the following systems is applicable: (a) Al_2O_3-V_2O_5, (b) Al_2O_3-ZrO_2, (c) Al_2O_3-LiF, (d) Al_2O_3-Si_3N_4 and (e) Al_2O_3-cryolite?

7.64 Plot the curves for total interfacial energy versus shrinkage during the early stage of liquid phase sintering for different dihedral angles. What is the significance of the initial slope of the curve?

7.65 "If the dihedral angle is greater than zero, the total interfacial energy will be minimum at a finite neck size. The depths, but not the positions, of the minima depend on the number of contacts." Elaborate on this statement.

7.66 In the parlance of ceramic technology, what is meant by vitrification?

7.67 Surface energies in ceramics tend to be more anisotropic than those in metals. Give reasons.

7.68 A small addition of MgO in alumina controls the mobility of the grain boundary migration during sintering. On what factors dose the selection of the additive depend?

7.69 Why do the powders of halides, KBr, KCl and KI, and metals such as Pb achieve full densification even during cold pressing?

7.70 Name two systems—one metallic and the other ceramic—which may be hot isostatic pressed at relatively low external pressure because of the presence of some liquid phase.

7.71 (a) It is noticed that the sintering of crushed glass powder is much faster than that based on spherical powders. Why?

(b) Is the presence of water vapour in the sintering atmosphere useful?

(c) What are the reasons for bloated glass?

7.72 Why is the role of vapour transport mechanism in pressure-assisted sintering negligible?

7.73 Referring to Fig. 7.31 comment on why the curves pertaining to (b) and (c) mechanisms for hot pressing crossing over in the early stage, while curve (d) prevails over the later stage.

7.74 An encapsulated powder in a metal can was subjected to HIP at an elevated temperature and applied pressure of 150 MPa. At the early stage, a density of 90% of the theoretical was noticed. Further densification rate was slow. What can be inferred about the plastic behaviour of this powder?

7.75 Success in nanopowder consolidation is closely related to the control of competition between densification and grain coarsening. Comment on the statement.

7.76 Write an expression for the approximate area fraction of short circuit paths (i.e., grain boundaries), if d is the grain boundary width and G is the grain size.

7.77 Referring to Fig. Ex. 4.1b (W-Ni binary phase diagram) describe the microstructural evolution in W-1 at% Ni premix isothermally sintered at (a) 1400°C, (b) 1450°C and (c) 1500°C, for 1 hour. What difference would you envisage in case the milling was done in an attritor mill? Do you expect mere solid-state sintering being enough to attain the desired end properties? Comment.

7.78 Nickel addition causes activated sintering of tungsten. Suggest an additive which might successfully activate the sintering of copper and Fe-Si systems. Justify your answer.

7.79 The dihedral angle in a certain crystalline or amorphous solid has the same value. Give reasons.

7.80 Although the dihedral angle of a crystalline material tends to be reported as a single value, it is not a single value. Comment on the statement.

7.81 The compositions in wt% of the following binary powder premixes subjected to liquid phase sintering are as follows: (i) WC-8Co, (ii) Fe-50Cu, (iii) W-7Ni. Calculate the vol% of the melts in the very beginning of the liquid phase sintering process.

7.82 During the early stages of liquid phase sintering, do you expect the microstructural parameters to be sensitive to sintering time?

7.83 Liquid phase sintering of powder pre-mixes of the following binary systems were carried out.

(a) low melt solubility in solid, but reverse for solid solubility;

(b) high melt solubility in solid, but reverse for solid solubility.

What should be the densification features during sintering of the above systems?

7.84 On what factors does the radius of curvature of the grain boundary between contacting grains depend?

7.85 In two liquid phase sintered alloy systems, rounded grains and flat-faced grains were observed. Indentify the mechanism controlling the evolution of such shapes.

8

Sintering Technology

LEARNING OBJECTIVES

- Various types of sintering furnaces and their significance

- The need for various sintering zones

- Rapid sintering methods—induction and microwave sinterings

- Neutral, reducing and carburising atmospheres

- The need for analysis and control of sintering atmosphere

- The role of sintering temperature/time and atmosphere on sintering

- The role of powder characteristics in sintering and resultant dimensional changes

- When and where to adopt the infiltration process

In the previous chapter, sintering theory was described. However, for successful production of PM parts, an appreciation of the technological aspect is essential. What types of sintering furnaces are to be selected and how many effective number of zones must they have? Can the process be modified for better productivity? What should be the sintering atmosphere? What is its interaction with the green part material in the range of elevated temperature? For successful sintering, the analysis and control of the furnace atmosphere is equally important. Both the process and the chemistry of the green part material could be modified for successful economical production. During sintering, the gross dimension and the microstructure of the products change. It is important to control these factors. During plant operation, the handling of sintered parts is also to be considered. The present chapter discusses the above factors.

8.1 DEBINDING

In the chapter related to powder treatment (Chapter 4), the description of binders and their role in powder shaping is given. It must be borne in mind that before actual sintering, the binder is to be eliminated from the compact. In many cases, in continuous sintering furnaces, this is done in the preheat zone of the furnace. In case of injection moulded parts, more recent developments have demonstrated shorter debinding times and better dimensional control using solvent debinding techniques. Failure to remove the binder before sintering results in compact cracking. The important feature is that the binder must be extracted from the pores as a fluid (liquid or vapour) without distorting or contaminating the compact.

The batch process of debinding has greater flexibility but lower productivity. Thermal debinding is carried out in either vacuum or in an atmosphere of air, hydrogen, nitrogen–hydrogen, hydrogen–argon or nitrogen–hydrogen–water. Debinding in air is cheap and efficient but will oxidise a metallic part and create subsequent problems in sintering. A protective atmosphere with proper control of dew point and carbon potential is therefore desirable.

As already indicated, if debinding is not carried out properly, a number of defects and problems can arise. These include the loss of the compact shape through distortion, slumping or warping, as well as the formation of surface pits, discoloured spots and internal cracks. Figure 8.1 shows several such defects in a compact.

8.2 LOOSE SINTERING

This method is quite widely used for the manufacture of highly porous parts like filters. Basically, metal powder is poured or vibrated into a mould which is then heated to the sintering temperature in an appropriate atmosphere. The form and complexity of the shapes which can be made by this method depend to a large extent on the flow characteristics of the powder. Since shrinkage usually takes place during sintering, only shapes where this can

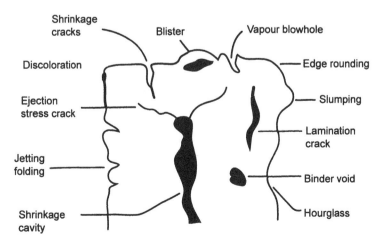

Fig. 8.1 Possible defects in a PIM compact that might be observed after debinding

occur without mould constraint causing cracking can be used. For loose sintering, the mould material should:

- be easily machined or formed into the required shape
- withstand the sintering temperature without appreciable deformation
- not weld to the powder during sintering.

For most applications, machined or welded metal moulds are used, although graphite may also be employed where no reaction with the powder is likely, and even here a refractory mould wash may suffice to prevent reaction. Since no pressure is applied to the powder, any unreducible oxide skin on the particle will prevent metal-to-metal contact and inhibit sintering. For this reason it is virtually impossible to loose sinter aluminium powder.

8.3 SINTERING FURNACES

Sintering furnaces may be classified as batch-type or continuous furnaces. Invariably they have a gas-tight furnace shell or gas-tight muffle to maintain a reducing atmosphere.

8.3.1 Batch-type furnace

These furnaces are used for sintering in protective atmospheres when the quantities produced do not warrant the installation of continuous furnaces. A typical box-type sintering furnace is shown in Fig. 8.2, which is similar to pusher-type continuous furnaces, except that the boats are stoked through the furnace by hand rather than by a mechanical stoker.

Small pilot plants or laboratory muffle furnaces are similar to box furnaces. Heating through a muffle is less efficient than exposing the electric heating element to the furnace

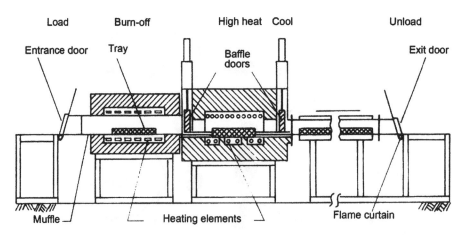

Fig. 8.2 Box-type manually operated sintering furnace

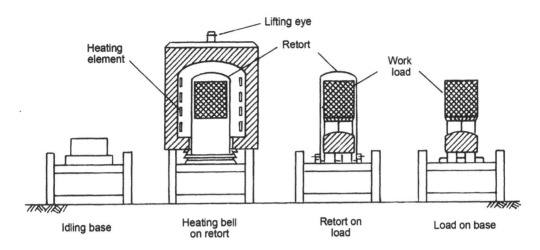

Fig. 8.3 Bell-type sintering furnace showing stationary bases, retorts and heating bell

atmosphere and furnace chamber. A full muffle furnace reduces the need for the flow of atmosphere purging gas, particularly while using expensive gases like hydrogen or dissociated ammonia.

In addition to box-type furnaces, bell-type furnaces are also used for batch-type sintering of metal powder compacts, particularly in cases when very good atmospheric control is required. Figure 8.3 shows a typical bell-type furnace. The compacts are arranged on a load supporting base with a removable sealed retort over the load to contain the protective atmosphere. A portable heating bell is lowered over the load and retort for the heating cycle.

8.3.2 Continuous furnace

Most sintering furnaces are used for large volume production work and are of the continuous-type. Here, a continuous movement of compacts through the three zones of the furnace takes place. Three main methods are used to convey the compacts. They are mesh belt conveyor–type, roller hearth–type, pusher-type and walking beam–type.

Mesh belt conveyor furnaces

It is shown schematically in Fig. 8.4; it has an endless woven belt, generally made of a nickel-based heat resistant alloy wire which runs over large motor-driven drums at the ends of the furnace. Inside the furnace it slides over the refractory hearth in the hot zone. The green compacts to be sintered can be placed directly on the belt or they can be placed on heat resistant trays. The doors in such furnaces are open, and therefore, the furnace requires large amounts of protective atmosphere. A humpback furnace is a modified mesh belt furnace (Fig. 8.5). These have inclined entrance and exit zones with the heating and cooling chamber at an elevated level. They are used for sintering materials, such as aluminium alloys and

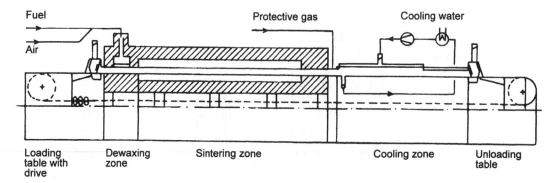

Fig. 8.4 Schematic of a conveyer belt sintering furnace

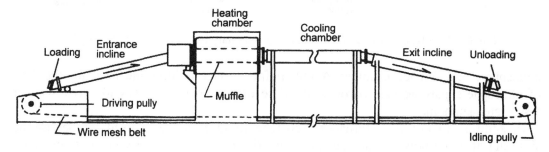

Fig. 8.5 Section through a humpback mesh belt furnace, belt drive and take up mechanisms

stainless steels, which require a particularly dry atmosphere. Sintering atmospheres high in hydrogen have a tendency to rise because of their low density; their contamination in such types of furnaces is naturally prevented.

The increasing demand for sintered products has resulted in a demand for higher mesh belt furnace performance, in conjunction with the requirement for cost efficient sintering. Conventional conveyor belt furnaces which can transport greater quantities without increasing their surface heat losses have a lower specific energy requirement. Assuming that the dwell time at the required sintering temperature must not drop below a specified value, such improvements can only be achieved by increasing the heat-up speed and the belt load. The overall furnace length is determined by the section lengths required for heat-up and for holding at sintering temperature. If it is possible to accelerate heat-up while achieving complete lubricant burn-off, the length of this furnace section can be reduced. In one of the furnaces this aspect has been introduced, where the green compacts are heated with a high energy density by vertically impacting high velocity protective gas jets distributed evenly over the conveyor belt width. The intensive contact of the part with the heating gas provides for a rapid heating, dewaxing and burn-off of the lubricant. The temperature in the rapid heating zone is kept constant by continuously adjusting the burner output.

The sintering zone proper is heated electrically by free radiating resistance heaters. These heaters are installed above and below the conveyor belt at right angles to the transport direction. This ensures uniform temperature distribution over the belt width. They can be replaced individually during belt operation. Several heating zones ensure that the desired temperature profile is maintained over the entire furnace length. The built-in gas generation in the furnace leads to lower investment and operating costs and less space is required. The composition of the protective gas can be adjusted between that of exo-gas and endo-gas. In the cooling zone, the sintered parts are cooled by circulating gas. The protective gas passes in a closed cycle via a water cooler. Regardless of the conveyor belt loading, the quantity of protective gas at the furnace outlet is kept constant via a control loop. The flow of protective gas from the muffle of the sintering zone into the dewaxing zone makes it possible to have a different protective gas atmosphere in the dewaxing and sintering zones. The details of sintering atmospheres are given in Section 8.6.

Roller hearth furnaces

There is a hearth in the hot zone of the furnace which consists of a series of parallel rollers made of a heat resistant alloy. The green compacts are placed on trays that are conveyed through the three zones by riding on driven rolls. For a given length the roller hearth method can allow much higher loading than the mesh belt technique, but the maximum sintering temperature is still 1150°C. The charge and discharge doors are operated automatically and open only when a tray of work is charged or discharged. This reduces the quantity of atmosphere consumed and minimises heat losses.

Pusher-type sintering furnace

They are suitable for sintering metal parts that are too heavy for the mesh belt–type, and where the production rate does not warrant the roller hearth furnace. In case the sintering temperature is high, say up to 1650°C, this furnace is suitable, because mesh belt and rollers are ineffective. Mechanical or hydraulic pusher furnaces are available for high outputs. Generally, two types of pushing mechanisms are used: the intermittent pusher mechanism and the continuous stoker pusher-type. A typical pusher furnace is shown in Fig. 8.6. The intermittent pusher-type is generally used for most common metals, while the continuous stoker pusher-type is more often used for cemented carbides. Smaller furnaces are often manually stoked.

Walking beam furnaces

This type of continuous sintering furnaces were first developed in Germany for use at high temperatures such as 1800°C, depending on the compatibility of the atmosphere and the heating element used. At high temperatures (>1150°C), mesh belt furnaces commonly used in industry cannot operate due to limitations of both the alloy wire belt and length of the heating chamber. In the walking beam furnace, the amount of production weight that can be conveyed safely is practically unlimited. Boats or trays are conveyed by a mechanism which periodically lifts them, advances them a short distance and then lets them settle back on a ceramic ledge. The part of the lifting mechanism which gets heated to the furnace temperature is made of refractory material. A typical furnace is schematically shown in Fig. 8.7. It illustrates a furnace with cross transfers and return conveyor that provide a fully automatic operation. This particular system consists of an entry lock chamber with an intermediate door, a preheat or

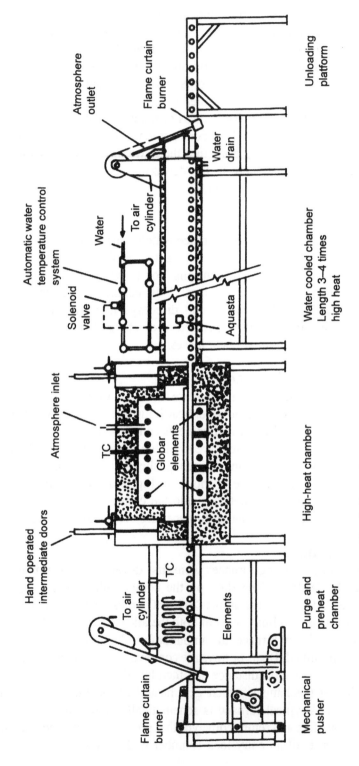

Fig. 8.6 Longitudinal section of a mechanical pusher furnace

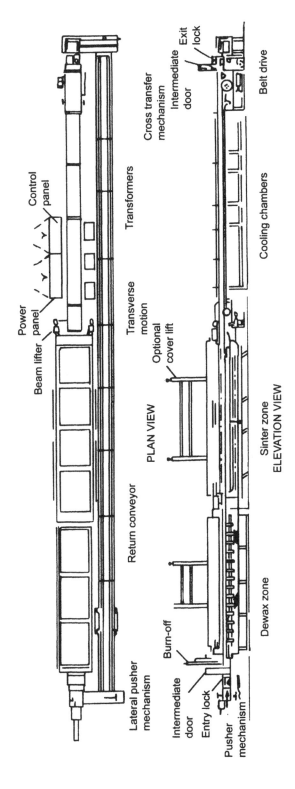

Fig. 8.7 Diagram of a walking beam sintering furnace illustrating cross transfers and return conveyer for automated operation

dewaxing zone, a sintering zone, cooling chambers, an exit lock chamber with intermediate doors, cross transfer mechanisms, and a return conveyor system. In this system, three means of transport are provided:

- **Pusher mechanism** which pushes the workload tray through a muffled preheat or dewaxing zone,
- **Walking beam mechanism** to transport the trays throughout the high heat sintering zone,
- **Cooling chamber** consisting of a conveyer belt drive system.

The three methods of conveying the workload throughout the furnace is used to provide a closed system, so that consumption of the atmosphere will be considerably reduced.

8.3.3 Vacuum furnace

Vacuum furnaces are generally of batch-type, but new developments in continuous vacuum furnaces have also been made, in which they incorporate burn-off chambers, vacuum locks and cooling zones through which the compacts are conveyed continuously. However, it may be mentioned that the batch-type furnaces offer better control of the sintering cycle than continuous furnaces.

A typical vacuum sintering furnace for cemented carbide parts is shown in Fig. 8.8. Since it is relatively easy to obtain high temperatures and control the sintering environment in vacuum furnaces, they have almost completely replaced atmosphere sintering furnaces for hard metal processing. Resistance furnaces are frequently used instead of induction furnaces, on a first cost basis, particularly when the work zone size is increased. The vacuum pumping system remains connected to the furnace chamber during the entire cycle (except during cooling) to pump away the outgassing or products of reaction. Because excessive paraffin may destroy the lubricating effect of the vacuum pump oil, effective condensing systems are necessary

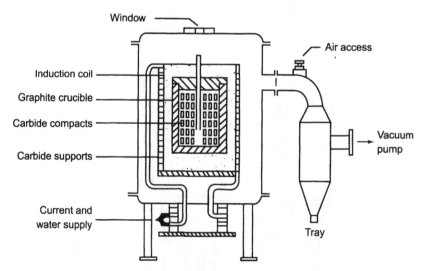

Fig. 8.8 Schematic of vacuum furnace for sintering cemented carbides

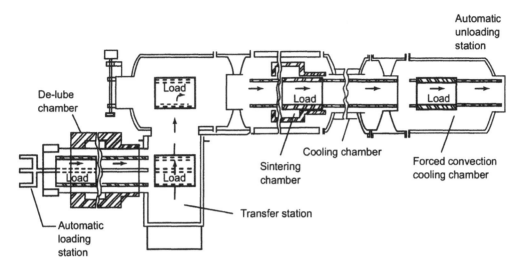

Fig. 8.9 A typical continuous vacuum sintering furnace

to trap the vapour before it reaches the pump. Pumping equipment usually includes a Roots blower, backed by a mechanical vacuum pump. The operating pressures range from 1 torr to 10 millitorr.

A typical continuous vacuum sintering furnace is shown in Fig. 8.9. It consists of an external loading table followed by an atmosphere pre-sinter chamber, a transfer tunnel, an 'atmosphere-to-vacuum' vestibule section followed by the heating chamber, vacuum cooling chamber, combination fan cooling and vacuum-to-atmosphere vestibule, and a two-tray loading table. A typical furnace of this type will accommodate trays that are 450 mm wide and 600 mm long, with up to 230 mm working height. The heating element is manufactured from solid graphite in the vacuum section and metallic alloy in the pre-sinter section. The vacuum section is capable of operating up to 1400°C at a vacuum level of 500 μm. The gross output of a typical unit is 250 kg h^{-1}.

8.3.4 Electric furnace heating elements

Electric heating elements in sintering furnaces may be:

1. Base-metal nickel–chromium alloys
2. Non-metallic heating elements: SiC, MoSi$_2$ or graphite
3. Refractory metal heating elements, Mo or W.

Graphite and refractory metal heating elements are used almost exclusively in vacuum furnaces. The wrong selection of heating element, particularly with reference to furnace atmosphere, can cause very high furnace maintenance. A mere comparison in hours for the same resistor in commercial furnaces operated at the same temperature would not give a true picture of the action of different atmospheres. Table 8.1 gives a summary of maximum temperature and atmospheres where each type of element can be used. The life of a resistance material depends

Table 8.1 Furnace heating elements and uses

Element	Max. temperature, °C	Atmosphere
Nichrome (Ni-Ci)	1150	1, 2, 3, 4, 5, 6
Kanthal (Fe-Cr-Al-Co)	1300	1, 3, 4, 6
Super kanthal ($MoSi_2$)	1800	1, 3
Silicon carbide (SiC)	1250	1, 2, 3, 5, 6
	1600	1, 3
Platinum	1500	1, 3, 4
Molybdenum	1800	2, 3, 4, 5, 6
	2200	4
Tantalum	1900	3, 4
	2400	4
Tungsten	1900	2, 3, 4
	2600	4
Graphite	2000	3, 4, 5
	3000	3, 4

Atmosphere key: 1: oxidising, 2: reducing, 3: inert, 4: vacuum, 5: carburising, 6: decarburising.

not only on the furnace atmosphere but also on many other working conditions, i.e., watt-density loading, cross-sectional area, operating temperature, frequency of switching the current on and off, support of the resistor in the furnace, physical shape and design of the resistor.

Resistance temperatures are always higher than furnace control temperatures. When the furnace is operated close to the maximum temperature, the watt-density loading must be lower and more conservative.

8.3.5 Furnace connectivity and control

The control of a sintering furnace can range in complexity from the very simple stand alone controllers to advanced computer systems. In the age of the Internet and mobile phones, furnace manufacturers and users are beginning to incorporate these technologies into sintering furnaces.

Real-time remote monitoring systems are being used to record information for both quality control and preventative maintenance. Information about power draw can be automatically sent to the furnace manufacturers on a routine basis for their review and maintenance recommendations.

The Internet is being used hand-in-hand with sintering furnaces. In the event of an alarm on the furnace, the furnace PLC can use an Internet connection to send a text message to

---

Example 8.1: Compare and contrast the heating elements (i) Ni-Cr-Al and (ii) SiC, which are commonly used in electric sintering furnaces.

Solution

(a) The maximum element temporarily in air is 1400°C for (i) and 1850°C for (ii); maximum use temperature is 1300°C and 1550°C, respectively, for (i) and (ii).

(b) (i) is metallic, while (ii) is a ceramic resister.

(c) Electrical resistance with time is constant for (i), and increases 4 times for (ii).

(d) Electrical resistance with temperature is constant for (i), and decreases to 800°C and then slowly increases for (ii).

(e) Type of control is on/off, for (i), and SCR (silicon control rectifiers) for (ii). SCR control is more expensive than on/off control, but can handle the increased voltage as the elements age and can also limit the current during the negative portion of the resistance curve.

(f) Major limitation is poor hot strength for (i) (which necessitates element support) and ageing for (ii).

---

a maintenance person's mobile phone or an email alerting them of the problem. A simple subscription to a service offering PC-to-PC remote connectivity allows remote access to the furnace controls and computer system.

8.4 SINTERING ZONES

8.4.1 Burn-off and entrance zone

This zone of the furnace is designated to heat the green compacts rather slowly to a moderate temperature in the order of 450°C. The main function of the burn-off zone is the volatilising and elimination of the admixed lubricant. A slow heating rate is necessary to avoid excessive pressures within the compact and possible expansion and fracture. The length of this zone must be sufficient to allow complete elimination of the lubricant before the compacts enter the high-temperature zone. The metal (for example, zinc from zinc stearate) and carbon from the breakdown of the hydrocarbons resulting from the volatilising of the lubricant can deposit on the furnace heating element and promote premature failure. Such deposits on refractory walls and cooling zones lead to poor heat transfer. On the other hand the compacts may be subjected to discoloration and possible undesirable chemical reactions.

The flow of the atmosphere is important in expelling the lubricant vapours. For this, sufficient atmosphere gas must be provided, and the flow should be directed such that the vapours are discharged towards the furnace entrance and not into the high-heat zone. The burn-off zone is sometimes separated with an air gap, using a flame curtain before the high-heat zone.

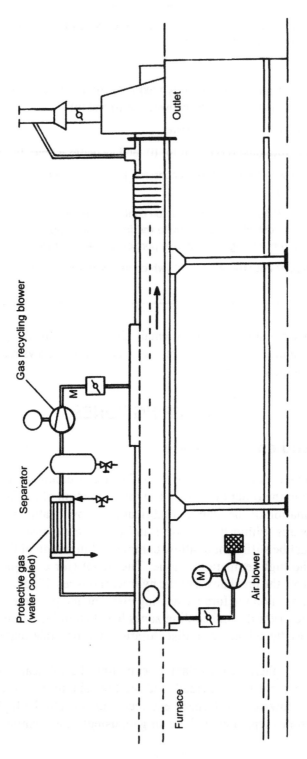

Fig. 8.10 Schematic of an accelerated cooling device (Courtesy: Mahler Gmbh, Germany)

During vacuum sintering metallic stearates are not used; stearic acid or a wax compound is instead used, as otherwise, the furnace and vacuum pumps get contaminated. For batch-type vacuum furnaces, the burn-off is best accomplished in a separate protective atmosphere oven at about 450°C.

In order to efficiently control the delubrication, recently, lubricant burners have been introduced just in front of the preheat zone. This controls emissisons, speeds up delubbing by eliminating 30%–40% of the lubricant before the parts enter the preheat zone, and preheats PM parts to 260–370°C. The burner assembly supplies oxygen-containing gas to the chamber at a temperature and velocity sufficient to mix and react oxygen with combustibles to generate a flame just before the parts enter the furnace. The advantage of such a unit is that it can be retrofitted to existing furnaces or installed in new sintering furnaces.

8.4.2 High-temperature zone

The actual sintering of the compacts takes place in this zone. It must be properly heated so that the desired temperature is reached. It should also be of sufficient length, so that the parts spend enough time in this temperature to achieve the necessary end properties in the sintered parts. The lengths of the burn-off and high temperature zones are usually about equal. Because of the necessity of having a reducing atmosphere present during sintering a gas, a tight muffle is utilised. However, heating through a muffle is less efficient than having the heating elements exposed to the atmosphere and furnace chamber.

8.4.3 Cooling zone

The cooling zone often consists of a short insulated section and a relatively long water-jacketed section. The former cools down the parts from the high sintering temperature to a lower one at a slow rate so as to avoid thermal shock in the compacts and the furnace. The latter provides cooling to a temperature low enough to prevent oxidation of the material on exposure to air. Since the cooling rate is rather slow at the low-temperature range, fans are used in the cooling chambers to circulate the atmosphere over the work. However, care must be taken to prevent air being sucked into the cooling chamber when the doors are opened.

Automatic water temperature control on cooling chambers is desirable to ensure that it does not get too cold or too hot. In case it is below the dew point of the atmosphere, water condenses on the chamber walls and the parts may get oxidised. To compensate automatically for varying loads, automatic flow control of the water is introduced.

In order to reduce the size of the furnace, a rapid cooling zone is thought of. Once the parts have passed through the sintering zone, they arrive at the first part of the cooling zone, which consists of a water-cooled duct. Here the importance should be attached to slow cooling in order to avoid distortion and unwanted hardening. The second part (end part) of the cooling zone is equipped with a gas circulation system (Fig. 8.10) to accelerate the final cooling process. In this convective cooling zone, the protective gas is circulated via a water-cooled heat exchanger. The other function of the system is to pressurise the furnace output to roughly the same pressure as that created by the fuel gas of the protecting zone at the furnace inlet, in order to ensure a constant distribution of the protective gas throughout the furnace.

In case a more enhanced cooling rate is required in the zone—for example, during sinter hardening of alloy steels—the zone can be equipped with a gas chiller device. The sintered parts with a wall thickness of up to 10 mm can be cooled at a rate up to 5°C per second over the temperature range 900–400°C. The elimination of subsequent hardening in a special hardening furnace saves considerable amount of time, and plant and energy costs.

8.5 RAPID SINTERING PROCESSES

The sintering furnaces described in the earlier section are related to heating the particulate compact externally by radiant and convective heat transfer. In order to achieve a reduction in the sintering period for equivalent densifications, two methods have been tried of late. They are induction sintering and microwave sintering.

8.5.1 Induction sintering

Induction sintering is based on eddy current heating which is caused by the skin effect. Eddy current heating produces heat in a layer of the specimen by the action of an alternating electromagnetic field. Most of the transferred energy is transformed in the sample material within a 'penetration depth, δ'. The relationship with the temperature and porosity-dependent material characteristics, electrical conductivity $K_{eff}(T, \theta)$, relative permeability $\mu_{eff}(T, \theta)$, and frequency f, are given by:

$$\delta = \sqrt{\frac{1}{K_{eff}(T,\theta)\mu_{eff}(T,\theta)f}} \tag{8.1}$$

Considering the special properties of sintering systems, appropriate conditions of induction sintering for materials of a given geometry may be derived on the basis of known physical and technological pre-conditions of eddy current heating.

The applicability of induction sintering is determined by efficient electrical material properties, the degree of symmetry of sintered part as well as by the efficiency of energy transfer. Efficient energy transfer is mainly accomplished on the basis of the distance between the sintered part (diameter d) and coil being as small as possible, and by applying a geometry-dependent minimum frequency.

$$f > f_{min} = \frac{8}{K_{eff}\mu_{eff}d^2} \quad \text{with } 2\delta < d \tag{8.2}$$

Strong irregularities in the geometry of a sintered part make induction sintering unsuitable and require complicated coil designs.

In industry this process is not found to be as versatile as conventional sintering. However, in sinter-forging operations, where the sintering parameters are more stringent, induction heating is very common. The parts pass individually or batchwise in the vertical or horizontal mode.

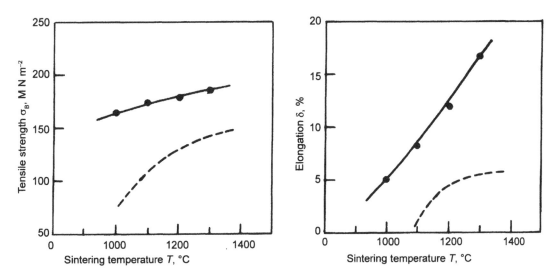

Fig. 8.11 Ultimate tensile strength and per cent elongation of conventional sintered (dotted curve) and induction sintered (full curve) iron powder green compacts (10% porosity) with respect to sintering temperature (sintering period: 1 min) (Reproduced with permission from Upadhyaya GS, (ed.), *Sintered Metal-Ceramic Composites*, Elsevier, Amsterdam, 1984)

Figure 8.11 presents a comparative presentation of conventional and induction sintering of iron PM parts from the viewpoint of mechanical properties.

8.5.2 Microwave sintering

Modern studies have confirmed that many ceramic and even metallic systems can be sintered by microwave energy. Microwaves are electromagnetic energy with a wavelength in the range of 10–1 mm (frequencies of 300–300 GHz). Ceramics with a moderate to high dielectric loss couple with the microwaves and convert the microwave energy to heat energy. Since the microwave energy penetrates deeply into the ceramic, thick sections can be heated uniformly and rapidly to the sintering temperature. In case the parts are small in size, complete sintering can be achieved within a few minutes, provided the part is insulated at the surface to avoid heat loss from the surface by infrared radiation.

Another rapid sintering method is *plasma sintering*. A plasma is high temperature, electrically ionised conducting gas. The temperature in a plasma is typically in the range 4000–10,000°C. Many oxide ceramics like Al_2O_3, ZrO_2, TiO_2 and MgO, including SiC, have been plasma sintered.

8.6 SINTERING ATMOSPHERE

Nearly all metals of technical importance react with the gas of their surrounding atmosphere even at room temperature, but more so when treated at higher temperatures. The most important reason for using special sintering atmospheres is to provide protection against oxidation and re-oxidation of the sintered metal powders. There are many other ways in which a sintering atmosphere can influence the basic sintering process. By reducing the oxides, the atmosphere may create highly mobile metal atoms. Gas atoms of the sintering atmosphere can enter the sintering compact via interconnected pores. They may later get trapped in closed pores, thus hindering their shrinkage. Gas atoms of the sintering atmosphere may also diffuse into the metal. At times these atoms might also alloy with the metal. In the following sections the details of different sintering atmospheres are given.

8.6.1 Hydrogen

Reducing atmospheres are most commonly used for sintering metal parts. Pure hydrogen is an excellent reducing gas, but it is not economical, except in case of high valued products. Hydrogen is highly flammable, having an extremely high rate of flame propagation. Because of its high flame propagation, hydrogen burns with a short, hot flame immediately on contact with air. The flame is an almost colourless blue. Hydrogen is the lightest element; its specific gravity is only 0.069, whereas for air it is 1.0. It is easily displaced by air, and rushes out rapidly from the top of the furnace door openings when free to do so. The thermal conductivity of the gas is seven times greater than air. Because of this, it accelerates both the heating and cooling rates of work in furnaces. The thermal losses in furnaces are higher with hydrogen than when using heavier, less conductive gases. Typical applications of hydrogen are in the reduction of oxides of iron, molybdenum, tungsten, cobalt, nickel and 18-8 stainless steel, annealing of electrolytic and carbonyl iron powders, and carburising of tungsten powders with lamp black to form tungsten carbide.

8.6.2 Reformed hydrocarbon gases

These are low-cost gaseous mixtures made by reforming hydrocarbon gases. They can be widely classified into 2 categories: exothermic and endothermic.

Exothermic gas

Exothermic gas is produced in specially designed generators, where fuel gas and air are mixed in such a ratio that incomplete combustion takes place. However, the heat generated in the combustion chamber is sufficient to support the reaction. Figure 8.12 shows the common range of analysis for an exothermic gas. This graph is based on the assumption that the fuel gas is pure methane, which corresponds reasonably well to a natural gas. In all cases, nitrogen is actually the largest single constituent. At an air-to-fuel ratio of 10.25:1, there is practically complete combustion. This gaseous mixture is relatively inert to metals such as hot copper, tin or silver. It will however oxidise hot iron and the reactive metals, because of the high proportion of CO_2 and water vapour as opposed to extremely low percentages of reducing components such as hydrogen and carbon monoxide. This gas is known as 'lean exothermic gas' and has very little application in powder metallurgy.

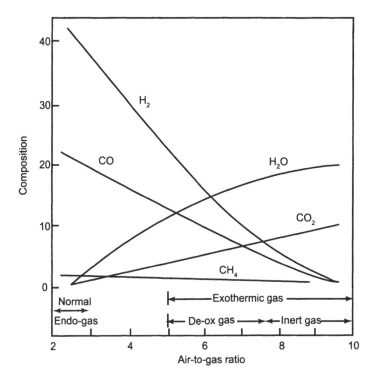

Fig. 8.12 Influence of air–gas ratio on exo-gas and endo-gas, assuming the fuel is pure methane

The exothermic gas generator is a refractory-lined, water-jacketed combustion chamber that operates at 1100–1425°C. The internal temperature of the chamber depends on:

1. Ratio of air to fuel gas
2. Volume of gas burned
3. Effectiveness of the heat transfer from the inside of the chamber to the water-jacketed exterior.

The water content in the gas mixture is very important for sintering. On leaving the combustion chamber, the gas has a water vapour content of approximately 5%–15%. This makes the atmosphere oxidising on most types of sintered products. Therefore, the gas has to be dried to at least below 1% H_2O. This may be accomplished, for example, in a refrigerant cooler by lowering the gas temperature to approximately 10°C. In many cases when a stronger reducing atmosphere is needed, a chemical desiccant agent is used, by which the dew point of the gas is lowered to –40°C, or approximately 0.1% H_2O. This dry gas is especially useful in a continuous belt furnace where air may enter from both ends of the furnace. Table 8.2 shows the compositions of some of the principal furnace atmospheres, according to the classification of the American Gas Association (ACA).

The correlation between the dew point of the gas and water vapour content is shown in Fig. 8.13. Figure 8.14 illustrates the amount of water vapour which can be present in a pure mixture of water vapour and hydrogen, without oxidising iron. The slope of the curve in this

Table 8.2 Composition of common sintering atmospheres

Factor	Endothermic	Exothermic	Dissociated ammonia	Nitrogen-based
Nitrogen, %	39	70–98	25	75–97
Hydrogen, %	39	2–20	75	20–2
Carbon monoxide, %	21	2–10		
Carbon dioxide, %	0.2	1–6		
Oxygen, ppm	10–150	10–150	10–35	5
Dew point, °C	−16 to 10	−25 to −45	−30 to −50	−50 to −75

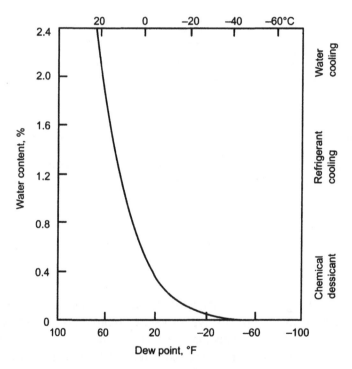

Fig. 8.13 Water content (vol%) versus dew point

figure shows that water vapour is more oxidising at low temperature than at high temperature. This means that even a fairly low water vapour content—which may not be dangerous at the actual maximum temperature in the furnace—might very well oxidise the compact in the cooling or in the preheating zone. Figure 8.15 shows the equilibrium constant (K) for the water gas reaction as a function of temperature. High temperatures favour the creation of CO and water vapour, whereas low temperatures stabilise CO_2 and H_2.

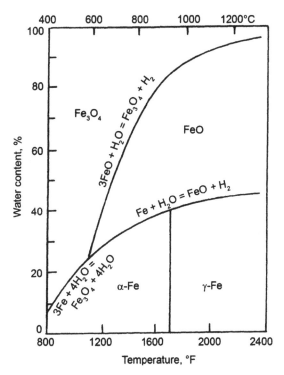

Fig. 8.14 Equilibrium diagram for the system Fe-O-H: per cent H_2O in H_2/H_2O mixture (100%) versus temperature

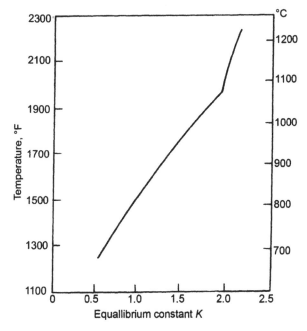

Fig. 8.15 Equilibrium constant K of water gas reaction $H_2 + CO_2 \rightarrow CO + H_2O$ versus temperature

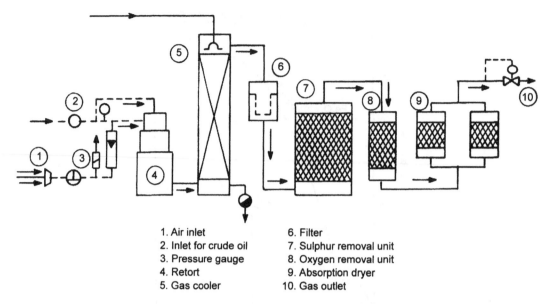

1. Air inlet
2. Inlet for crude oil
3. Pressure gauge
4. Retort
5. Gas cooler
6. Filter
7. Sulphur removal unit
8. Oxygen removal unit
9. Absorption dryer
10. Gas outlet

Fig. 8.16 Schematic flow diagram of an exothermic gas generator operating with crude oil

Production of exo-gases from hydrocarbon (petrol, crude oil, LPG) has become widely used. From crude oil, an exo-gas containing 20%–30% CO and H_2 can be obtained. Figure 8.16 shows a schematic diagram of an exothermic gas generator operating with crude oil. The air and crude oil enter the system at points 1 and 2, respectively, and then pass to retort 4 through flowmeters. A pressure gauge (3) is provided to check the air pressure. The oil is burned in the retort and the generated gas passes to the cooler 5 and filter 6. The crude oil contains about 1% sulphur in the form of H_2S or SO_2 and is removed to a level of 1 ppm. The gas then passes to the O_2 removal unit 8 and hence to the dryer 9. The gas finally leaves the system at point 10 and is ready to use.

Endothermic gas

Endothermic gas is produced at a lower air-to-gas ratio, and heat must be supplied to support the reaction. It is richer than exo-gas in CO, H_2 and CH_4. This gas is not only strongly reducing, but is also carburising. Endothermic gas is most suitable for sintering compacts of mixtures of iron–carbon and other alloy steels. The composition of the endo-gas is controlled, so that it is in equilibrium with the carbon potential of the steel to be sintered. The gas formation is facilitated over a clean catalyst such as nickel oxide in an externally heated chamber. Typically the generator operates at temperatures of about 1060–1100°C. The factors that influence the composition of the produced endo-gas are:

- Temperature in the cracking zone
- Air-to-gas ratio
- Efficiency of the catalyst
- Time in the cracking unit.

The temperature of the cracker should be kept close to the maximum temperature in the sintering furnace. Otherwise, the gas may not be stable in the sintering furnace. This could cause sooting or uneven properties of the sintered steel.

In Fig. 2.1 (refer Chapter 2), it has been illustrated how at lower temperatures the CO is strongly carburising and reducing. At higher temperatures, however, the action of the gas is weaker. The two almost parallel inverted S-shaped curves show the producer gas reaction:

$$CO_2 + C \rightarrow 2CO \tag{8.3}$$

The carburising effect of CO on iron is:

$$3Fe + 2CO \rightarrow Fe_3C + CO_2 \tag{8.4}$$

At lower temperatures, the producer gas reaction is generally the most prevalent and results in soot deposition on the sintered parts. However, at higher temperatures, say at 800°C, the carburising action is dominant. Soot deposition can be suppressed by fast heating and cooling in the furnace. The effect of methane is different from that of CO. Methane, with increasing temperature, increases the reducing action and the rate of carbon deposition. Figure 8.17

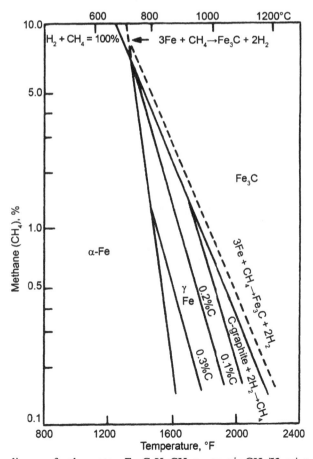

Fig. 8.17 Equilibrium diagram for the system Fe-C-H: CH_4 content in CH_4/H_2 mixture versus temperature

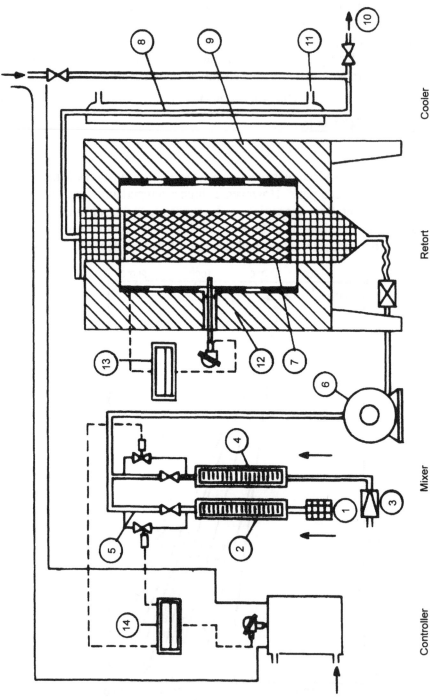

Fig. 8.18 Schematic flow diagram of an endothermic generator: 1 air filter, 3 valve, 2 and 4 flow meters, 5 mixer, 6 compressor, 7 retort, 8 gas cooler, 9 heating equipment, 10 endo-gas outlet, 11 cooling water, 12 thermocouples, 13 dew point analyser

presents the equilibrium between a methane–hydrogen atmosphere and iron with different carbon contents, at various temperatures.

The extent of decarburisation and recarburisation determines the finished carbon content at or near the surface of the sintered steel. Carbon content can be controlled by the dew point or the carbon dioxide content in the endo-gas. Most sintering under endothermic atmosphere is done at dew point ranging from –5 to 15°C. Above 260°C, endo-gas is reducing to iron, because its typical 40% H_2 and 0.8% H_2O (dew point 5°C) contents provide a hydrogen-to-water ratio of 50 to 1. Although this ratio achieves acceptable oxide reduction, other atmospheres— notably those based on hydrogen or dissociated ammonia or synthetic nitrogen-based atmospheres—can provide higher ratios.

Figure 8.18 is a schematic diagram of an endothermic generator. The air and gas enter the flowmeters (2 and 4) and then the mixing unit 5 through filter 1 and valve 3, respectively. The mixture is then passed in to the catalyst-filled reaction chamber of the generator by means of the compressor 6. The retort 7 is heated to the desired temperature by heating equipment 9. The gas leaving the retort is cooled to about 200–300°C in the cooler 8 and leaves the system at point 10.

8.6.3 Nitrogen and nitrogen-based atmospheres

Nitrogen is inert to most of the common metals and alloys. Because it is non-flammable, it is also used as a safety purge for flammable atmospheres.

The main constituent of the nitrogen-based system is molecular nitrogen. It is obtained from air, which consists of approximately 78% nitrogen, 21% oxygen, 0.93% argon, 0.03% carbon dioxide and a small amount of rare gases such as neon and helium. Nitrogen is most commonly produced through air separation, i.e., liquefaction and fractional distillation. Air is filtered, purified, compressed to drive it through the rest of the system, and cooled to remove H_2O and CO_2. After being liquefied it is distilled into the major constituents, the most abundant being nitrogen. In addition to being plentiful and independent of natural gas, the nitrogen thus produced has the following characteristics:

- It is very dry, with a dew point of less than –65°C
- It is very pure, with less than 10 ppm of oxygen
- It is essentially inert to materials most commonly sintered and to furnace components such as muffles, conveyer belt, heating element, radiant tube, fixtures, etc.

In actual practice, furnaces are not very tight and some air does get in. Nitrogen by itself does not control the resulting surface oxidising and decarburising. Chemically 'active' gases are therefore added to the nitrogen, when the atmosphere has to perform functions requiring transfer of some element, such as carbon, from the atmosphere to the component being treated, or such as oxygen, from the oxide to the atmosphere. These active ingredients can be divided into three categories discussed on the next page.

Oxide reducing agents

The most desirable ingredient for reducing surface oxides is hydrogen. It can be derived from liquid hydrogen storage tanks or dissociated ammonia. It can also be derived from endo-gas or dissociated alcohols. Methanol dissociates or cracks at temperatures above 815°C to produce H_2 and CO in the same ratio (2:1) as is normally found in the endo-gas generated from natural gas:

$$CH_3OH \rightarrow CO + 2H_2 + small\ amount\ of\ CO_2,\ H_2O\ and\ CH_4$$

Dissociated or cracked methanol when mixed with appropriate amounts of nitrogen can produce an atmosphere almost identical to endo-gas.

Carburising agents

The most desirable ingredient to carburise is CO. It can be derived from endo-gas or dissociated alcohols such as methanol, or by reacting a hydrocarbon such as natural gas or propane and an oxidant such as water in the carburising zone. It is generally found that in order to maintain effective carburising rates, the CO level in the carburising zone should not be less than 10%.

Oxidant

Small but controlled amounts of oxidants such as CO_2, H_2O, O_2 or some combination of them can be added to nitrogen at selected sections of the furnace to provide decarburising, oxidising or burning of lubricants just before sintering, PM parts.

Manufacturing the nitrogen consumes less energy than producing an equivalent volume of endo-gas: 44% less energy in the liquid storage process and 80% less energy in the 'on site' separator process. Nitrogen allows smaller volumes of atmosphere to be used because it permits reduced flow rates, increase in production rates, lower part rejection and increased furnace utilisation.

With the proper choice of enrichment gas, nitrogen-based atmospheres can be used to sinter and infiltrate iron, carbon steel and other ferrous and non-ferrous alloys. Stainless steel and refractory metals can be sintered in nitrogen-based atmospheres when nitriding is not critical.

8.6.4 Dissociated ammonia

Dissociated ammonia consists of 75% hydrogen and 25% nitrogen by volume. Liquid ammonia from the tank enters a vaporiser at high pressure where heat converts the liquid in vapour. The pressure of the vapour is then reduced in an expansion valve and the low pressure vapour passes through a dissociator element filled with catalyst. When heated at 900–1010°C, ammonia dissociates into hydrogen and nitrogen. Figure 8.19 illustrates the flow diagram of an ammonia dissociator. Normally the gas contains only a trace (0.05% or less) of uncracked ammonia, which can be eliminated by passing it through either water or activated alumina. The dissociated gas at an elevated temperature is highly flammable. Its specific gravity is 0.295 and thermal conductivity is 5.507,

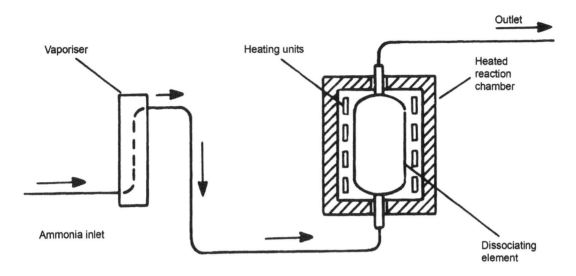

Fig. 8.19 Flow diagram of an ammonia dissociator

which is 1 for air. Typical applications of dissociated ammonia include sintering of brass, copper, iron–copper, tungsten and tungsten alloys, aluminium and its alloys and stainless steel.

8.6.5 Argon and helium

Argon and helium are non-flammable and are inert to all application. They are used for sintering refractory and reactive metals and also as a back fill in vacuum furnaces. Argon is cryogenically produced from air. Its purity is very high, less than 0.0005% oxygen and the dew point is lower than –68°C. Its specific gravity is 1.379×10^{-3} kg m^{-3} while the relative thermal conductivity is 0.745.

8.6.6 Vacuum

Vacuum retains the proper chemistry of the parts during sintering. It is often more economical than atmosphere gases, particularly bottled gas. The only operating costs involved in producing the vacuum are for electrical energy and oil for the pumps. The commonly used vacuum pumps are mechanical pumps and oil vapour pumps. Vacua are generally classified with four ranges; rough (>1 to 1 torr), medium ($1-10^{-3}$ torr), high ($10^{-3}-10^{-7}$ torr) and ultimate vacuum (<10^{-7} torr). Most sintering furnaces are medium or high vacuum furnaces and such vacua are readily created by oil vapour pumps backed by a mechanical rotary pump.

All metallic oxides have a so-called dissociation pressure which is equal to the partial pressure of the oxygen present in the gas atmosphere at equilibrium with the oxide. If the partial pressure of the oxygen is lower than this, the compound will be transformed into a lower value oxide or metal and oxygen. If the partial pressure of the oxygen is higher than the dissociation

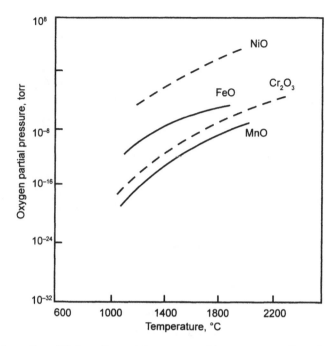

Fig. 8.20 Variation of equilibrium O_2 partial pressure with temperature for selected metal oxides

pressure, the reverse is true. Figure 8.20 shows the dissociation pressure of some of the metal oxides as a function of temperature. While sintering an alloy, the selective evaporation of some alloying elements, due to the different vapour pressures of the individual metals, must also be taken into account. This naturally depends on the duration and temperature of vacuum sintering. Table 8.3 gives the vapour pressures of some common metals.

Table 8.4 illustrates the dew point existing at different degrees of vacuum for different levels of water vapour in the gas. From the table it can be seen that for a vacuum of 10^{-1} torr the dew point is between –40 and –55°C, depending on the water vapour content.

Most of the vacuum sintering is on reactive metals which are highly susceptible to the formation of hydrides, nitrides or oxides in gaseous atmospheres. Refractory metal carbides, stainless steels, beryllium, titanium, zirconium, tantalum, niobium, vanadium, thorium, uranium and cermets are best examples for vacuum sintering.

8.7 SINTERING ATMOSPHERE ANALYSIS AND CONTROL

8.7.1 Gas analysis

Gas analysis is done either by an Orsat-type analyser or infrared analyser. In the Orsat-type, the complete analysis of a gaseous mixture is divided into the absorption phase and the explosion or burning phase. The amount of carbon dioxide, oxygen and carbon monoxide are determined by chemical absorption. These units are portable and permit relatively rapid analyses.

Table 8.3 Vapour pressures of some common metals

Vapour pressure, torr	Temperature, °C						Melting point, °C
	10^{-5}	10^{-4}	10^{-3}	10^{-2}	10^{-1}	1	
Aluminium				1210			660
Carbon		2100	2250	2430			3730
Chromium	1060	1160	1270				1875
Cobalt	1160	1260					1495
Copper		1030	1140	1270	1430	1620	1083
Iron	1110	1210	1320				1537
Lead	490	550	630	720	830	980	327
Manganese	700	770	850	950	1070	1230	1245
Molybdenum	1990	2170					2610
Nickel	1140	1250	1360				1455
Niobium	2190	2360	2500				2470
Silicon	1180	1280	1360	1550	1720	1930	1410
Silver	760	830	920	1030			960
Tantalum	2400	2590	2810				3000
Tin			1090	1230	1400		231
Tungsten	2550	2760	3010	3330	3650		3410
Vanadium	1430	1551					1900
Zinc		250	290	340			420

Table 8.4 Dew point as a function of water vapour content for different vacuum levels

Vacuum, torr	Dew point for different vol% of water vapour in the gas, °C		
	20%	70%	100%
1	−35	−21	−17
10^{-1}	−55	−43	−40
10^{-2}	−70	−61	−58
10^{-3}	−86	−77	−74
10^{-4}	−92	−90	−89

The infrared analyser depends on the principles that different gases absorb infrared energy at characteristic wave lengths. Because of this property, changes in the concentration of a single component in a mixture produce corresponding changes in the total energy remaining in an infrared beam passed through the mixture. Proper selection of apparatus permits accurate, rapid analyses for such constituents as carbon monoxide, carbon dioxide and methane. Such analysers are not suitable for measuring oxygen, hydrogen and nitrogen, which have no infrared absorption band. They are highly sensitive and are successfully applied for analysing gases with high carbon potential, such as purified exothermic gas or dry endothermic gas, and those with high purity such as dry hydrogen or argon.

8.7.2 Specific gravity analysis

The specific gravity of gases can be measured and compared against that of air. Since carbon dioxide is much heavier than the other sintering atmosphere constituents, this analysis is especially sensitive to changes in carbon dioxide. In case carbon dioxide appears in the influent or effluent furnace atmosphere, the specific gravity analyser could be used for corrective measures. This analysis is useful, for example, in checking sintering furnace atmospheres of purified rich exothermic or rich endothermic gas which are supposed to be free from carbon dioxide. By continuously measuring the specific gravity of the effluent gas from the furnace, the completeness of purge can be established.

Example 8.2 What are the major pollution problems in the sintering shop of a powder metallurgy plant? How best can they be mitigated?

Solution: The biggest problem is with the protective atmosphere used in the sintering furnace. Endo-gas is very common in sintering ferrous products. This is prepared in a generator with 2.5 to 1 air-to-natural gas volume ratio. The details of gas composition are already highlighted in the appropriate section of this chapter. CO (~20 v/o) gas constituent is the main problem, if the non-combusted gas is allowed to vent or leak into the surroundings. However, with the correct furnace operation the CO gets burnt. Some of the solutions to prevent pollution in the shop are:

- Proper operation with the right equipment
- Correct air-to-gas ratio in the gas generator
- Clean catalyst in gas generator
- Correct temperatures
- Consistent gas compositions
- Proper calibration of the instruments
- Proper furnace conditions.

Another problem in the sintering furnace is the inadequate burning of the lubricants from the green parts. If lubricants are not burned completely, the un-burnt lubricant vapours escape into the environment as volatile organic compounds. Some metal-based stearates, such as zinc stearate, produce zinc oxides which are hazardous.

8.7.3 Moisture determination

The moisture content of the sintering atmosphere has significant effect on sintering. One of the simplest methods for determining moisture content is by checking the dew point. The device used to determine this is called a 'dew cup', and contains a glass thermometer. The cup is placed directly in the gas stream so that the entering gas impinges on its polished surface. To begin the test, the atmosphere gas is flown through the container, and acetone is poured into the cup. After about 5 minutes, small amounts of crushed dry ice are added to the acetone while stirring constantly with the thermometer. At the first sign of dew or moisture on the polished surface of the cup, the temperature is read from the thermometer, which happens to be the dew point of the gas.

In a more instrumented type of indicator, the sample of the gas is compressed, and then quickly expanded. In case the gas cools down to below its dew point due to rapid expansion, a fog is noticed in the expansion chamber. The pressure ratio of the gas is converted to dew point.

The dew point measurements of a gas, either entering or leaving a furnace, give an excellent indication of the carbon potential. For example, excess moisture in a supposedly non-decarburising atmosphere can cause decarburisation. The measurement is also an indicator whether the water gas reaction occurs within the atmosphere by reaction of carbon dioxide with hydrogen at elevated temperatures. The dew point determination is a convenient quick check to see whether conditions are right throughout a furnace system.

8.7.4 Carbon potential control

Carbon potential control of the sintering atmosphere is important during sintering of steels, since it should be in equilibrium with the carbon potential of the steel concerned. Such a control would prevent the steel from either carburising or decarburising. Either dew point or carbon dioxide analysis of the atmosphere can be used as a measure of the carbon potential. The gas sample taken for measurement of the carbon potential should be from the high-heat zone of the sintering furnace. As the furnace is usually not under positive pressure, a pump is required to pump a gas sample from the furnace through the instrument. Dew point analysers are not as sensitive, and are slower in response than carbon dioxide infrared analysers. Because of the fast response, infrared analysers are used for multipoint control systems. Dew point on the other hand is not satisfactory to use for multipoint control because water molecules are difficult to purge from an instrument in a short time.

Before controlling the process, it is necessary to refer to the equilibrium data of dew point vs carbon content, or carbon dioxide vs carbon content, at the operating temperature of sintering. Such data (Fig. 8.21) have been determined by actual measurement and are fairly accurate. However, they should be used only as a guide and the final adjustment should be made after the carbon analysis of the steel concerned.

The sintering furnace has different zones, each requiring a unique combination of temperature, time and atmosphere composition. To achieve this the atmosphere is either

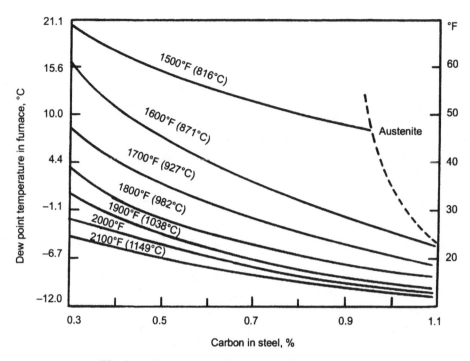

Fig. 8.21 Dew point and carbon equilibrium diagram

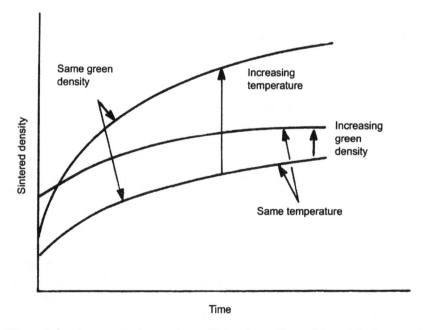

Fig. 8.22 Sintered density versus time curves, illustrating effect of increasing green density and sintering temperature on densification

Example 8.3: A FLC-4608-70 HT grade steel (density 6.85 g cm^{-3}) was sinter hardened in an endothermic atmosphere at 1120°C for 15 min with a CH$_4$ spike to achieve 0.8% C potential. After sintering, many parts revealed melting on the surface. What could be the reasons? What is the effect of such faulty sintering on the microstructure? What remedial measures should be taken? [For full details of composition refer Section 12.15 (Chapter 12)]

Solution: The reasons for melting may be two-fold: (i) High sintering temperature and (ii) High localised carbon.

The corrective measures should be:

· Burn-off and clean the furnace

· Replace the thermocouple in the sintering zone.

The microstructure of the melted region will show a quarter fraction of carbide. In addition the pores in the melted region will be rounded. The microstructure is shown in Fig. Ex. 8.3 (Reproduced with permission from Haas MR, *International Journal of Powder Metallurgy*, 42(5):29, 2006).

Fig. Ex. 8.3

diluted or enriched by hydrocarbons. To achieve a specific carbon content in the sintered steel, it is important to understand the effect of temperature of each zone on the final carbon equilibrium of the concerned steel.

8.8 PROCESS VARIABLES

The most important factors involved during the sintering process are temperature, time and furnace atmosphere. The influence of these factors on the sintering process is described below:

Sintering temperature: Increasing the sintering temperature greatly increases the rate and magnitude of any change in properties occurring during the sintering. Figure 8.22 illustrates the effect of increasing temperature and green density on sintered density.

Figure 8.23 schematically represents the variation in sintered properties of compacts with respect to homologous temperature. It is evident that electrical conductivity improves

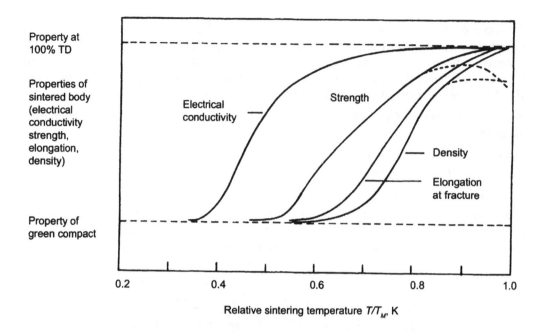

Fig. 8.23 Schematics of variation in various properties of metal powder compacts with respect to sintering temperature (powder characteristics, compaction pressure and sintering period kept constant)

at very low sintering temperatures, while strength is developed after higher sintering temperatures, reaching a saturation earlier than the elongation at fracture. The strength may decrease at the highest sintering temperature due to grain coarsening. The sintered density increases only at the higher temperatures, except in case of fine powders. The influence of porosity is more important in fatigue tests than in other mechanical tests. As electrical conductivity is directly related to particle-to-particle contact, it is affected by several factors including oxide films, lubricant, compacting pressure and sintering parameters as well as powder characteristics.

Sintering time: Although the degree of sintering increases with increasing time, the effect is small in comparison to the temperature dependence. The loss of driving force with increasing time at any temperature is one of the reasons why it is so difficult to remove all porosity by sintering. An attempt should be made to achieve the desired properties of the sintered parts by shorter sintering times and correspondingly higher temperatures. However, the maintenance costs and energy consumption of a furnace increase when its operating temperature is raised.

Sintering atmosphere: The proper production, use and control of sintering atmospheres which are essential for the optimum use of the powder metallurgy process have already been described in Section 8.6.

8.9 MATERIAL VARIABLES

The variables described briefly below have been adequately elaborated in Chapter 7. However, for the sake of recapitulation they are briefly described here.

Particle size: In terms of the basic stages of sintering, decreasing particle size leads to increased sintering. The smaller particle size has a greater pore/solid interfacial area producing a greater driving force for sintering. It promotes all types of diffusion transport, e.g., greater surface area leads to more surface diffusion; small grain-size promotes grain boundary diffusion and a larger interparticle contact area to volume diffusion.

Particle shape: The factors that lead to greater intimate contact between particles and increased internal surface area promote sintering. These factors include decreasing sphericity and increasing macro- or micro-surface roughness.

Particle structure: A fine grain structure within the original particles can promote sintering because of the favourable effect on several material transport mechanisms.

Particle composition: Alloying additions or impurities within a metal can affect the sintering kinetics. The effect can either be deleterious or beneficial, depending on the distribution and reaction of the impurity. Surface contamination, such as oxidation, is usually undesirable. Dispersed phases within the matrix may promote sintering by inhibiting grain boundary motion. Reaction between impurities and either the base metal or alloying additions at the relatively high sintering temperature may be undesirable.

Green density: A decreasing green density signifies an increasing amount of internal surface area and consequently, a greater driving force for sintering. Although the percentage change in density increases with decreasing green density, the absolute value of the sintered density remains highest for the higher green density material.

8.10 DIMENSIONAL CHANGES

Changes in dimensions resulting from sintering represent an important field in powder metallurgy, especially with respect to large-scale production of parts with small dimensional tolerance. The fundamental process of sintering leads to a reduction in volume because of pore shrinkage and elimination. In this regard the following factors may be considered.

Entrapped gases: The expansion of gas in closed porosity has been postulated as producing compact expansion.

Chemical reactions: Hydrogen is a common component of sintering atmospheres and can often diffuse through the metal to isolated portions of the compact where it reacts with oxygen to form water vapour. The pressure of the water vapour can lead to the expansion of the entire mass. It is also possible to have reactions that lead to the loss of some element from the sinter mass to the atmosphere, such as volatilising, and result in a shrinkage of the material.

Alloying: Alloying that may take place between two or more elemental powders very often leads to compact expansion. This effect, which is due to the formation of a solid solution, is often offset by shrinkage of the original porosity. Dimensional changes may also occur in a binary system where the rate of diffusion of each metal into the other is different. The theoretical basis has been described in Chapter 7.

Dilatometric study is a convenient method to study dimensional changes (shrinkage or growth) in green compacts during sintering. Figure 8.24 shows typical curves for 90/10 tin bronze alloy, when the powders are in premix form or pre-alloyed. It is evident that in the case of a premix compact the dimensional changes are more than in the pre-alloyed one during sintering. This is due to the fact that in the former case, during alloying, the dilation effect is more.

Shape changes: Green parts invariably contain variations in green density that can lead to substantial changes in shape because of the strong dependence of sintering—especially shrinkage—on green density. Low green density regions will exhibit a greater amount of shrinkage during sintering. For example, a cylinder with a relatively high *L/D* ratio compacted by a single action method would have a gradually decreasing green density from one end to the other. Such a cylinder would change during sintering into a truncated cone, as shown in Fig. 8.25. On the other hand, a cylinder prepared by a double action method, would likely achieve an 'hour-glass' shape (Fig. 8.25). Often the shrinkage (or expansion) is different in two directions—axial and radial—and must be taken into account while designing the compacting tools. Parts with complex and

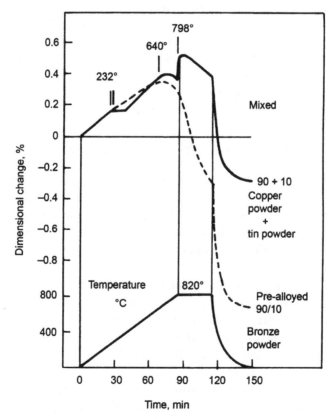

Fig. 8.24 Dilatometric plot of 90/10 tin bronze compacts during sintering cycle, based on premix and pre-alloyed powders

Original green   Single-end compacted   Double-end
cylinder   after sintering   compacted

Fig. 8.25 Illustrations of changes in the shape of cylindrical compact after sintering for single- and double-end green compaction

asymmetrical shapes exhibit uneven shrinkage during sintering and loss of the desired tolerances. To restore lost tolerance, sizing operation is generally carried out. This will be described later in Chapter 10. From the users' point of view it is easier to maintain narrow tolerances on the contours than on the axial dimensions.

8.11 MICROSTRUCTURAL CHANGES

The basic scientific features related to microstructural changes during solid state and liquid phase sinterings have been described in the previous chapter. In this section, a brief qualitative summary is given.

When a green compact is sintered, the original particle boundaries can no longer be observed. Instead, the structure becomes similar to that of the metal in wrought and annealed conditions, except that it contains pores. Pores are of two types: open or closed. With progression of sintering the pores continue to shrink. At about 5% total porosity, the formation of closed pores commences. Typical microstructures in PM parts are caused by porosity and by the blends of elemental powders that constitute many alloys. These blends do not always result in homogeneous, well diffused structures. Such heterogeneity is not necessarily detrimental, and in certain nickel steels and diffusion alloy steels, it may be advantageous.

Grain growth is an important aspect of sintering. There is normally a large driving force for grain growth. Grain growth in a sinter mass can be somewhat different from that in conventional metal, since the former can be treated as a two phase material—the second phase being pore. Porosity in green compacts and in the developing sinter mass represents a very effective hindrance to grain growth. The addition of other components in the powder blend may drastically hinder bonding between adjacent particles and the formation of grain boundaries. Grain boundary grooves also tend to inhibit grain growth, since the movement of grain boundary away from its groove leads to an increase in the area and energy of the grain boundary.

The control of grain growth so that grain boundaries do not pull away from pores is essential in sintering to zero porosity. The more non-uniform the grain structure—particularly if it develops discontinuous grain growth—earlier (in the sintering process) is the separation of pores from grain boundaries likely to occur. Therefore, to reduce this tendency, it is important to have a starting powder that is uniform in size. Often, a proper additive also plays a vital role. An ideal example is the sintering of MgO (0.05%)-doped alumina, where the additive is preferentially concentrated at the grain boundaries. Such a control has been shown to impart full densification to alumina.

Many types of phase transformations may occur in the solid state during sintering—at constant temperature or during the cooling of the metal from the sintering temperature. Porosity and fine grain structure influence these transformations. The best example of a phase transformation associated with sintering is the production of sintered steels. Since the porosity reduces the thermal conductivity of the mass, for a given set of cooling conditions the actual cooling rate throughout the sintered metal is considerably less than for the conventional solid one. Such an effect would have bearing on martensitic transformation, which is dependent on the cooling rate. Precipitation from solid solution is also a common type of transformation associated with sintering.

During liquid phase sintering the porosity level falls, while grain size increases. The shape of the pores varies rapidly during liquid phase sintering. In the first stage, the pores are irregular; later they form a cylindrical network, and finally attain a spherical shape. The interfacial energies can change during liquid phase sintering, as they depend on solubility, surface contamination and temperature. Hence, the microstructural parameters keep shifting with time during liquid phase sintering. A proper development of the microstructure during liquid phase sintering is very important for attaining mechanical properties, particularly ductility.

8.12 INFILTRATION

Infiltration is a variant of liquid phase sintering, in which a liquid is formed outside the green compact during the very early stage of sintering. Although simultaneous infiltration and sintering appear to be dominant, infiltration of a previously sintered material is also practised.

Infiltration of an iron PM compact by copper at a temperature higher than the melting point of the latter is common for densifying the porous skeleton. Through capillary action, the molten copper is drawn into the interconnected pores and fills the entire pore volume. The resultant alloys have high tensile strength and hardness, in addition to improved impact and fatigue strengths. The infiltration process is generally classified as single-step or double-step. The former method consists of infiltration of the green iron part by contact with a green copper compact in conjunction with sintering. The double-step method consists of two full cycle runs through the furnace. The first is the sintering cycle and the second the actual infiltration cycle in which the green infiltrant compact is placed in contact with the sintered part and the full furnace cycle is repeated. The infiltrant constitutes copper and about 5% iron to minimise 'erosion' or localised dissolving of iron beneath the contact area as infiltration proceeds. Other ingredients may include lubricant, graphite, nickel and manganese, which tend to favour non-adhering residuals. A typical composition of such a premix is Fe 5%, Ni 0.6%, Mn 1.5%, Al 0.3%, C 0.5% and the balance copper. The compacting characteristics of infiltrants may differ depending on their composition and method of manufacture. The shape of the infiltrating compact must be compatible with the part to be infiltrated and is also dependeat on the characteristics of the infiltrant. If possible, the contact area per unit weight of infiltrant should always be maximised to avoid erosion. Common protective sintering atmospheres suffice for infiltration; however, certain types of infiltrants require atmosphere dew points around 0°C or higher to prevent tenacious adherence of constituent residuals. The sintering cycle used for the infiltrating alloy is essentially that used for sintering standard parts. Excessive outgassing due to gaseous reactions occurring at temperatures prior to reaching the infiltrant melting point could inhibit the inward flow of the infiltrant and hence effective infiltration cannot occur. The difficulties encountered due to outgassing can be minimised by double-step infiltration or by prolonged sintering prior to reaching the infiltration temperature during the single-step process.

There has been a recent introduction of copper infiltrant in the form of pre-cut and pre-weighed form as a wire ring, coil or slug. Such infiltrants eliminate the need to press an infiltrant preform and the associated costs. This new development claims to leave no residue or erosion on the surface of the PM iron skeleton.

8.13 SINTERED PARTS MATERIAL HANDLING

This category of material handling comprises sintering furnace loading and furnace unloading. During loading, the simplest design is to pass the green parts directly from the transport carrier on to the belt. Usually, the parts travel via gravity down a smooth chute on to the belt of the sintering furnace. Sometimes, the chute is vibrated to assist the motion. However, there are limitations for this method:

- Part-to-part contact on the sinter belt
- Random pattern while sintering
- Potential for parts to tip during transfer to sintering furnace belt leading to bottom-edge damage.

To eliminate the above limitations, more sophisticated automation is employed. Row-building automation is used to organise the parts with a predetermined gap and then release one row at a time using metering devices. For PM parts with delicate bottom-side flatness, a pick-and-place motion can be used to handle from one part to a trayful.

In case of furnace unloading, the simplest method is to directly dump the parts into a container or on a take away conveyor. This is viable for those parts which travel directly on the belt and can withstand part-to-part contact. For parts that travel on sintering plates, an auxiliary handling device is used to gather the sintering plates and then send them back to the front of the sintering furnace. Vision and mechanical gauging methods are used to verify the quality of the sintering plate before returning it to the front of the sintering furnace.

SUMMARY

- Prior to sintering, removal of binders, if any, in the green compact is necessary.
- The form and complexity of part shapes during loose sintering are dependent to a large extent on the flow behaviour of the powder.
- The main methods to convey the parts in a continuous furnace are mesh belt, roller hearth, pusher-type and walking beam types.
- A sintering furnace should have burn-off, high temperature and cooling zones.
- The sintering atmosphere is generally analysed by an Orsat-type or infrared analyser.
- The dew point measurements of a sintering atmosphere either entering or leaving the furnace give an indication of the carbon potential.
- In a given PM steel, it is necessary to understand the effect of temperature on each furnace zone to achieve the final carbon equilibrium.
- To reduce discontinuous grain growth during sintering, it is necessary to have a uniformly-sized starting powder.
- Infiltration is a variant of liquid phase sintering, in which a liquid is formed outside the green compact during the very early stage of sintering.

Further Reading

Bose A, *Advances in Particulate Materials*, Butterworth–Heinemann, London, 1995.

Bradbury S, (ed.), *Powder Metallurgy Equipment Manual*, 3rd edn, Metal Powder Industries Federation, Princeton, 1980.

German RM, *Sintering Theory and Practice*, John Wiley and Sons, New York, 1996.

German RM, *Powder Metallurgy and Particulate Materials Processing*, Metal Powder Industries Federation, Princeton, NJ, USA, 2005.

German RM and Bose A, *Injection Molding of Metals and Ceramics*, Metal Powder Industries Federation, Princeton, NJ, USA, 1997.

Rahaman MN, *Ceramic Processing and Sintering*, 2nd edn, Taylor and Francis, London, 2003.

Reed JS, *Principles of Ceramic Processing*, John Wiley and Sons, New York, 1995.

Remmy Jr, GB, *Firing Ceramics*, World Scientific Co., Singapore, 1994.

Rice RW, *Ceramic Fabrication Technology*, Marcel Dekker, New York, 2003.

Sanderow H, (ed.), *High Temperature Sintering*, Metal Powder Industries Federation, Princeton, 1990.

EXERCISES

8.1   Explain why the temperature for complete binder burnout is commonly higher when products of a larger size or finer particle size are sintered.

8.2 Binder burnout may produce considerable gas. Calculate the volume of gas produced from 1 cm^3 of an alumina compact containing 2 wt% polyvinyl alcohol binder. Assume that the compact is 60% dense and the gas is at 400°C and contains H_2O and CO_2.

8.3 On what factors does the burn-off period in the sintering furnace depend?

8.4 What are the disadvantages of too fast and too slow burn-off in the continuous sintering furnaces?

8.5 Explain why in both burn-off and high-heat zones, the heat-up rate is rapid at the start and slow at the finish.

8.6 What is the purpose of mounting exhaust hoods on the sintering furnace ends at the door openings? What ideal proposition do you have against exhaust hoods?

8.7 Explain the significance of multizone control in each chamber of a continuous mesh belt sintering furnace.

8.8 Suggest what heating elemeat is to be used in the electric furnace for the following materials produced through pressing/sintering route:

(a) Low-alloy steel

(b) Stainless steel

(c) Pure oxide ceramic, say Al_2O_3

(d) Cemented carbides

(e) W-Re alloys.

Name the type of furnace(s) used industrially.

8.9 Zirconia heating element can be used upto 2000°C in air. This is not achievable by any other ceramic heating elements. However, its use is limited to laboratory size furnaces only. What is the reason?

8.10 Describe why producer gas, which is produced from cheap coal, is not preferred as a fuel for any ceramic kiln.

8.11 Because of erratic electric supply, a PM plant management decided to switch over to combustion heating. Describe what changes in furnace design have to be made.

8.12 UO_2 oxide ceramic nuclear fuel pellets are sintered in a reducing atmosphere at 1700°C. Suggest the type of sintering furnace. Can you use an alternative furnace?

8.13 Which metal is more stable in water vapour (super-heated steam) at 1000°C: Cr or Ni? (Consult Ellingham diagram).

8.14 Will a gas mixture containing 97 v/o H_2O vapour and 2 v/o $H_2(g)$ deoxidise nickel oxide over pure nickel at 1000 K? Assume 1 atm pressure (Use Ellingham diagram).

8.15 Using thermodynamic data calculate the equilibrium P_{O_2} over nickel at 1200°C. Determine the corresponding vacuum below which NiO will begin to dissociate.

8.16 Calculate the equilibrium ratio P_{H_2}/P_{H_2O} for the oxidation of Cr in water vapour at 1000°C. Also determine it using a nomograph (Fig. 2.2).

8.17 Establish the relative thermal stability of Si_3N_4 and BN in a mixture of Si_3N_4 and B at 1 atm. pressure. Assume all components are pure (use the nitride Ellingham diagram).

8.18 Al-8Mn alloy was prepared through PM route using two routes: premix and mechanical alloying. The starting manganese powders were fine and coarse. Identify the routes in the table below which show their mechanical properties.

	Material	UTS	Young's modulus	Elongation
		MPa	GPa	%
A	Premix Al-8Mn (Fine)	114	68	29
B	-----------------	338	84	11
C	-----------------	104	68	14
D	-----------------	338	85	8

8.19 Is it possible to guess the maximum achievable strengths and ductilities for a given set of sintered premixed and pre-alloyed powders from the respective dilatometric plots? If not, why?

8.20 For plain carbon sintered steel (0.4% C) that has been cooled slowly in a sintering furnace from 1000°C to 200°C temperature, determine the fractional amounts by weight of pro-eutectoid ferrite and pearlite.

8.21 A reduced iron powder (Höganäs make ASC 100.29 grade) of following characteristics was mixed with 3% copper powder (−200 and −100 mesh size respectively), 0.5% graphite (ultra-fine powder) and 0.9% lubricant. The green compacts were sintered in a reducing atmosphere at different temperatures and time.

Particle size range   20–180 μm

Apparent density   2.96 g cm^{-3}

Flow   24 s/50 g

H$_2$ loss   0.08%

C   0.002%

Compressibility   7.21 g cm^{-3} at 600 MPa

Green strength   38 MPa at 600 MPa

Discuss the qualitative order of dimensional change by various material and processing factors. Note the advantage that reduced dimensional change will not necessitate sizing operation.

8.22 If the ASC 100-29 grade powder (refer problem 8.21) is replaced by water-atomised powder of similar particle size, what significant changes in the powder characteristics of the latter do you anticipate?

8.23 For producing a PM telephone token, a 75Cu-25Ni pre-alloyed powder was selected, followed by double pressing and double sintering.

(a) Why was double pressing/double sintering selected?

(b) Would you prefer a fine or coarse size powder? Why?

(c) Why were cast and wrought tokens not preferred?

(d) What should be the sintering temperature range? Can the continuous mesh furnace be used?

8.24 Figure P. 8.24 shows the pseudo-binary phase diagram of TiB$_2$-Fe system. Suggest the selection of sintering temperature and atmosphere for a TiB$_2$-20 vol% Fe composition. Would you prefer single-step or double-step sintering? Why? TiB$_2$ was produced by carbothermic reduction, while the iron was carbonyl iron powder.

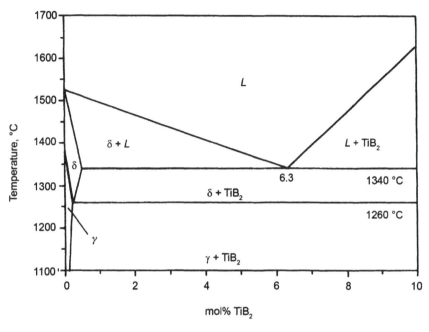

Fig. P. 8.24

8.25 An 80 vol% TiB$_2$-20 vol% Fe PM cermet is subjected to solid-state sintering in vacuum. The commercial TiB$_2$ powder had the following other species: 0.85% B$_4$C, 0.02% C, 2.6% TiO$_2$ and 1.5% B$_2$O$_3$. Iron powder on the other hand had 0.2% O$_2$ and 0.05% carbon impurities. Suggest the solution, reaction and compatibility aspects, if any, of these constituents among each other during sintering.

8.26 PZT is synthesised from original oxide powder mixtures. PbO has high vapour pressure at elevated temperatures and is therefore added sligtly in excess of required amount. Fine grain size microstructure gives rise to better quality piezoelectric devices; therefore, the fast sintering cycle was applied. It is found that this method is beneficial in case of specimens without excess PbO, but in case of excess PbO the end properties were inadequate. Give reasons.

8.27 Explain why 'bloating' defect is common while sintering ceramic whiteware, but is not present in sintered pure oxide ceramic. [Hint: Refer composition and state of the raw material.]

8.28 In order to develop a tungsten/copper-graded material by infiltration, what processing strategy should be adopted? [Hint: A number of grades of tungsten powder can be selected.]

8.29 Suggest a simple but effective test procedure to confirm whether the oxidation taking place in the cooling zone of the continuous sintering furnace is due to air or water leak.

8.30 For preparing Ti-6Al-4V PM alloy through the premix route, Al-40V master alloy powder of three different types were used: (a) gas atomised 26 μm size, (b) coarse powder irregular 20 μm size and (c) fine powder irregular 6.4 μm size. The green compacts were sintered at various temperatures (1300–1600 K). Suggest in which cases the sintered density will be maximum and minimum. Give reasons.

8.31 In preparing the hyper-stoichiometric uranium–plutonium mixed oxide nuclear fuel, CO/CO_2 gas atmosphere is preferred in comparison to H_2/H_2O in the sintering atmosphere. Why? What atmosphere would you prefer for stoichiometric composition?

8.32 A firm producing bonechina-ware operated a 150 kW electrically heated kiln for firing. The production cycle was 15 to 20 h and power consumption averaged 1300 kWh per cycle. If you are asked to reduce the cycle to 8 hours by faster heating and cooling of the same kiln, what modification would you suggest in the construction material of the kiln?

8.33 PM processing of a nanocomposite is far less problematic than for a nano-monolith. Comment on the statement.

8.34 How would you control cobalt loss in a cemented carbide during vacuum sintering, without any change in the chemistry of the premix?

8.35 What are the advantages and disadvantages of vacuum sintering?

9

Full Density Consolidation

LEARNING OBJECTIVES

- The significance of hot pressing
- The advantages of hot isostatic pressing over hot pressing
- The need for sinter-HIP or overpressure sintering
- How to achieve consolidation during powder hot extrusion and hot forging
- Spark sintering

For full density consolidation, conventional pressing and sintering of green powder compacts is not sufficient and further introduction of mechanical working over them becomes a desirable aspect. This makes the products stronger and more dutiful. The operations are mainly dynamic compaction, hot pressing, hot isostatic pressing, hot extrusion, forging and rolling. Recently, spark sintering has been introduced in a big way as an improved method for full density consolidation.

9.1 DYNAMIC POWDER COMPACTION

Dynamic powder compaction is a single-operation consolidation brought about by the impact of a high-speed punch on powder. This produces a discrete shock wave in the powder, which under optimised conditions, results in metallurgical bonding and sometimes fusion of the surface of particles. The work of compaction produced by inter-particle shear is transferred on the surface of the powders in such a short time (microseconds or less) that there is no possibility for heat to be conducted away from the surface and thus localised melting or welding occurs. The welding is similar to that which occurs during the seizure of a bearing or during friction or explosive welding. The production of greater than 99% theoretical density compacts is facilitated by the inter-particle lubrication that the melted surface of the particle provides. The rise in temperature of the interior of each particle is small. Dynamic powder compaction differs from such high-speed techniques as Petro Forge or Dynapak. The latter processes are closely related to crank presses and drop forges. They involve a large mass travelling at low velocity, while dynamic powder compaction involves a very light punch travelling at very high velocity. The actual compaction press consists of a high-pressure reservoir, fast action valve, guide tube, compaction chamber and ejector unit. Compressed air is usually used as the drive system. At very high velocities helium may be used. It has been noticed that in dynamic powder compaction, the liquid zone between the particles resolidifies in the same time-scale as its formation. This is in the range of 10^6–10^8 °C s^{-1}. These cooling rates result in a rapidly solidified material with an extremely fine structure or even an amorphous glassy structure. Rapid solidification imparts unique properties to the compacts. Depending on the material, the weld zone may have a hardness above or below that of the work-hardened particles. Heat treatment may be given to bring out a specific property in either the weld zone or the particles. For instance, the rapidly-quenched weld zone may be given an ageing treatment or the heavily work-hardened particles may be made to recrystallise.

Among the possible applications, the process allows non-equilibrium powder or powder mixtures to be consolidated with either chemical reactions or a degeneration of metallurgical structures. For example, aluminium compacts with very good wear and seizure resistance have been made from Al-steel mixed powders. Conventional sintering of these would result in the formation of a brittle intermetallic. It is also possible to compact ad-mixed carbides in steels.

In an extreme case, it has been possible to consolidate amorphous materials without the occurance of crystallisation. The finer grain size, higher solute contents, uniform alloy distribution and cleanliness of the powder make rapidly-solidified powders highly desirable for many demanding applications, especially with regard to superalloys and aluminium alloys for aircraft. Dynamic powder compaction can also consolidate non-metals, and again, several interesting possibilities exist in this area.

9.2 HOT PRESSING

The basic mechanisms for pressure-assisted sintering—hot pressing and hot isostatic pressing—have been described in Chapter 7. Hot pressing is a suitable method for densifying materials with poor sintering behaviour. This technique, which combines powder pressing and sintering into one single operation, offers many advantages over conventional powder consolidation. By simultaneous application of temperature and pressure, it is feasible to achieve near theoretical density in a wide range of hard-to-work materials. As the resistance of metal particles to plastic deformation decreases rapidly with increase in temperature, much lower pressures are required for consolidation by hot pressing. Further, densification by hot pressing is relatively less sensitive to powder characteristics—shape, size and size distribution which are important in cold pressing and sintering. Hot pressing parameters—pressure, temperature, time and the working atmosphere—largely control the properties of compacts. High-speed tool steels, superalloys, beryllium and the refractory metals are particularly amenable to hot pressing. Hot pressing is perhaps the only method of producing dense and fine-grained shapes of materials such as pure carbides, nitrides and borides which are otherwise difficult to sinter due to the lack of adequate atomic mobility at the sintering temperature.

9.2.1 Process and equipment

The various steps involved in the hot pressing procedure are as follows:

1. Powder or a cold compacted preform is placed in the die mould.
2. The mould is heated either by resistance or by induction method to a predetermined temperature.
3. The powder in the die cavity is then pressurised.
4. The temperature is steadily increased during compacting until a maximum required temperature is reached.
5. Compacting pressure and temperature are maintained for a dwell time.
6. The mould is cooled slowly under pressure to a temperature at which oxidation of the material will not occur.

There are many variations on the general procedure given above. In many cases it is preferable to apply a nominal pressure, or even the maximum required compacting pressure, before the initiation of the consolidation cycle. This is particularly true in the case of reactive hot pressing.

Many oxide and carbide ceramics can be hot pressed using graphite dies in the open atmosphere. However, the die life is limited due to severe oxidation at temperatures above 500°C. Hot compacting of refractory and reactive metal powders demands an inert atmosphere or vacuum. Using vacuum in place of inert gas for hot pressing offers additional advantages of removing air from the powder body (thus eliminating the possibility of air entrapment), and also degassing it during the initial heating up of the pressing cycle. This would enhance the activity of the powder and thus help in sintering. Figure 9.1 gives the schematic design of a hot pressing unit.

While graphite is the most common die material, other ceramics and refractory metals have also been used on a limited scale.

Graphite is available in many grades, differing in density, strength, thermal and electrical properties. The strength increases with temperature, up to about 2500°C, beyond which it falls. The strength at 2500°C is almost twice that at room temperature. A low thermal expansion coefficient, ease of machining and low cost make it a nearly ideal choice for die material.

Oxide ceramic dies are promising for hot pressing in an oxidising atmosphere. High-density alumina can be operated up to 40 hot pressing cycles, provided extreme care is taken in alignment and rate of loading and heating.

Carbides have been tried in hot pressing die materials on a limited scale. The machining of these materials is difficult and expensive.

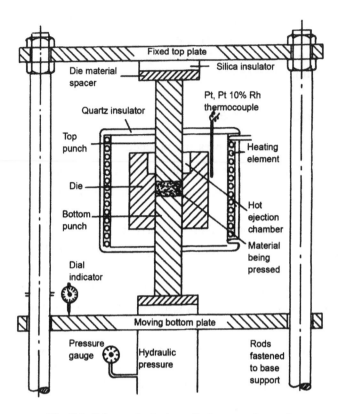

Fig. 9.1 Schematic design of a hot pressing unit

Titanium boride, although showing excellent high temperature strength and increase in strength with temperature (flexural strength of 240 MPa at RT and 414 MPa at 1600°C), is used only to a limited extent because of difficulty in machining.

Tungsten and ***molybdenum*** have been used as die materials but creep of the metals at high temperatures limits their usefulness.

Different kinds of die washes, coatings, spacers, liners and sleeves have been employed to overcome or decrease the die compact interaction, and thus decrease contamination and facilitate ejection of the final compact. Die washers such as B_4N and Al_2O_3, liners or spacers of foliated graphite and punch coatings such as pyrolytic graphite have been used to prevent interaction between the materials being pressed and the graphite die parts. Foils of noble or refractory metal (e.g., Pt, W, Mo, Ta) and spacers of refractory compounds, such as SiC, and refractory oxides, such as ZrO_2, have also been used for the same purpose.

Induction heating is most commonly employed for heating the graphite die tooling. Heating by radiation from resistance heating elements facilitates closer control of temperature. However, the maximum temperature attainable by indirect resistance heating is limited to about 1800°C.

Earlier hot pressing was primarily employed to improve the densification of some metal and metal-bonded carbide cutting tools. Currently, hot pressing as a fabrication process is being used increasingly in the ceramic industry for the preparation of materials with improved properties through the control of composition, microstructure and density. Hot pressing has also been employed in the fabrication of various high-temperature components, multiphase ceramics, ceramic–metal systems and pressure bonding. Tungsten, tungsten-based alloys, dispersion-strengthened aluminium and copper-based alloys, superalloys and beryllium metal have also been consolidated by vacuum hot pressing for many years.

9.3 HOT ISOSTATIC PRESSING (HIP)

The aim of the process is to compact materials that would normally compact only under considerably higher pressures, or to combine the pressing and sintering operations in one step. The pressure medium used is an inert gas. Heat is applied by an internal furnace in the pressure vessel. In most cases the powder or workpiece is encapsulated in a gas-tight material, which can withstand high pressure and temperature and does not react with the powder. Typical temperatures are 1000–1750°C and typical pressures are 100–320 MPa or 1000–3200 bar.

9.3.1 Equipment and process variables

A HIP system usually consists of five major components: pressure vessel, internal furnace, gas handling, electrical and auxiliary systems (Fig. 9.2).

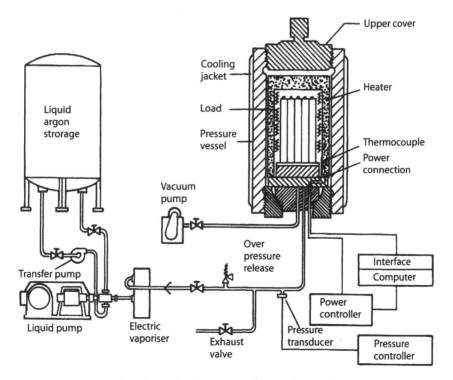

Fig. 9.2 Schematic of hot isostatic pressing equipment

Pressure vessel: The pressure vessel in a HIP system contains the high-temperature furnace and retains the high-pressure gas. Most HIP vessels have threaded closures. The sealing of the gas in the vessel is done with elastomer O-rings since the vessel temperature is kept below 250°C. The ASME Code Section VIII, Division 2, requires a minimum 3:1 safety factor of stress versus tensile strength. The quality assurance of a vessel after installation is very important, regardless of the type, size, design and stress analysis.

Furnace: There are different types of furnaces: radiation, natural convection and forced convection. The workpiece is not exposed to direct radiation from heater elements. Heating elements are not susceptible to damage by the load/unload process. The common heating elements are graphite, molybdenum and nickel/chrome. As the pressure vessel is designed as a 'cold-wall vessel', a thermal barrier, which prevents penetration of hot gas into the inner vessel wall, is used.

Gas handling: Most systems use argon as the pressurising medium. Gas pressures can be achieved with a compressor by thermal expansion. Gas purity is very important. Gas with 50 ppm total impurities is acceptable to prevent premature furnace failure, but can be harmful to superalloys or titanium which require less than 5 ppm total impurities.

Controls: This can be done at three levels: relay, programmable (Level I), mini and microcomputer (Level II) and total computer supervision (Level III).

Auxiliary systems: These include a cooling system, vacuum system, material handling with workpiece fixtures and facilities subsystems, including exhaust fans, oxygen monitoring equipment and cranes.

The variables of a HIP system are given below:

- Workpiece configuration (powder, compact, bonding, etc.)
- Shape, size, density, thermal characteristic, and number of workpieces processed per year
- Hot or cold loading/unloading
- Type of fixturing: manual or automatic workpiece loading/unloading
- Method of loading/unloading into and from vessel with crane or automatic jib
- Maximum pressure and temperature
- Maximum tolerable temperature uniformity during steady-state heating and cooling
- Maximum heating and cooling rates during transients. Is rapid quenching of workpiece required for optimum property recovery?
- Type of cycle: (i) pressurise cold, then heat; (ii) pressurise and heat simultaneously; (iii) heat first at vacuum or low pressure, then pressurise
- Maximum time at steady-state
- Desired cycle time
- Minimum required gas purity for workpiece

- Electrical control of Levels I, II or III
- Utility requirements: electrical, water, air or gas
- Shutdown modes: preventive maintenance, quality assurance, power failure, etc
- Room size for equipment pit for vessel (water table) and barricading, crane for loading/unloading, etc.
- Amortisation time of equipment (life of vessel and other components)
- Applicable codes ASME, ASA, ASTM, ASM, IEEE, NEC, NEMA, OSHA and their codes
- Requirements of insurance carrier, local safety codes, seismic, etc.
- Cost effectiveness: cycle cost versus loading efficiency.

Hot isostatic pressing PM parts has the demonstrated potential to produce 100% dense high-performance alloy parts on a consistent basis. Thus, the applications of PM are now possible where residual porosity cannot be tolerated because of its adverse effect on key properties. However, HIP is used for preparing porous parts as well, e.g., filters, grinding wheels and porous electrodes. In this case, the process is capsule-free HIP on powder compacts or pre-sintered bodies, but under high gas pressure. The high gas pressure prevents the open pores from closing. Hot isostatic pressed porous materials have larger neck growth between particles than when conventional pressureless sintering is used. The former have better mechanical properties.

9.3.2 Sinter-HIP or overpressure sintering

The widespread application of HIP is restricted by two factors:

- Lack of HIP equipment capable of necessary high temperature
- Difficulty in containerising the green powder compact.

By introducing containerless HIP a higher throughput of as much as 50%–90% can be achieved when compared to containerised parts. Furthermore, it is difficult to be assured that containers are fully sealed with no pinholes. The major benefit is the combination of sintering with the introduction of gas at the end of the sinter cycle for HIP. This eliminates one major piece of equipment, reduces materials handling, provides greater control of the process, and saves energy and labour costs. This process is particularly useful for cemented carbides, superalloys and ceramic products where HIP is used extensively. Sinter-HIP eliminates the need for reheating the product to solidus temperature a second time, which risks abnormal grain growth. Sinter-HIP pressure is rather low, being in the range 6–10 MPa in comparison to that for classical HIP, since the material structures are soft in the sintered condition.

Figure 9.3 illustrates the processing profile of various HIP variants and their effect on ultimate densification.

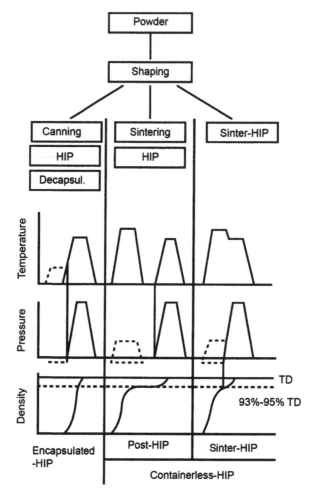

Fig. 9.3 Process variants of hot isostatic pressing

9.4 POWDER HOT EXTRUSION

The basics of powder working processes like forging, extrusion or rolling are similar, being the result of direct compression (Fig. 9.4). Other types of stresses like indirect compression (drawing and deep drawing) and biaxial tension (stretch forming) are not of direct concern to powder metallurgists. Hot working processes are related to stress and temperature. The temperature is maintained above the recrystallisation temperature of the material concerned. Figure 9.5 illustrates the effects of pressure and strain rate on temperature. It is evident that the increase in applied pressure increases the deformation range, while an increase in strain rate has a reverse effect. The stress conditions in mechanical working processes are invariably complex, particularly due to the presence of friction. This leads to inhomogeneous plastic flow. The triaxial stress system set up causes the formation of the friction hill, which leads to increased deformation load, and this increases power requirement. It is worth mentioning that

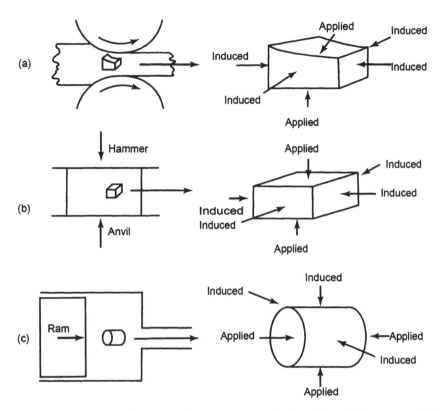

Fig. 9.4 General schematics and stress application patterns in (a) rolling, (b) forging and (c) extrusion processes

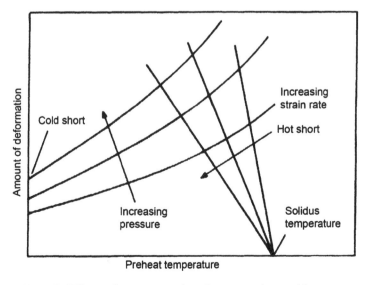

Fig. 9.5 Effects of pressure and strain rate on the working range

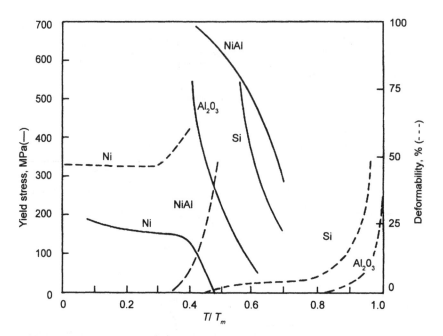

Fig. 9.6 Yield stress (—) and deformability (- - -) for various materials as a function of homologous temperature T/T_m (T_m = melting temperature)

hot rolling of a very thin workpiece can cause extremely high working loads, and thus, require extra power.

The ideal or theoretical pressure required for extrusion of a fully dense material is expressed as

$$P = \sigma_0 \ln r \qquad (9.1)$$

where, σ_0 is the yield stress and r is the extrusion ratio of the material.

This expression does not take into account the effect of friction or redundant work. These factors make the extrusion pattern rather inhomogeneous. In order to get the actual pressure, one has to place a multiplying factor α in the equation. Thus,

$$P' = \alpha \sigma_0 \ln r \qquad (9.2)$$

Figure 9.6 illustrates the yield stress and plastic deformability of pure metals (e.g., Ni), intermetallic compounds (e.g., NiAl), ceramics (e.g., Al_2O_3), and covalently bonded materials (e.g., silicon) as a function of homologous temperature. Note that NiAl is significantly stronger than Al_2O_3, Si and Ni, and that it softens and shows significant ductility at temperatures above approximately $0.8T_m$, whereas Si and Al_2O_3 soften in a similar temperature range but show significant ductility only above ~$0.8T_m$. The difference in these materials is obviously related to the type of chemical bonds.

9.4.1 Process

Figure 9.7 shows the three basic methods of hot extrusion of metal powders.

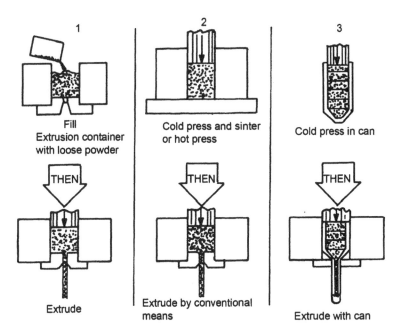

Fig. 9.7 Three hot extrusion methods for metal powders

1. Loose powder is placed in the heated extrusion container and extruded directly through the die. This method has been developed for the extrusion of certain magnesium-alloy powders.

2. The powder is cold compacted and then hot pressed. The hot pressed compact is then extruded as per the conventional method. Aluminium-alloy powder billets are extruded by this method.

3. The metal powders are placed in a metallic capsule or 'can', heated and extruded with the can. A green metal powder compact may be canned or the powder may be cold pressed into a metal can under moderate pressure. The can is outgassed by evacuation at room or elevated temperature and sealed off before the can and powder are heated for extrusion.

To prevent turbulent flow during extrusion, the end of the can has a conical shape to fit into an extrusion die with a conical opening. To avoid folding when the powder in the can is not packed very densely, a penetrator ram may be used as shown in Fig. 9.8. In such cases, the material of the can should:

- have the same stiffness at the extrusion temperature as the powder to be extruded
- not react with the powder
- lend itself to removal by etching or mechanical stripping.

Copper and low carbon steels are most commonly used as can material. Hot extrusion of powders encapsulated in cans was first developed for consolidating beryllium powder and powders of dispersions of fissile material in a matrix of zirconium and stainless steel.

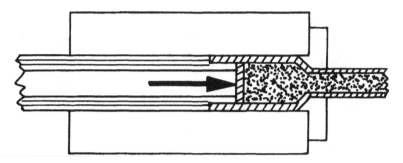

Fig. 9.8 Penetrator technique in powder extrusion to avoid folding

The method has been almost universally used for copper- and nickel-based dispersion-strengthened alloys. A hot extrusion process for producing seamless tubing from stainless steel powder was developed in Sweden.

9.5 POWDER HOT FORGING

There are two types of forging processes: impact forging and press forging. In the former, the load is applied by impact, and deformation takes place over a very short time. On the other hand, press forging involves the gradual build-up of pressure to make the metal yield. Powder forging can be broadly classified into:

1. Conventional powder forging
2. Preform powder forging.

Conventional forging as applied to original compacting of loose powder consists of 'canning' the powder in some type of metal container. Once the powder is within such an enclosure it can be evacuated, and eventually, the assembly is treated in a conventional forging press. After forging the can material can be removed by chemical or mechanical means. The resulting compact can have a very high density and may require no sintering if the powder was heated prior to the forging step.

Powder preform forging is a combination of powder metallurgy and forging. Powders used for preform forging are essentially those used for conventional powder metallurgy practice. Powder preform forging processes have developed along two distinct lines:

1. Compression of a preform which is very similar in cross section to the final part (Fig. 9.9a). This is known as hot repressing.
2. Compression of a preform of relatively simple shape into a part with a much different shape (Fig. 9.9b). This is similar to true forging.

The first process involves densification with little or no lateral flow, while the second achieves densification and shape change simultaneously through a large degree of plastic deformation and lateral flow. In true forging the associated lateral flow subjects the pores to a combination of normal pressure and shear (Fig. 9.10). This shearing action aids the normal pressure in closing up pores and lower pressures are required for densification of

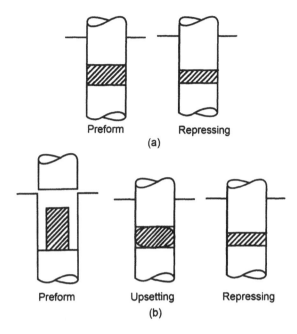

Fig. 9.9 Schematic diagrams of powder preform forging involving (a) hot pressing and (b) forging in a confined die showing upsetting and repressing modes

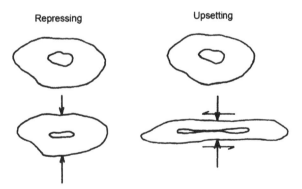

Fig. 9.10 Pore closure by repressing and upsetting (lateral flow)

the material. In addition, the shearing action increases the bond strength between opposite sides of collapsed pores, which enhances the soundness of the metallurgical structure. The lateral flow during forging involves a period of deformation in which the expanding vertical surfaces are free and tensile stresses can develop, leading to fracture. This is designated in Fig. 9.10 as the upsetting stage. The likelihood of fracture increases as the amount of lateral flow before reaching the die side walls increases. In case of hot repressing, fracture is prevented; the preform is initially in contact with the die side walls and tensile stresses do not occur. Rational design of preforms involves specification of the shape and dimensions such that the deformation during forging is sufficient to achieve full density and a sound metallurgical structure, but less than the amount at which fracture occurs.

The advantages of forging powder preforms are:

- High production rates
- No waste of materials as is usually associated with more conventional production methods
- Elimination of much or all finish machining
- Good surface finish
- Ability to form quite complex components in one finishing operation
- Randomly oriented fine grain structure
- Equiaxed orientation of physical properties
- Forging costs are lower than in conventional forging
- Lower forging load.

The powder forging process is suitable for the manufacture of a wide range of parts and competes with sintering in the area of longer sintered parts and with forgings over the small to medium-large range. While comparing the route by which traditionally forged and powder preform forged components are made, it is clear that whilst the basic number of production stages are similar, the number of actual forging steps varies significantly. In conventional forging the heated bar or billet stock is subjected to a number of forging blows in a series of dies to develop the final shape with transfer from one die to the next. With preform forging, however, a fully formed component can be produced with one forging stroke only—in one set of closed dies with an attendant reduction in actual forging cost and a greater improvement in press utilisation. In its relatively low density condition at forging temperature, the preform starts to plastically deform under relatively low forging loads. As the density approaches the 100% value, the required forging load increases. A great deal of the final forged shape is developed during the early stages of forming when the loads required are relatively low. By the end of the forging stroke when the load requirement is at its maximum, most of the shape detail has been achieved and the last stages are mainly hot compacting involving relatively little plastic flow. In cases where the preform has a form which corresponds closely to the finished part, hot working is essentially simple compacting, involving very little lateral metal flow.

9.5.1 Preform manufacturing and forging

Powder blending

In the conventional method of powder blending, the alloy ingredients and the lubricant are mixed. The alloy ingredients introduce chemical heterogeneity into the finished part because the powders can never be mixed ideally and sintering is not sufficient to permit complete inter-diffusion of the different elements. The lubricant decreases die wear, makes ejection of the part easier and acts to distribute the pressure throughout the part while it is being pressed. After pressing the part, the lubricant must be removed, which poses many problems. If the green parts are larger or dense, complete lubricant removal becomes extremely difficult and the entrapped lubricant either remains as an impurity in the part or erupts during sintering to cause defects in the parts.

Preform compacting
The design of the powder preform depends on the process. The preform design, i.e., its size, shape, weight, density and design tolerance, limits the methods that can be used to manufacture it. Preform weight control and speed of die cavity filling are critical production factors. Coarse, dense, regular-shaped powders would be easier to densify than fine, porous and irregular powders, and provided their lower green strength and sintering activity can be tolerated, they would be more desirable for both the preform compacting and preform forging operations. Preform density and density distribution become more important as the complexity of the part increases. On complex parts it is necessary to have the correct mass distribution in the preform to ensure proper metal flow during forming. Improper mass distribution could result in low density in some areas and overloading of tooling in other areas.

Preform sintering
Preform sintering is carried out after preform compacting and before forging operation. The sintered preforms are then normally cooled to room temperature and reheated to the forging temperature. In case of alloy steels, the prevention of oxidation is a serious matter. An alloying element in dilute solution is more difficult to oxidise than if it existed as the pure metal or as a rich master alloy. It would therefore be easier to prevent oxidation in the pre-alloyed powders than in mixtures of elemental powders. The details and precautions described in Chapter 8, Sintering Technology, are valid for preform sintering as well.

Preform forging
The deformability and oxidation characteristics of the preform are important for the forging operation. The deformability of a preform is a function of many variables—forging, temperature, preform design and the composition and characteristics of the powder constituting the preform. In general, the highly alloyed preforms would be more difficult to deform. As far as oxidation control of the preform is concerned, carbon-containing preform coatings are very common. Lower the density of the preform, the more serious the oxidation problem. In general, oxidation increases with increasing time and temperature of exposure, although the oxides become less stable and the carbon becomes more protective towards the metal, as the temperature is raised.

Figure 9.11 illustrates the influence of processing parameters like forging temperature, and speed, tool temperature and tool lubrication on the surface porosity of forged structural parts. Surface porosity originates as a result of rapid cooling of the surface of the preform when it comes in contact with the forging tool, and the resulting reduced forgeability in those zones. Only the optimisation and strict control of all these sintering parameters yield flawless powder forgings.

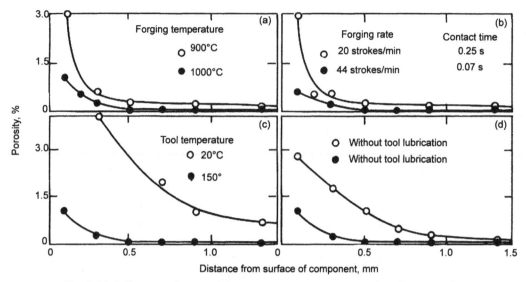

Fig. 9.11 Influence of processing parameters on components' surface porosity

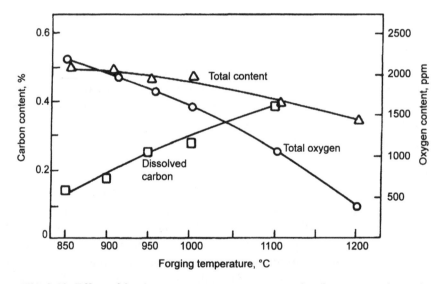

Fig. 9.12 Effect of forging temperature on oxygen and carbon content in steel

Figure 9.12 shows the effect of forging temperature on total carbon, dissolved carbon and total oxygen content in steel. It is evident that a temperature above 1120°C is required to dissolve admixed graphite into solution as well as to decrease the oxygen content.

As has already been indicated earlier, the forging temperature has an adverse effect on surface decarburisation and surface oxidation.

Once the part is hot forged, the ejection of the part from the die is important. The force required to eject the part from the forge die is dependent on both the preform temperature

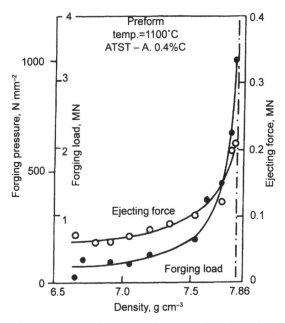

Fig. 9.13 Forging pressure and ejecting force as a function of residual porosity

and the density of the part. In general, the ejection force decreases with elevated temperature. Figure 9.13 shows the dramatic increase in forging tonnage applied and die ejection force required as full density is achieved. The usability of the forging die can be prolonged by adjusting both the forging pressure and the ejection force to be as low as possible. Since these two variables increase exponentially as zero porosity is reached, it is worthwhile to consider whether a totally pore-free material is essential for the application in question.

Hot forging of preforms often involves not only upsetting, but also extrusion. A typical example is hub extrusion (Fig. 9.14). The top surface of the hub is a free surface which undergoes bulging and tensile strains. Free surface fracture occurs on this surface when the strains reach the critical amount for fracture. These strains can be altered by changing the draft angle of the hub or by using a preform that partially fills the hub section of the die. Both free surface fractures and contact surface fractures may occur during extrusion forging of a hub on a cylinder. Kuhn and Ferguson made an in-depth analysis of the conditions under which such cracks are found and how they may be eliminated by design changes. A typical example of preform design could be

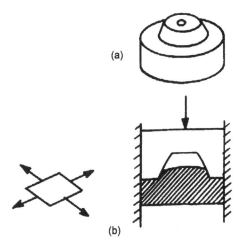

Fig. 9.14 Differential pinion gear: (a) Sintered preform and (b) Part under forging process

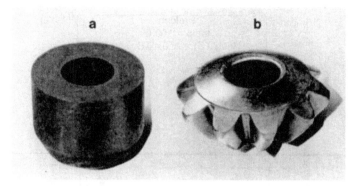

Fig. 9.15 Differential pinion gear: (a) Sintered preform and (b) Finished forged part

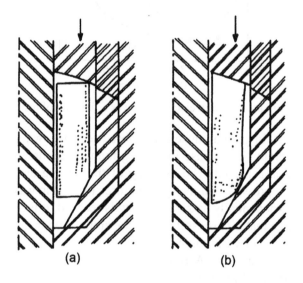

Fig. 9.16 Longitudinal section view of preforms and tooling for forging pinion gear: (a) Original preform which leads to crack formation at teeth and (b) Modified preform which leads to defect-free gears

understood from the case study for differential pinion gears (Fig. 9.15) in which the preform is flat-topped with a bevelled bottom surface. When this preform is forged, cracks develop in the partially formed gear during tooth formation. These cracks close up in the final stages of the forging stroke as the material is pressed against the die surface. However, oxidation of the crack surface and trapping of lubricant in the cracks may lead to structural weakness of the tooth surfaces. Kuhn proposed a modification of the preform contour involving a slightly tapered top surface and a rounded instead of a beveled bottom surface (Fig. 9.16).

ASTM standard B-797 is related to surface finger oxide penetration depth and presence of inter-particle oxide networks in forged steel parts of low alloy powder. Surface finger oxides are surface oxides that follow prior particle boundaries into a powder-forged part from the surface and cannot be removed by physical means such as rotary tumbling.

Example 9.1: Give the advantages and limitations of sinter-HIP versus HIP powder consolidation in a tabular from.

Solution: See Table Ex. 9.1.

Table Ex. 9.1

Sinter-HIP processing	HIP powder consolidation
Advantages	
No canning required	No pre-HIP of sinter
Sintered parts easily handled	Lower additive concentrations
Higher packing density in pressure vessel	Minimal grain growth at full density
Limitations	
Requires sintering to state of closed porosity	Requires coating or canning
Sintering aid addition can degrade properties	Lower packing density in pressure vessel
May develop coarse grains during sintering, reducing product strength	Preforms may be difficult to handle
Surface connected defects will not heal during HIPing	
Potential volatilisation of constituents or decomposition effects	

On the other hand, the inter-particle oxide networks are continuous or discontinuous. They follow particle boundaries in powder-forged parts. The testing is done with an optical microscope. Critical areas of forging are defined by the drawing of the applicable part on the purchase order. Specimens are taken from the powder-forged part in the condition in which it is to be supplied. The polished surface of the specimens must be parallel to the forging direction.

9.6 HOT ROLLING OF PM PREFORMS

Rolling of PM preforms must be seen in conjunction with powder roll compaction, already described in Chapter 5. PM hot rolling is a continuous process, where powder rolling and sintering operations are initial steps. As the sintered flat sheet is around 95% dense, the mechanism of hot rolling is akin to hot rolling of cast products. The two-high rolling mill is the most commonly used one. The four-high rolling mill is a special case of the two-high mill. The middle–low diameter rolls cause the lowering of rolling load which is given by the relation

$$\text{Rolling load} = \sigma_o w (R\Delta h)^{0.5} \tag{9.3}$$

where, σ_o is the yield strength of the metal, w is the width of metal workpiece, R is the radius of the roll and Δh is the reduction in thickness, also known as *draft*.

Example 9.2: Compare the stages involved in conventional drop forging and powder forging of a connecting rod. Give a historical perspective.

Solution: Currently, the vast majority of connecting rods are produced by drop forging; pearlitic malleable cast iron connecting rods are used in some applications. Powder forging competes with these processes. Such rods are being used by major automobile producers like Ford, Chrysler, General Motors, Toyota and BMW. Although work on PM forging started in the 1960s in UK, it was only in 1976 that the first powder-forged connecting rod was produced commercially for the Porsche 928V8 engine. The PM-forged rod for Porsche engine was made from a water-atomised, pre-alloyed, low-alloy steel powder (0.3 to 0.4 Mn, 0.1 to 0.25 Cr, 0.2 to 0.3 Ni and 0.25 to 0.35% Mo – Astaloy D) to which graphite was added to give a forged carbon content of 0.35% to 0.45%. The forgings were oil quenched and tempered to a core hardness of 28 HRC (UTS 835-960 MPa), followed by shot peening. In conventional forging, if the billet weight is 1.2 kg, the same for PM forging is only 0.7 kg and requires little machining.

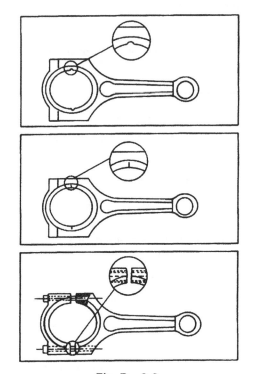

Good dimensional tolerance and precise control of part mass (± 0.5%) require fewer machining

Fig. Ex. 9.2

operations for PM-forged connecting rods as compared to the drop-forged ones. An additional development in PM-forged rods has been the elimination of additional expensive machining steps. This development involves fracture splitting of the crank end cap from the rest of the connecting rod. This leads to a low investment for new machining lines and to savings in production cost. In the patented process of Sinter Metallwerke Krebsoge (now Sintermetal), a V-shaped groove is formed during compaction of the forging preforms and this groove is closed to form a short crack during the forging operation. This controlled defect initiates splitting of the end cap during the fracture splitting process (Fig. Ex. 9.2). A comparison of conventional and PM-forged (fracture split) methods is given in Table Ex. 9.2.(Continued on next page)

Draft is sometimes expressed as a function of the starting stock thickness, called the *reduction*, i.e.,

Reduction = draft/original thickness of the metal

In addition to thickness reduction, rolling usually increases workpiece width. This is known as *spreading*. It tends to be most pronounced with low width thickness ratios and low coefficient of friction.

Table Ex. 9.2

Conventional method	Fracture splitting method
Alignment	Obsolete
Machining of outer contours	Obsolete
Deburring of outer contours	Obsolete
Preliminary grinding of side faces	Obsolete
Broaching of small and big end bearings	Obsolete
Machining of cap bolt bores	Machining of cap bolt bores
Drilling of oil bore	Drilling of oil bore
Machining of bolt head and nut seats	Obsolete
Separation of cap from rod	Fracture splitting
Grinding of rod/cap contact surfaces	Obsolete
Insertion of bolt sleeves	Obsolete
Assembly of cap and rod	Assembly of cap and rod
Finish-grinding of side faces	Finish grinding of side faces
Final drilling of small and big end bearings	Final drilling of small and big end bearings
Milling of bearing shell positioning grooves	Milling of bearing shell positioning grooves
Honing of big end bearing	Honing of big end bearing
Insertion of small end bearing bush	Insertion of small end bearing bush
Insertion of big end bearing shells	Insertion of big end bearing shells
Machining of balance pads	Obsolete
Sorting according to weight	Obsolete

It is not necessary that only fully dense PM sheets are produced. Porous powder-rolled strip, in particular stainless steel strip for filters and nickel strip for electrodes, are also produced. In order to obtain a completely dense strip, additional cold rolling and annealing steps are incorporated in the total cycle. The main problem with this type of mill is to synchronise the rate of rolling with that of sintering and annealing.

9.7 SPARK SINTERING

Spark sintering combines electrical energy and mechanical pressure to convert metal powder into a solid part of desired configuration and density. It is performed in air, vacuum or inert atmosphere using graphite or other specially developed materials for tooling. Powder is fed into the cavity of a punch and die assembly. The powders used are usually electrically conductive; however, non-conductive mixtures can also be used. High-density electrical energy of moderately high frequency AC and DC combined is passed through the powder while compacting it at relatively low pressures. Using both AC and DC energy simultaneously, accelerates and augments particle-to-particle bonding, which in turn promotes more uniform part density. At the end of the densification cycle, power is turned off, and pressure is maintained for some time while the part cools. The die stays relatively cold during this type of hot pressing. The time taken for the entire operation ranges from seconds to minutes depending on the material, part size, configuration, tooling and equipment capacity. This method was originally introduced by Lockheed Missiles and Space Company for the production of several beryllium components. It requires considerably less material and less subsequent machining than machining from hot-pressed blocks.

In spark sintering, pulsing the current helps removal of contaminants. Very high frequencies and high current densities induce rapid densification. Pressure up to 50 MPa is applied. The process time is short and thus the method can be categorised as rapid hot pressing method. In case of functionally gradient products this method of consolidations is very useful, as different powders can be placed in different locations and sintered as a single operation. In case of laminate metal-intermetallic composites too this method can be applied. Table 9.1 shows how pulse spark sintering is useful not only in shortening the processing period, but also in enhancing the properties. Pulse plasma method is very convenient for consolidating nanocrystalline powders. Figure 9.17 illustrates the schematics of this method. The electric current results in local Joule heating as well as heating from the creation of localised plasma between nearby powder particles. Since the amount of Joule heating and plasma heating are highest in less dense areas of the compact, regions that would not sinter well under only radiative heat and pressure experience enhanced sintering rates.

Table 9.1 Processing routes and mechanical properties of Ti-aluminides reinforced Ti-matrix laminate composites

Process	Heating rate, K s^{-1}	Holding temp., K	Holding time, ks	Aluminide phases	Tensile strength, MPa	Elongation, %
Vacuum hot pressing	0.33	1233	7.2	TiAl, Ti$_3$Al, Al$_3$Ti	543	1.8
Pulse spark plasma sintering	1.7	1173	0.6	Al$_2$Ti, TiAl, Ti$_3$Al	810	3.64

(Adapted from Mizuuchi K et al, *J. Jpn. Soc. Powder Powder Met.*, 54(8):560, 2007)

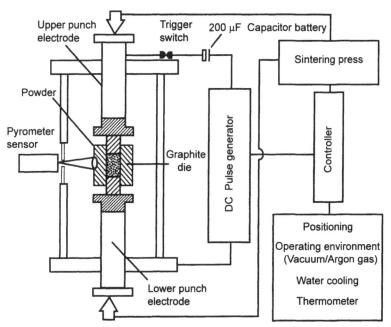

Fig. 9.17 Schematic representation of the equipment for pulse plasma sintering

SUMMARY

- Dynamic powder compaction is the consolidation process brought about by the impact of high-speed punch on powder.
- Hot pressing and hot isostatic pressing are suitable methods for densifying materials with poor sinterability, e.g., ceramics, cermets, etc.
- Sinter-HIP is a containerless consolidation method, which eliminates a major piece of equipment, reduces materials handling, provides greater control of the process and saves energy and labour costs.
- Powder hot forging is of two types: impact forging and press forging. The latter are classified as conventional powder forging and preform powder forging.
- The stages involved in powder preform forging are powder blending, preform compaction, preform sintering and preform forging.
- Spark sintering combines electrical energy and mechanical pressure to convert metal powder into a solid part of desired configuration and density.

Further Reading

Arunachalam VS and Roman OV, (eds), *Powder Metallurgy: Recent Advances*, Oxford and IBH Publishing Co., New Delhi, 1989.

Bose A, *Advances in Particulate Materials*, Butterworth-Heinemann, London, 1995.

Bose A and Eisen WB, *Hot Consolidation*, Metal Powder Industries Federation, Princeton, 2003.

Kuhn HA and Ferguson BL, *Powder Forging*, Metal Powder Industries Federation, Princeton, 1990.

Kuhn HA and Lawley A, (eds), *Powder Metallurgy Processing*, Academic Press, New York, 1978.

EXERCISES

9.1 PM Ni-base superalloy MA-6000 was hot isostatically pressed under different combinations of maximum pressure, holding period and maximum temperature.

Table P. 9.1

Max. temp, °C	Hold time, h	Max. pressure, MPa	Density, g cm^{-3}
950	4	-	8.20
1100	-	130	8.20
1150	0.2	-	8.17
1205	3	-	8.11

(a) Fill in the blanks in the table as best as you can and justify your answer.

(b) Give the full composition of this alloy. (Refer *Metals Handbook*).

(c) What type of starting powder was selected and why?

9.2 A PM Ni-base superalloy aerospace component which was hot isostatically pressed, was found to contain the following defects:

(a) Ceramic inclusions

(b) Metallic inclusions

(c) Voids and pores

(d) Prior particle boundary contamination.

Enumerate methods to minimise the above defects.

9.3 For high-performance liquid chromatography and fuel injection analysis, high-performance porous platinum electrodes are required. A simple powder compact is not preferred because of poor reproducibility. Suggest a method to produce such porous electrodes.

9.4 You are given a high purity oxide ceramic to get a fully dense compact by HIP. What atmospheres can you use?

9.5 An Al-8 Mg powder premix was subjected to attritor milling under argon for producing intermetallic strengthened alloys. The additives were PbO, SnO_2 and GeO_2 respectively. The resultant powders were then subjected to vacuum hot pressing under 100 MPa at 400°C for 1 hour. The billets were subsequently hot extruded at 400°C with a reduction ratio of 1:25. After hot pressing it was obtained that the oxides decomposed and formed intermetallics.

(a) Do you think all these oxides will be stable after attritor milling?

(b) Will the intermetallics be Al-based or Mg-based and what are they?

[Refer to the respective binary phase diagrams].

9.6 Kadam et al reported powder extrusion (extrusion ratio 16:1) of air-atomised lead, Pb-Zn and Pb-Al_2O_3 composites (*Powder Metallurgy Int.*, 15(1):20, 1983). Room temperature powder compacting

(384 MPa) followed by vacuum sintering at 280°C for 100 h gave almost full density. The products were then tensile loaded at room and elevated temperatures. The yield stress dependence is shown in Fig. P. 9.6. Curves 1 and 2 represent cast extruded and PM extruded lead, respectively. Label the remaining curves for Pb-Zn and Pb-Al$_2$O$_3$ composites. Why is the curve corresponding to cast/extruded lead much below that for the PM extruded one? Relevant property data are given in Table P. 9.6.

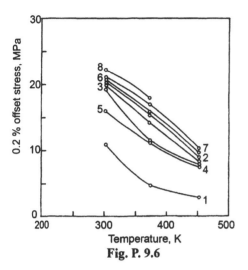

Fig. P. 9.6

Table P. 9.6

Property	Pb	Zn	Al$_2$O$_3$
Density, g cm^{-3}	11.4	7.1	3.98
Melting point, °C	328	420	2050
Surface energy, J m^{-2}	0.6	0.8	~0.9

9.7 Whitehouse and Clyne at the University of Cambridge presented experimental results and predicted ductility values of PM hot extruded Al-Al$_2$O$_3$ composites (Fig. P. 9.7) pertaining to spherical, angular and short fibre shapes. The curves describe the predicted values, while the specific points pertain to the experimental ones. For a pure matrix, an unalloyed aluminium powder was hot extruded under identical conditions. Identify the shapes of the reinforcing particulate.

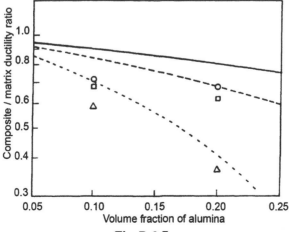

Fig. P. 9.7

9.8   A medium-carbon steel powder produced after water atomising was hot forged to nearly full density. It was noticed that inclusion shape control increased the transverse ductility and toughness but had little effect on the longitudinal properties. Explain this observation.

9.9   Prepare the flow diagrams for (a) the full density PM processing of Ni-base superalloy and (b) an oxide dispersion-strengthened (ODS) alloy of the same base.

9.10 Figure P. 9.10 shows the effect of HIP, hot die forging and isothermal forging methods on low cycle fatigue life of the Ni-base PM superalloy. Identify the methods.

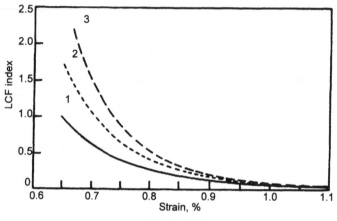

Fig. P. 9.10

9.11 Spray forming is described in Chapter 6. Describe the advantages of hot forged spray deposited preform in comparison to conventional hot forging of a cast product.

9.12 How can the force/power required to roll a PM strip of a given width and work metal be reduced?

9.13 Figure P. 9.13, shows the schematics of preparation of a hexagonal monofilamentory strand for fine superconductor NbTi-based filamentary material. In case the NbTi rod was proposed by PM route, describe the details starting from the powder stage. Do you think the PM rod will have premium over the cast and wrought one?

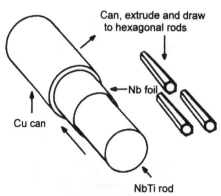

Fig. P. 9.13

9.14 What are the advantages and disadvantages of hydraulic hot die and isothermal forging presses for gas turbine discs and wheels?

9.15 It is observed that a thicker hot-rolled PM strip made of atomised nickel powder is possible than when made from hydro-metallurgical nickel powder. Comment on this statement.

9.16 What are the advantages of spray forming compared to powder hot forming?

9.17 Rapidly solidified alloy particulates containing precipitates are generally not hot consolidated to achieve full densification. Explain.

9.18 A WC-12Co cemented carbide with an initial WC crystallite size of 60 nm was subjected to pulse plasma sintering. What is the required approximate sintering temperature? If you increase the sintering temperature, will it be useful from the properties point of view? Why is conventional sintering not suited for such powders?

9.19 PM ferritic oxide dispersion-strengthened superalloy after elevated temperature treatment was found to promote pore formation. The superalloy was prepared through mechanical alloying under argon followed by hot extrusion. Give reasons.

9.20 You are asked to prepare a titanium intermetallic dense laminate composite based on titanium and aluminium foils through the hot pressing method. Describe various steps to be followed. Will the thickness of the foils of both the materials be similar or different? Why?

9.21 Which of the following is not a major advantage of uniaxial hot pressing?

(a) Decreased densification temperature

(b) Complex shape capability

(c) Fine grain size

(d) Minimum porosity.

9.22 Arrange the following powder consolidation methods in increasing order of the shear component of the applied stress: hot pressing, hot isostatic pressing, extrusion, quasi-isostatic pressing and sinter-forging.

9.23 What are the approximate magnitudes of heating rate and sintering time in spark plasma sintering? What is the current pulse time?

10

Secondary Treatments

LEARNING OBJECTIVES

- Why a sintered part should be sized or coined
- When and where to introduce machining operation in sintered parts
- How to impregnate porous parts with oil or resins
- Surface engineering of PM parts by steam treatment, coatings and shot peening
- Various heat treatments like hardening/tempering, case hardening and age hardening
- Joining of PM parts

In practice, PM components require closer tolerances, increased mechanical properties and features not possible by simply sintering. Most of the operations that accomplish these processes are performed on PM components in the same manner as on cast or wrought components. However, porosity frequently imposes limitations on some secondary operations.

In this chapter various types of secondary post consolidation treatments are described.

10.1 SIZING AND COINING

Sizing is done to refine dimensional accuracy, or to compensate for warpage or other defects which may occur during sintering. The sized parts will be straighter, dimensional tolerances will be closer, and the surface finish will be improved. Generally, little or no increase in density is achieved since pressures used are usually no more than the initial compacting pressure.

Coining, because of the use of higher pressures, increases the part density, in addition to improving dimensional accuracy. During the process, considerable plastic flow takes place, as a result of relatively soft sintered parts, where heating is involved. This process, therefore, increases the mechanical properties of the sintered product.

10.2 MACHINING

Sintered parts do have some porosity, and therefore, it is necessary to differentiate their machining behaviour from those of fully dense wrought products. The major differences are:

1. When a porous metal is machined, the depth of work hardening is more important than for the wrought metal; in the case of the former, pores create stress concentration.
2. The temperature at the tool end causes oxidation of the pore surface.
3. The surface porosity enhances tool vibration, which is submitted to fatigue.

Although PM parts are often specified because machining operations can be eliminated, in some cases it is more economical to leave certain part details desired for machining, rather than to incorporate them in the pressed configuration. Parts made by PM techniques generally require machining methods that differ from those used for wrought or cast parts of similar composition, because of the inherent porosity of PM parts. The following are some of the causes of machining problems in PM parts:

- Pores may be closed by smearing the metal surface
- Cutting fluids may cause problems by entering the parts
- Parts may become charged with abrasives when ground, honed or lapped.

Variations in machinability may be caused by differences in sintering conditions and by different alloy compositions. In steels the degree of surface carburisation or decarburisation would have significant effect on machinability. Machinability will also be affected if the parts are either over- or under-sintered. As parts increase in density, they begin to approach the same machinability as wrought and cast parts. This is due to the fact that porosity simulates a series of interrupted cuts. To partially eliminate or minimise this, many ferrous parts are copper infiltrated to fill the pores, or are impregnated with a polyester resin or wax. Various additives like sulphur, copper, lead, bronze, phosphorus and moly disulphide have been introduced in iron powder to obtain better machinability. Any addition to an iron mix which will have a hardening effect on the matrix will improve machinability. Copper improves the surface finish and also gives longer tool life. Sulphur in the form of iron sulphide improves tool life. Manganese in the form of manganese sulphide acts as a chip breaker and gives a smooth surface finish, but at the same time tends to decrease tool life. With quantities between 0.3% and 0.5% MnS, the wear is reduced by a factor between 5 and 20, and hence the solid lubricant addition can be decreased by 0.2%–0.3%. Phosphorus has the general effect of decreasing tool life, although smoother surface finishes can be obtained at faster machining rates.

The machining of sintered stainless steel parts is more difficult than machining wrought stainless steel parts. Poor machinability of stainless steel can be attributed to carbide precipitation resulting from improper sintering.

Aluminium sintered parts of more than 95% theoretical density do not pose much machining problems. However, improper lubricant removal before sintering, or excessive aluminium oxide in the powder may deteriorate machinability.

In case of sintered bronze—widely used as porous bearings—it is essential to maintain a good surface finish without closing the surface porosity. Care must be taken to minimise the number of machining operations. Extremely sharp cutting tools must be used and it must be seen that the tool gives a good surface finish free from the 'saw tooth' effect. Light cuts, no greater than 0.38 mm, may be used. It is difficult to mill, drill or thread-sinter bronze and maintain surface porosity. Grinding is definitely not recommended because of the extreme tendency to close the surface pores.

As already described in Chapter 5, there are geometric features in PM parts, e.g., threads, cross-holes, undercut, etc., which cannot be pressed but have to be machined. Sometimes it may be more economical or technically easier to introduce some geometric features into sintered parts through machining rather than pressing, even if the latter would be technically possible. Figure 10.1 shows the factors affecting the machinability of PM materials. Generally the machinability is influenced by three interacting systems: workpiece, tool and machining conditions. The system 'workpiece' will be described in more detail. The system 'tool' consists of the material and the tool geometry. The substrate and coating, if any, have to be adjusted to each other and to the geometry. The third factor— machining conditions—dictates the selection of machine system. It can restructure the choice of tools and cutting conditions. Cutting parameters, feed, speed and depth of cut and other factors

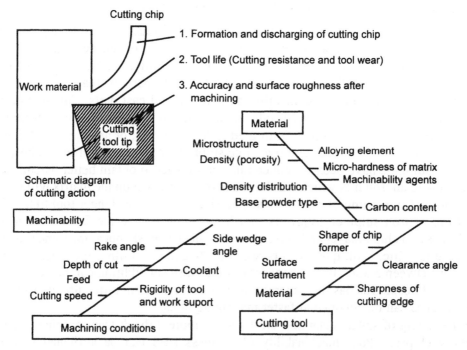

Fig. 10.1 Fundamentel problems in machining, and factors affecting the machinability of PM materials

have to be selected with respect to the application demands. Both cost efficiency and product quality have to be balanced for the optimum machining results.

10.2.1 Machining practice

Turning and boring: Carbide tools with a sharp nose point are best suited for such operations.
Drilling: Carbide or high-speed steel drills with a low right-hand helix angle prevent the drill from digging in. The cutting edges should also be dubbed to reduce the axial rake.
Tapping: Spiral pointed taps are most desirable because they throw the chips ahead and prevent them from driving into the pores.
Milling: Milling is generally difficult because of the tendency of the material to smear. To minimise smearing, it is recommended that dead sharp helical tool cutters with an axial rake be used so that the chips are sheared on an angle.
Shaping: Shaping is very similar to turning or boring, where a single point tool is employed.
Reaming: It is satisfactory for structural parts. However, it is not recommended if porosity is to be maintained, because of the tendency to smear the bearing surface.
Burnishing: Ball sizing or burnishing of holes maintains open pores, providing proper sintered dimensions are held.
Broaching: Broaching is not recommended if porosity is to be maintained. Standard draw broaching is recommended to obtain best tolerances and finish.
Grinding, Honing and *Lapping:* These operations are usually a finishing operation for heat treated materials; and only a small degree of pore closure occurs.

The standard machining charts for cast or wrought parts of the same composition can be referred and the recommended speed can be increased by 10% for PM materials.

Another solution to easy machining is to machine the parts after a pre-sintering operation at a low temperature (850–950°C for steels). The parts are then sintered and eventually sized.

Table 10.1 gives the approximate range of recommended cutting speeds for the turning operation for a number of ferrous and non-ferrous alloys.

10.3 IMPREGNATION

Controlled porosity permits powder metallurgical parts to be impregnated with oils or resins. The lubricant oil is filled in the interconnected pores, which imparts lubricating function to bearings and mechanical parts subjected to wear such as cams, gears and

---

Example 10.1 What are the special features of machining a PM part when compared to a wrought one of the same chemistry?

Solution: In PM parts, the cutting speeds may be significantly faster than those used for cutting wrought materials. However, the cutting feed may have to be reduced due to the difficulty in holding the part in the chucking device or to distortion of the PM part.

Table 10.1 Approximate ranges of recommended cutting speeds for turning operations

Workpiece material	Cutting speed, m min^{-1}
Aluminium alloys	200–1000
Copper alloys	50–700
High-temperature alloys	20–400
Steels	50–500
Stainless steels	50–300
Titanium alloys	10–100
Tungsten alloys	60–150

Note:

(a) The speeds given in this table are for carbides and ceramic cutting tools. Speeds for high-speed steel tools are lower than indicated. The higher ranges are for coated carbides and cermets. Speeds for diamond tools are significantly higher than any of the values indicated in the table.

(b) Depths of cut (d) are generally in the range of 0.5–12 mm.

(c) Feeds (f) are generally in the range of 0.15–1 mm rev^{-1}.

connecting rods. The choice of lubricant is important, as it affects the functionality of the system. The parts to be impregnated may be simply submerged in an oil bath for several hours. Reasonably good impregnation can be obtained if the oil temperature is maintained at about 80°C. The best results are obtained with vacuum impregnation. In this process, the parts to be impregnated with oil are placed in a basket in a vacuum chamber. After the air in the chamber and the parts have been evacuated, the chamber is flooded with oil. Next, the chamber is returned to atmospheric pressure, and the oil is drained from the chamber and the surfaces of the parts. Formation of gas bubbles can pose a serious problem in complete pore filling. The formation of such gas bubbles may be due to:

- Dissolved gases in the oil
- Gases formed by chemical reaction
- Dissolved gases in the sintered material.

In case of resin impregnation, most of the interconnected pores are filled with resins, usually polyesters of the thermosetting type. This process greatly improves the machinability of PM parts by filling the pores and increasing the density. A drawback of this process is that machined plastic impregnated parts cannot be heat treated. The process takes place in a suitable vacuum chamber. PM parts have a lot of potential for making pressure-tight components such as valves, pumps, meters, compressors, brake pistons and hydraulic systems.

The problem posed by the presence of pores in parts to be plated is eliminated by resin impregnation. Some PM parts which are subjected to such treatment include gun components, pole pieces, decorative automotive parts and outboard motor parts. Filling the parts with hardened resin prevents entrapment of other fluids which could later leach out and ruin the surface finish. Improved structural strength and the virtual elimination of internal corrosion are related benefits.

10.4 SURFACE ENGINEERING

The first step in selecting a surface modification treatment for a PM part, irrespective of its density, is to determine the surface and substrate engineering property requirements, such as:

- Abrasion wear resistance under conditions of low or high compression loading
- Resistance to scuffing and seizure
- Bending or torsional fatigue strength
- Rolling contact fatigue
- Resistance to case cracking (surface collapse)
- Resistance to corrosion.

The thickness of the engineered surface can vary from several millimetres for weld overlays to a few micrometres for physical or chemical vapour deposited coatings, while the depth of surface modification induced by ion implantation is < 0.1 μm (Fig. 10.2).

Table 10.2 summarises some of the typical surface treatment methods applied on a wide range of PM parts that are either fully dense or porous.

In the following sections some major surface modification methods are discussed. Thermochemical methods have been described separately in Section 10.5

10.4.1 Steam treatment

It is commonly used to improve the wear properties of ferrous PM components. It also provides improved corrosion resistance. During the process, all exposed surfaces, interiors and exteriors, are coated with hard black iron oxide, Fe_3O_4. The parts are first made free of oil or grease and subsequently placed in a forced convection furnace. The parts are first heated to 370°C to drive off moisture. Steam is then introduced into the furnace to purge the air from the furnace. The temperature is then raised to 510–540°C, when

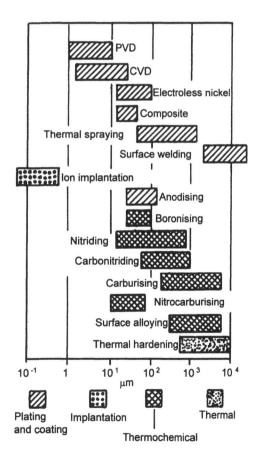

Fig. 10.2 Typical thicknesses of surface-engineered layers

Table 10.2 Typical PM components surface engineered to improve specific properties

Component	Material	Key property	Surface treatment
Injection-moulded computer parts	2% Ni/Fe	High load carrying capacity High cycle fatigue	Carburising
Cold forming tool	ASP23	Abrasive wear resistance	TiN-TiC CVD coating
Heavy duty transmission components	Cr/Mn steel	Rolling contact fatigue	Vacuum carburised
Shock absorbers	Sintered Fe	Frictions	Steam treatment
Gearbox synchromesh ring	Sintered Fe	Localised friction	Induction hardening
Precision gears	Sintered low-alloy iron	Good tribological properties; fatigue strength	Plasma nitriding
Automotive lock assembly fasteners	Alloy steel	400–800 h 5% NaCl salt spray corrosion resistance	Zn coatings
Bevel gears	Powder-forged alloy steels	Increased fatigue life	Shot peening

iron combines with the oxygen in the steam to form magnetite. The hydrogen from water is given off to the furnace atmosphere. Care should be taken that steam does not come in contact with the parts before the temperature reaches 100°C, as the parts will rust. After steam treatment the parts are immersed in oil to help both corrosion and wear resistance. The thickness of the oxide layer is limited and does not exceed 10 μm.

Steam treatment increases resistance to compression stresses and there is considerable increase in hardness. Dimensional changes are minimal, when compared to traditional hardening or case hardening heat treatment, and warpage is much reduced because of the low temperature of the treatment. However, the toughness gets lowered, particularly in parts with low relative density.

For iron–copper and iron–copper–carbon PM alloys, the steam treatment is always combined with some age hardening.

Figure 10.3 illustrates the variation in weight increase with respect to steam treatment period. At the beginning the slope is high, because the oxide formation happens on extended surfaces. When the smallest pore sections start to close, the slope changes to that typical for oxidation of non-porous materials. The rate of layer growth decreases with time, as a consequence of the continuous increase offered by the formed oxide layer to the migration of iron and oxygen atoms.

10.4.2 Coating

Coatings are useful because they can impart surface properties which cannot be produced in the body of a structure. Because of interconnected porosity in sintered parts, surfaces are exposed to the environment. Therefore, sintered products are more susceptible to corrosion damage than their cast or machined counterparts. A common method is to plate a sacrificial metal, e.g., zinc, to a part which is first sealed with a resin or some other filler.

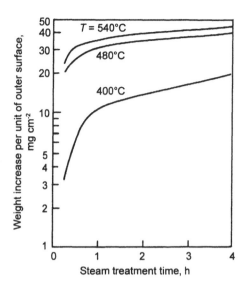

Fig. 10.3 Effect of steam treatment time on weight gain of a ferrous PM part at different temperatures

Powder metallurgy components can be electroplated with the same metals as cast and wrought components. For this, the PM component should have porosity sealed to avoid entrapment of plating solutions in the pores. Copper, nickel, chromium, zinc and cadmium plating are commonly used. Electroless nickel plating can also be used. Acid copper plating generally gives the best result on PM components because of the excellent throwing power. Nickel, applied either alone or over a copper base, is used to increase wear resistance and to provide a bright corrosion-resistant surface. Chromium plating is done for either decorative or wear-resistant applications. Cadmium plates faster and has greater throwing power than zinc. To plate a part with more than 10% porosity, it is important to first impregnate the pores with a resin or close them by peening the surface to exclude the plating salts from the pores of the part.

The organic metallic zinc coatings, commercially available in the UK under the trade names of Deltatone and Dacromet, involve the application of a coating of bichromated zinc flakes. The components are dipped in a mixture of zinc flakes with an epoxy resin and subsequently dried at 140°C. Such coatings can resist corrosion for up to 800 h in a salt spray environment. This being an electroless process, it is impossible for the coatings to incur hydrogen embrittlement effects of the substrate material.

Vapour deposition processes

There are two main methods in this group: chemical vapour deposition (CVD) and physical vapour deposition (PVD). Two important parameters of the coating process—process temperature and working pressure—largely determine the production equipment and the properties of the end products.

Chemical vapour deposition (CVD): It can be broadly defined as the process whereby a chemical reaction between gaseous reactants at or near the heated surfaces is used to deposit material on such surfaces. Table 10.3 gives the ranges of substrate temperature required for

Table 10.3 Hard materials produced by CVD and their temperature of formation

Material	Hardness* (kgf mm^{-2})	Formation reagents	Substrate temperature, °C
VC	2000–3000	$VCl_4 + C_6H_5CH_3 + H_2$	1500–2000
Si_3N_4	2500–3000	$SiCl_4 + NH_3$	1200–1600
SiC	2500–4000	$CH_3SiCl_3 + H_2$	~1400
B_4C	3000–3500	$BCl_3 + CH_4 + H_2$	~1300
HfC	1800–2500	$HfCl_4 + CH_4 + H_2$	1000–1300
Al_2O_3	2000–2500	$AlCl_3 + H_2 + CO_2$	800–1300
TiN	2000–2700	$TiCl_4 + N_2 + H_2$	650–1700
TiC	>3200	$TiCl_4 + CH_4$	800–1100

* 1 kgf mm^{-2} = 9.81 MN m^{-2}.

various ceramic coatings. The requirements of a relatively high temperature (~1000°C) in hot wall reactors for CVD coatings presents grain growth and distortion problems. This has been overcome by using N_2H_2-amine-$TiCl_4$ mixture to get TiCN deposition at temperatures as low as 600°C. This modification is known as moderate temperature chemical vapour deposition (MTCVD). Plasma activated CVD is a variant of the CVD process, where it is carried out in the temperature range 400–600°C, that is, much lower than the conventional process. A schematic sketch of the unit is shown in Fig.10.4.

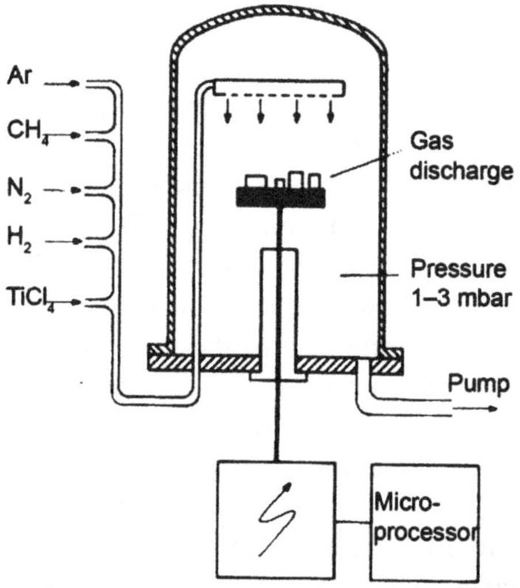

Physical vapour deposition (PVD): The coating material is evaporated from a solid source under a partial vacuum. The most advanced methods involve the creation of a glow discharge or plasma. Other

Fig. 10.4 Schematic of set up for plasma CVD

methods include sputtering, activated reactive evaporation and most recently, ion implantation. PVD coating on sintered cutting tool inserts provides a damage-free interface, as a consequence of which the total tool life is increased.

Of late, diamond coating has attracted much attention. Coatings with predominantly sp^3 bonds are termed 'diamond coating' (DC), while the term 'diamond-like coating' (DLC) is applied to films of carbon exhibiting a mixture of sp^3 and sp^2 bonds.

Thermal spraying

The spray material, which is fed into a heat source, is in the form of a wire or powder. The near molten particle is rapidly accelerated to impact a prepared substrate. On collision, the particle deforms to form a splat, cools rapidly and adheres to the substrate. Subsequent particle build-up gives rise to the coating. Although the term 'thermal' is prefixed with this type of coating, thermal spraying is a metallurgically cool process, as very little heat is carried over to the base material. With due precautions even polymers can be successfully thermal sprayed. Figure 10.5 illustrates various modes of thermal spraying. Traditionally, metals that could be easily drawn into wires are used as feedstock. The extensive use of powder mixtures has greatly expanded the choice of spray materials. The production of fully alloyed, smooth, spherical powders has been made possible due to improved spray rate and consistency, leading to high-quality coatings.

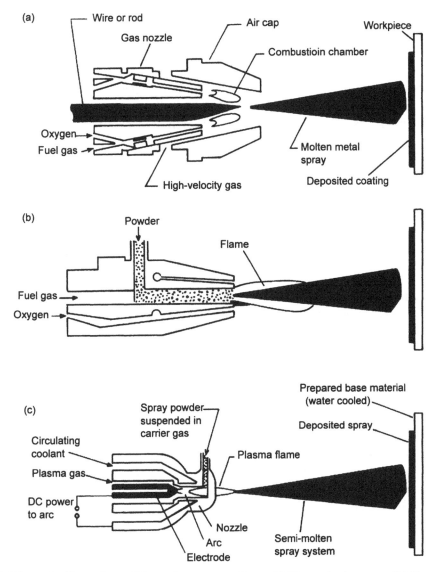

Fig. 10.5 Schematic illustrations of thermal spray operations: (a) Thermal wire spray, (b) Thermal metal powder spray and (c) Plasma spray (Reproduced with permission from Kalpakjian S and Schmid SR, *Manufacturing Engineering and Technology*, 4th edn, Pearson Education, Inc., NJ, 2000)

10.4.3 Shot peening

Significant improvement in fatigue properties can result from the formation of harder and stronger surfaces. This causes favourable compressive residual stresses on the surface which can be measured by X-ray methods. Shot peening is a convenient method to impart residual compressive stress on the surface. Here the sintered parts are tumbled in a medium of steel balls of varying diameters according to the size of the parts.

10.5 HEAT TREATMENT

Heat treatment of PM parts is an important secondary operation. With the emergence of high compressibility powders, the role of heat treatment in enhancing properties of PM alloys has been better appreciated. As the density approaches the theoretical value, as in PM forging, compressive residual stress patterns, similar to those in wrought steels, are created. This improves properties like fatigue strength, impact strength and hardenability.

10.5.1 Hardening and tempering

The technological aspects of heat treatment of steel is discussed in this section. The physical metallurgical aspects are discussed in Section 12.1.1.

The response of steel to heat treatment depends on its thermal conductivity, which in turn is dependent on the surface area. In wrought alloys, because of the high weight-to-volume ratio, heating and cooling rates are fast. In low-density PM steels, the slow removal of heat inhibits the hardenability and slack quenching or shallow hardening results.

As with conventional steels, the physical properties of sintered steels must exhibit a consistent response to heat treatment. A steel's response is measured by its hardenability, which is quantified using the Jominy end-quench test. Hardenability is expressed as the ideal diameter, Di, of a bar of steel which will under specified conditions harden to the centre to a condition corresponding to 50% martensite in the microstructure. Each alloying element in the steel contributes to hardenability, depending on the multiplying factor. Published data on the magnitudes of the multiplying factors for specific elements vary considerably. A number of investigations were reported by Ford Motors which indicated that the multiplying factor determined by Doane enabled the most consistent Di calculations, for both conventional and PM steels.

In the case of a homogeneous microstructure, when the master alloy additives are distributed uniformly, the PM steel's response to heat treatment is consistent and corresponds to the Di value calculated from the composition. The alloying efficiency N_{AF} can be defined as $N_{AF} = 100(Di_{exp}/Di_{cal})$. Clean, low-oxygen base powder, without a coarse size fraction, is desirable for good alloying.

With respect to high hardening of steels at a lower cost, manganese and chromium are the leading candidates. For some elements of interest, Table 10.4 shows the melting point, the Grossman factor at a concentration of 1 wt% and the diffusivity in austenite at 1000°C, related to the self-diffusivity of iron at the same temperature. As a rough guide, the table also shows the

Table 10.4 Important data for elements alloying with iron

Element	Melting point, °C	Grossman factor at 1% level	Diffusivity, $\left(D_x^{(\gamma)}/D_{Fe}^{(\gamma)}\right)$ 1000°C	ΔG (oxide), kJ mol⁻¹
Mn	1243	4.5	2.5	−586
Cr	1845	3.1	5	−544
Mo	2000	3.7	5	−314
Si	1423	1.7	10	−678
Zn	419	-	3	−418
Cu	1083	1.7	1	−155
Co	1495	(~2)	0.5	−272
Ni	1455	1.3	0.5	−251

free energy of oxidation for 1 mol of oxygen at 1000°C. It can also be seen that both manganese and chromium have favourable diffusivities, although their oxygen affinity is high. Among the elements with low oxygen affinity, nickel and cobalt have low diffusivity in austenite.

In steels sintered out of partially pre-alloyed powders, transformation during heat treatment is rather complex due to compositions varying from plain high-carbon steels to that of high alloying low-carbon steels. Consequently, incubation periods are short while transformation periods are long, and initial rates of transformation are high, but fall progressively as transformation spreads.

Figure 10.6 shows the hardenability response of SAE 1080 steel. This illustrates the poor thermal conductivity of low-density parts. As quenched surface hardness decreases with density, the depth of hardening drops off.

Various methods of heating may be used to austenitise sintered parts, including induction heating; the only limitation being in the use of salt baths. Quenching is usually done in oil. The disadvantage of oil for quenching is that it is less severe than brine or water, which means that one must quench from a higher temperature to get equivalent hardening. However, in one way it is advantageous, as it reduces the amount of distortion and possibility of cracking.

In sintered steels, the most common alloying elements are carbon and nickel. Addition of copper increases both hardness and tensile strength in the sintered condition. The effect of copper addition in relation to combined carbon content is shown in Fig. 10.7. Increasing nickel content significantly increases heat treated tensile strength. However, this benefit begins to taper off when the addition exceeds 2%.

The heating atmosphere is significant in controlling the end properties of PM steels.

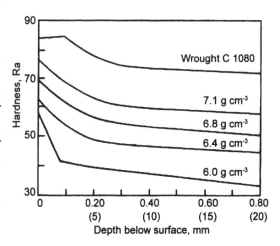

Fig. 10.6 Effect of density on the hardenability of SAE 1080 steel

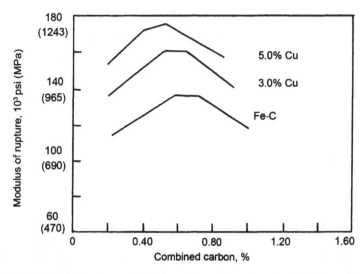

Fig. 10.7 Effect of carbon and copper on heat treated strength of sintered steel

Most heat treating is done in an endothermic atmosphere which contains approximately 20 CO, 35 H_2, 0.3 CH_4 and the balance N_2. Other gases are usually added to adjust the carbon potential to meet requirements for a part. Nitrogen-based atmospheres are also getting popular as they give better part uniformity.

A distinctive feature of sintered steel is that by varying the cooling rate of the steel in the cooling zone of the mesh belt continuous furnace, a wide range of microstructures can be achieved.

The specific surface of the iron powder used for developing sintered steels is also important, as it can be seen that both the strength and the martensite content are linear functions of the specific surface (Fig. 10.8). This infers that surface diffusion plays a prominent role in distributing the alloying elements in steel, particularly nickel.

Hardening is generally followed by tempering at an appropriate temperature to obtain the correct balance between hardness, strength and toughness. The normal tempering range is 150–650°C, depending on the properties desired.

The tempering process involves the decomposition of martensite to the stable

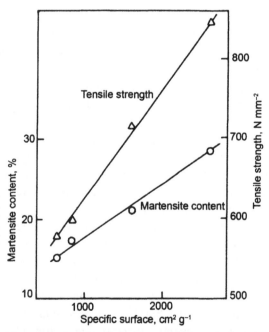

Fig. 10.8 Tensile strength and martensite/bainite content versus the specific surface of the base iron powder for a steel with composition: 1.75% Ni, 1.5% Cu, 0.5% Mo, 0.5% C at a density of 7 g cm^{-3}. Sintering temperature 1120°C; sintering period 1 h

phases of ferrite and Fe$_3$C. For low-carbon steels, since M_s and M_f temperatures are quite high, the martensite obtained is essentially a tempered one.

For fully hardened carbon and low-alloy steels containing 0.2% to 0.85% C and less than 5% total alloying elements, the tempering parameters (i.e., temperature and time at the temperature) can be estimated by using the *Holloman–Jaffe* (H–J) parameter, when the tempering temperature is in the range 345–650°C. The expression for the H–J parameter is

$$\text{H–J parameter} = T(\log t + 18) \times 10^{-3} \qquad (10.1)$$

where, T is the absolute temperature and t is the time in hours. Using the data given in Figs. 10.9 and 10.10, we can determine the possible combination of tempering time and temperature.

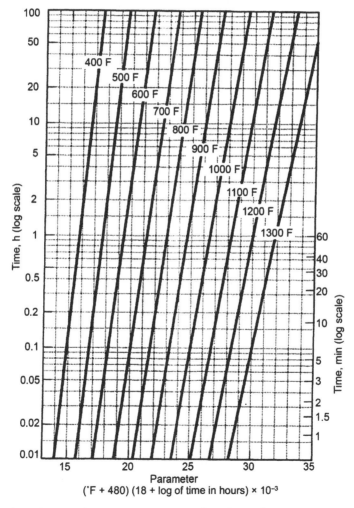

Fig. 10.9 Tempering time and temperature as a function of tempering parameter of steel (Reproduced with permission from Kern RF and Muess MC, *Steel Selection*, John Wiley and Sons, Hoboken, NJ, 1979)

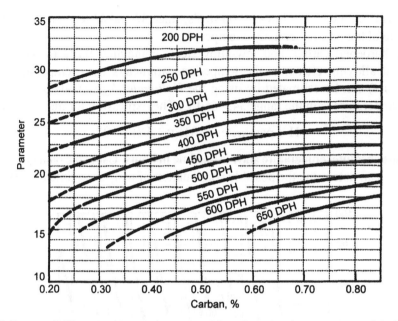

Fig. 10.10 Hollomon–Jaffe tempering parameters as a function of carbon content and desired hardness for plain carbon steels (Reproduced with permission from Kern RF and Muess MC, *Steel Selection*, John Wiley and Sons, Hoboken, NJ, 1979)

Example 10.2: In a PM plant, crank-shaft sprocket is produced from sintered steel of composition 0.8 Mo, 2 Ni, 0.6 C in heat treated (quenched from 840°C in oil followed by tempering) condition. Due to the high price of molybdenum, a plant reduced its content in steel to 0.3%, keeping other alloying additions the same. It was noticed that the fall in mechanical properties after heat treatment was marginal.

(a) For better productivity what change in heat treatment method do you suggest?

(b) Will the mechanical properties be the same for either heat treatment method? Comment.

(c) In case the addition of nickel is altogether avoided, will the mechanical properties be affected drastically? Comment.

(d) Do you think there will be any change in the mechanical properties of the heat treated sintered steel test pieces and the finished PM sprocket?

Solution

(a) Induction hardening followed by oil quenching.

(b) The induction hardened PM parts will have some inferior properties when compared to the conventional hardened and tempered ones. The reason is inadequacy in proper control of heat treatment parameters in the former.

(c) Yes, because the hardenability is lowered. Some bainite might be present.

(d) The mechanical properties of the PM part will be somewhat lower than those for test pieces. This is attributed to some variation in the sintered porosity in the PM part, which is generally absent in simple single-height test pieces.

10.5.2 Case hardening

Under similar carburising conditions, the diffusion depths in steels are much deeper when the porosity level is higher; they are far above those found in wrought steels. Carbon and nitrogen are the most common case hardening elements. For PM steels, classical gas carburising atmospheres for carbonitriding are adopted. Figure 10.11 illustrates a typical case of depth variation in C1018 steel. It is obvious that lower the sintered density, greater the case depth. This is due to the greater porosity, which allows for greater gas penetration. It can be concluded that it is difficult to achieve maximum physical properties in low density parts. With the deep carbon penetration in case of low densities, the core properties would be approximately the same as the case properties.

The case hardening operation is invariably followed by oil quenching and tempering. This is done in order to reduce internal stresses and improve the toughness. Because of the development of compressive processes in the surface layers during case hardening treatment, the fatigue strength of the steel is also increased. The most common form of case hardening is through case carburising. In PM, gases are affected at a temperature normally between 825°C and 925°C. The transport of carbon from the carburising medium invariably takes place via a gaseous phase, usually carbon monoxide. The methods are (Fig. 10.12):

- *Direct hardening:* The part is quenched straight from the carburising medium.
- *Single quenching*: Heating and quenching of the carburised parts is done after first allowing them to cool to room temperature from the carburising treatment.
- *Double quenching*: This consists of a direct quench and a requench from a lower temperature.

The lower temperature range 780–820°C relates to the range of hardening for the case. Apart from time and temperature, the depth of carburisation depends on the carbon potential of the

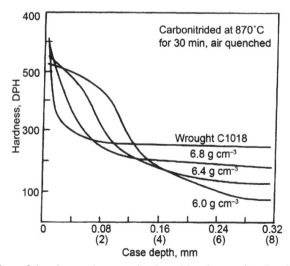

Fig. 10.11 Effect of density on the case depth and hardness of carbonitrided PM steel

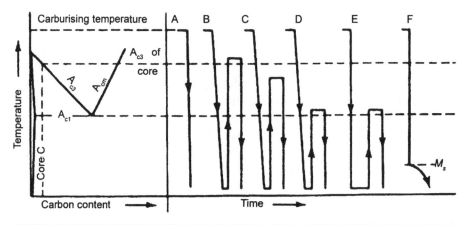

Fig. 10.12 Various heat treatment schedules followed for carburising steels

Treatment	Case	Core
A	Unrefined; solution of excess carbide favoured; austenite retained; distortion low to medium	Unrefined but hardened
B	Refined; solution of excess carbide favoured; austenite retention promoted in highly alloyed steels	Refined; maximum core strength and hardness
C	Refined; some solution of excess carbide	Partially refined; stronger and tougher than treatment D
D	Refined; excess carbide not dissolved	Unrefined; soft and machinable
E	Refined; some solution of excess carbide; austenite retention minimised	Low hardness; high toughness; machinable
F (Interrupted quench; marquenching)	Unrefined; solution of excess carbide favoured; austenite retained; distortion minimised	Fully hardened

carburising medium and on the composition of the steel. Higher the carbon potential of the medium, higher the carbon concentration at the surface of the steel, when equilibrium has been established, and deeper the carburising depth.

The validity of the simple diffusion equation $x = k \sqrt{t}$ (x is in mm and t in hours) has been established and the k values are 0.34, 0.41 and 0.52 at 875, 900 and 925°C, respectively. For example, a carburising treatment at 900°C for 12 h would give a case depth of 1.42 mm. Details of the carburising atmospheres and their thermodynamics can be found in the *Metals Handbook*.

The *carbonitriding process* is a modification of the straight gas carburising process, where anhydrous ammonia is added to the atmosphere of the furnace along with the hydrocarbon gas and endothermic gas. At the carburising temperature, the ammonia dissociates into hydrogen and nitrogen with some of the nitrogen combining with the iron and the hydrogen being given off to the furnace atmosphere. Higher the carburising temperature, lower the nitrogen taken into the part. In general, where light case depths of extremely hard surfaces are desired, carbonitriding is the most desired practice.

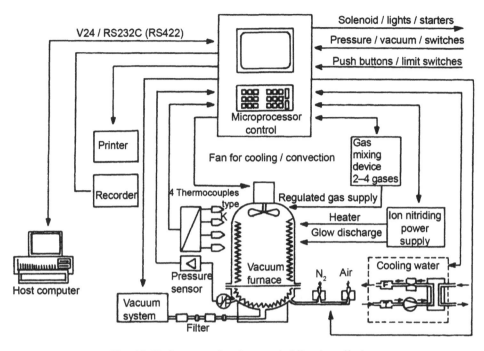

Fig. 10.13 Layout of a plasma nitriding installation

Nitriding of steel is a very attractive process. It is a ferritic thermochemical treatment involving the introduction of atomic nitrogen into the ferrite phase in the temperature range 500–590°C, and consequently no phase transformation occurs on cooling to room temperature. Nitrogen can be incorporated in liquid (tufftriding), gas (ammonia) or plasma (nitrogen–hydrogen) medium. In case of plasma nitriding (also called ion nitriding), N_2 and H_2 are fragmented to ions and radicals by the glow discharge plasma surrounding the part to be surface modified. The ions diffuse inside the surface because of high temperature. The part temperature is maintained by ion or neutral (argon) bombardment and by external heating. In case the reactor is small, external heating may not be required. Figure 10.13 illustrates the layout of a multifunctional plasma nitriding installation.

10.5.3 Age hardening

Age hardening or precipitation hardening is an important heat treatment for many aluminium alloys. Similarly, parts made from iron and copper or iron, carbon and copper can be age hardened, since the solubility of copper in iron decreases with temperature. In 85Fe-15Cu alloy, the tensile strength can be raised as much as 170 MPa by the precipitation hardening process. Sintered aluminium alloys, after solution hardening are invariably subjected to sizing, prior to ageing treatment. Most of the sintering furnaces for aluminium alloy parts also have a provision for quenching. This avoids additional solution treatment in a separate furnace, thus offering energy efficiency.

Example 10.3: Many automobile manufactures use piston rings made of martensitic stainless steel. Describe the production of such rings on a mass scale by PM route. Describe various surface treatment alternatives for such rings.

Solution: Piston rings in automobiles help:

- to seal hot gases
- as a heat transferring media
- to lubricate film metering device
- as a sliding support for piston

Generally there are three piston rings in a piston:

- Top compression ring for sealing
- Second compression ring for sealing and oil scraping
- Oil control ring for oil film control.

Some part of the material of the ring was generally SG Iron, but of late martensitic stainless steel is more common, being mainly developed in Japan.

For mass production of PM piston rings, the aim must be to produce hollow pipes and to then slit the rings out of them. The best solution for this is to cold isostatic press the powder in the form of hollow cylinders, followed by sintering and machining. The machined cylinder is then slit into slices.

The surface engineering of a piston ring can be done in any of the following ways:

- Plasma spraying
- Hard chromium plating
- Gas nitriding

In case of plasma spraying, molybdenum alloys are sprayed on steel either as in-lay or over-lay.

Advantages of plasma spraying

- Very high flame temperature upto 10,000°C; even ceramics can be melted and sprayed
- Scuff, wear and thermal resistant coatings
- Good control of process parameters

Disadvantages of plasma spraying

- Bonding is only mechanical; bond strength is less compared to metallurgical coatings (diffusion type coatings)
- Differential thermal property
- Thermo-chemical degradation
- Cost

Chromium coating is produced by electro-deposition from chromic acid bath (CrO_3). It provides hard wear resistant coating of minimum 800 $HV_{0.1}$ hardness. Piston rings can be selectively plated on ID, OD or all over the surface. (Continued on next page)

Advantages of chromium coating

- Good wear and corrosion resistance
- Very hard deposit
- Resists galling and abrasive wear

Disadvantages of chromium coating

- Poor throwing power
- Poor plating efficiency
- Hardness drops above 400°C
- High rejections during process
- Generation of chrome effluent
- Scuffing resistance less than plasma coating
- Fatigue strength of base metal is reduced about 50%

In case of gas nitriding, the process is carried out in gaseous ammonia or nitrogen. Nitriding of martensitic stainless steel requires special depassivation and activation techniques, e.g., dry honing and activation using NH_4Cl. An essential feature of the steel is that it must contain nitro-alloying elements, e.g., Cr, Al, Mo, V, etc.

Advantages of nitriding

- High hardness
- Good wear resistance
- High fatigue strength

Disadvantages of nitriding

- Formation of white layer (brittle layer) which needs to be removed
- Longer nitriding period

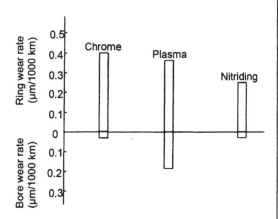

Fig. Ex. 10.3

Figure Ex. 10.3 shows the ring wear and bore wear data of some surface-engineered piston rings. The summary of these three surface engineering processes in relation to various properties of the ring is presented in Table Ex. 10.3.

Table Ex. 10.3 Coating comparison of piston rings

	Gas nitriding	Plasma coating	Chrome plating
Wear resistance	Best	Better	Good
Scuff resistance	Good	Best	Better
Corrosion resistance	Good	Better	Best
Fatigue strength	Best	Better	Good

10.6 JOINING

All major joining processes can be equally applied to sintered products. Powder metallurgy itself makes it possible to obtain complex shapes, and hence the joining operation is not often required. Sometimes, the very presence of pores in sintered parts may pose problems during joining, for example, brazing is unsuitable because the molten brazing alloy prefers to infiltrate the pores by capillary action. However, in case of necessity the parts to be joined should first be infiltrated using an alloy with melting point equal to or higher than that of the brazing alloy, or by using an excess of brazing material to infiltrate the pores during the joining process.

The combination of sintering and brazing for components has become attractive. A typical example of sinter-brazing is planetary carriers for automatic transmission in automobiles. The conventional route involves eight processes, including stamping, bending, welding and machining of hot-rolled sheet steel. In the powder metallurgy route only three stages—powder compaction, simultaneous sintering/brazing and finishing— are involved. Figure 10.14 illustrates the schematic configuration of the part. The PM route gave rise to 20% improvement in the torsional fatigue strength. In addition to this, reduced deformation during operation and improved wear characteristics between pinion shaft and gear were noticed.

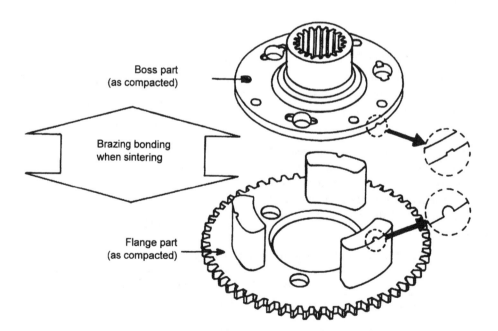

Fig. 10.14 Schematics of sinter brazing of planetary carrier for automatic transmission in an automobile (Courtesy: Hyundai-Kia Motors Corporation, Korea)

As far as welding of PM parts is concerned, most conventional welding methods (TIG, MIG, electron beam, resistance, projection and friction) are used. There are no special technological problems in joining sintered to wrought steel parts so long as the joints are correctly dimensioned and the operational parameters of the welding process are well selected. Best welding results for PM steels are achieved when the component density is 6.8 g cm^{-3} (88% of theoretical density) or higher, and the carbon content is held to the minimum practical for the application. Higher densities yield greater weld strengths, while higher carbon levels tend to make welds more brittle.

Another method for joining is equivalent to 'shrink' fit by using two materials with different growth characteristics during sintering. PM parts can also be joined during the infiltration process by assembling the component parts and infiltrating the assembly during sintering. Sinter welding consists of joining two metal parts by inserting a suitable powder of the same metal between the parts to be joined, and then hot pressing the assembly in a suitable atmosphere. The process can be modified to 'activated sinter welding', in which reactive metal hydride decomposition and hot pressing are combined.

Example 10.4: Generally cemented carbide cutting tools are brazed on steel shanks. A filler metal of typical composition 52Cu-38Mn-5.5Ni is used. Answer the following:

(a) What is the role of Ni in the filler metal?

(b) The use of a composite filler metal, consisting of a sandwich of two layers of a silver base filler metal and a copper shim is often recommended. Why?

(c) Why are the hydrogen-sintered cemented carbides poorly wetted by copper?

(d) What are the other reasons for weak brazing?

(e) Induction brazing is carried out for mass production. What limitation does it have?

(f) What composition/microstructure of cemented carbide is preferred for brazing?

Solution

(a) Increases bond strength.

(b) The ductile copper layer diffuses during brazing to decrease the residual stress generated in the joint.

(c) Alumina is often used as packing in the sintering boats. Cobalt binder in the cemented carbide gets contaminated by aluminium. Aluminium during brazing gets oxidised to alumina, which prevents brazing.

(d) Draining of cobalt from the interfaces and also of carbon cause η-phase formation.

(e) Since the method is fast, the relief of internal stresses is problematic.

(f) Coarse-grained carbides with a relatively larger proportion of cobalt binder are preferred, as they minimise brazing cracks.

SUMMARY

- Sizing of a PM part is performed to refine dimensional accuracy, or to compensate for warpage or other defects, which may occur during sintering.
- Porosity in PM parts makes the machining operation more challenging.
- Various additives like sulphur, copper, lead, bronze, phosphorus and molybdenum disulphide are introduced in iron powder to obtain better machinability.
- Impregnated PM parts have great potential as pressure-tight compacts such as valves, pumps, meters, compressors, brake pistons and hydraulic systems.
- Chemical and physical vapour deposition methods are in vogue. Other methods are sputtering, activated reactive evaporation and ion implantation.
- Shot peening operation improves fatigue properties.
- In low-density PM steels, the slow heat removal inhibits the hardenability and shallow hardening results.
- In PM steels, by varying the cooling rate in the cooling zone of the continuous mesh belt furnace, a wide range of microstructures can be achieved.
- The most conventional welding methods for PM parts are TIG, MIG, electron beam, resistance, projection and friction welding.

Further Reading

Bradbury S, (ed.), *Powder Metallurgy Equipment Manual*, 3rd edn, Metal Powder Industries Federation, Princeton, 1986.

Burekowski T and Wienshon T, *Surface Engineering of Metals: Principles, Equipments, Technologies,* CRC Press, Boca Raton, 1998.

Krauss G, *Steel: Heat-treatment and Processing Principles*, ASM International, Materials Park, OH, 1990.

Salak A, Selescka M and Danninger H, *Machinability of Powder Metallurgy Steels,* Cambridge International Science Publishing, Cambridge, 2005.

Thelning KE, *Steel and its Heat Treatment*, 2nd edn, Butterworth Heinemann, Oxford, UK, 1984.

EXERCISES

10.1 Explain why the specific energy in grinding operations is much greater than in conventional machining.

10.2 Unlike conventional PM parts, PIM parts cannot be oil impregnated or metal infiltrated. Why?

10.3 What current density is required to plate 20 μm of chromium on a surface in 20 minutes from a Cr^{2+} solution?

10.4 How long does it take to electroplate 225 g of nickel from a Ni^{2+} electrolyte with a current of 160 A?

10.5 An average current density of 100 A m^{-2} is used to build up an anodised coating on aluminium. The Al_2O_3 is to be 1 μm thick. How much time is required? Density of aluminium is 3.8 g cm^{-3}.

10.6 What makes electrodeposits hard?

10.7 What is the current efficiency in the deposition of 0.132 g of Cr from a CrO_3 solution ($n = 6$) when a current of 6.78 A is used for 20.8 min?

10.8 It is desired to electroplate a sintered ferrous part with 25.4 μm of nickel. The area of the part is 77.9 cm² and the current efficiency is 98.4%. How many minutes will be required if the current is 6.40 A? The density of nickel is 8.90 g cm⁻³.

10.9 Why are very thin plates of copper followed by nickel recommended as base coats for chromium electroplate on steel?

10.10 Sketch schematically the variation in hardness from the surface to the centre for cylindrical parts of the following steels that have been quenched from the austenite region:

(a) A low hardenability steel containing 0.6 wt% C.

(b) A medium hardenability steel containing 0.6 wt% C.

(c) A high hardenability steel containing 0.6 wt% C.

[Hint: Carbon content is constant, but other alloying addition elements are in different quantities.]

10.11 Refer to the *Metals Handbook* for the Jominy End quench hardenability data for Fe-1.8Ni and 0.5Mo-0.2C wrought steel. Position your curves for PM preforms of same grade steel prepared from pre-alloyed and premixed powders. Which type of powder offers better hardenability and why?

10.12 A tempering treatment for a low-alloy PM steel specifies 3 h at 300°C. What temperature would be required to achieve the same tempering effect in 1 hour.

10.13 A PM high-speed steel (18W-4Cr-0.36C, balance Fe) when quenched to form martensite retains about 5% austenite by volume. Describe the tempering treatment you will use to eliminate the retained austenite and describe the resulting microstructure.

10.14 Sketch schematically the variation in hardness from the surface to the centre of the carburised and quenched steel cylindrical part that has a surface composition of 1.0 wt% C and an interior composition of 0.2 wt% C.

10.15 Using thermodynamic data from the *Metals Handbook*, calculate the equilibrium P_{CO}/P_{CO_2} ratio, during carburisation of steel at 870°C in a mixture of CO and CO_2. Also, determine the same using Ellingham diagram. [Note: Surface oxidation of iron may prevent penetration of C if P_{CO_2} is too high.]

10.16 A piece of sintered steel containing 0.40% C was exposed in a carburising atmosphere for 10 min at 1000°C. What is the carbon concentration at a depth of 0.025 cm, if the carbon concentration at the surface is 1.00 wt%? $D_c = 3.1 \times 10^{-7}$ cm² s⁻¹ at 1000°C.

10.17 Calculate the depth or distance from the surface at which the concentration of carbon is 0.70% for the above specimen.

10.18 Wear resistant sintered steels are often subjected to case carburising or nitriding. Such treatment has a disadvantage of substantial dimensional changes. For proper tolerance, a subsequent grinding operation becomes necessary. What other alternative do you have in mind to avoid the heat treatment? [Hint: Manganese surface alloying.]

10.19 Compare plasma nitriding and chrome-plated surface modification processes for a steel from the following view points:

(a) Modified region

(b) Sharpness of the cutting edge

(c) Resistance against indentation

(d) Fatigue strength

(e) Environment friendliness

(f) Cost effectiveness.

10.20 Very thin coating (1.5–2.0 μm) of (TiAl)N and (Cr,Al)N was made by magnetron co-sputtering. Two nitrogen pressures, 0.4 and 0.96 m torr, were used. It was found that grain size increased with increase in nitrogen pressure, while the hardness and elastic modulus dropped. Which coating's mechanical properties are sensitive to grain size variation? Why? Why should one take recourse to nano-indentation measurement for hardness estimation?

10.21 It is proposed to sinter bond a composition graded WC–Co cemented carbide over a steel shaft. Indicate which end of the carbide tool—high- or low-cobalt—should be joined with the steel substrate. Give reasons.

10.22 Explain why the fusion welding of steam treated or quenched and tempered PM steel parts is not recommended.

10.23 A nickel base superalloy part prepared by the PM route is to be brazed. The filler metal has alloying addition like boride to lower the brazing temperature. This resulted in some embrittlement due to precipitation, for which a long time diffusion annealing was required. This however leads to grain coarsening. Suggest some remedial measures.

10.24 For better surface hardening of a ferrous structural PM part, if an engineer suggests boriding of the available case-carburised parts, is his advice correct? [Refer Fe-B binary phase diagram]

10.25 In what type of coating is the change of grain size most sensitive to mechanical properties? Why?

10.26 Complex-shaped high-speed steel milling cutters produced through PM method are to be hardened. Will you prefer oil quenching or gas quenching? Give reasons.

10.27 What is the advantage of bimodal structured coatings by thermal spraying in which one of the constituents is a nanopowder?

10.28 At times dry machining of sinter-hardened PM steel parts becomes necessary, but such an operation causes significant heat generation. What modifications (excluding wet machining) in machining operation do you suggest? In the absence of any modification, what should be the effect on steel microstructure?

10.29 What is the general carbon limit in sintered steel beyond which the secondary machining of the part is avoided?

10.30 Joining of metal/non-oxide ceramics is generally more problematic than the metal/oxide ceramics. Comment on the statement.

10.31 α-Al$_2$O$_3$ / Niobium seal is an excellent joint as it matches the coefficient of thermal expansion. The temperature is high—1700°C. What alternative would you propose to lower the joining temperature to 1400°C?

10.32 Which of the following is likely to exhibit the least wear during green machining?

(a) Cubic boron nitride

(b) Co-bonded tungsten carbide

(c) Diamond

(d) Tool steel

10.33 For a chemical process industry application, a seamless pipe of stainless steel prepared by PM extrusion route is to be joined with a similar pipe made of titanium. Describe the procedure to be adopted.

11

Testing and Quality Control of PM Materials and Products

LEARNING OBJECTIVES

- How to sample a PM material and product
- The physical and mechanical properties of PM parts
- Quality control of parts like filters, bearings, structural parts and cemented carbides
- Non-destructive testing of PM parts
- Statistical quality control

In Chapter 3, the description of various characterisation methods for metal/ceramic powders was given. Presently, we are concerned with the sintered materials and products based on them. The properties embrace physical, mechanical and electrochemical ones. Some physico-technical properties like wear resistance are also included. Microstructural analysis is important, as properties are related to the microstructure of the sintered material.

11.1 SAMPLING

Before testing any PM product, sampling is an important operation. The sampling of sintered products must be dealt with in the same way as any other industrial product. Care must be taken to choose sampling procedures and acceptable quality levels that take cost into account.

Before implementing a sampling plan into action, the supplier and user must agree on the following points:

- The properties to be inspected, their values and relative tolerance ranges
- The methods of measurement for each of the properties to be inspected
- Sampling plans and the acceptance or rejection criteria for the lots supplied.

When there is no interaction between different testing procedures, more than one type of test may be carried out on the same test specimen, for example, density, hardness, metallographic examination, etc.

11.2 DENSITY

In general, the density achieved in sintered products is between 70% and 95% of the fully dense wrought products, depending on the production technology in use and the type of application.

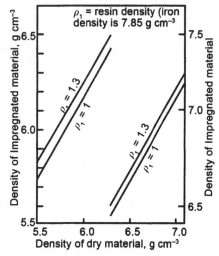

Fig. 11.1 True density of resin-impregnated parts

The sintered density determination is carried out following the procedures of ISO standard 2738. This is valid for both dry parts and parts that have been impregnated with oil. In case parts have been impregnated with thermosetting polymers, the true density is obtained graphically (Fig. 11.1). This allows the calculation of an unknown density by determining the density of the part examined and assuming a degree of impregnation equal to 0.8. As most impregnating resins have a density between 1 and 1.3, two straight lines have been drawn on the diagram, delineating the two limits of the band; intermediate points may be obtained by interpolation.

The density may be calculated by the oil impregnation method, the formula being:

$$D = \frac{A}{B-C}(\text{g cm}^{-3})\tag{11.1}$$

where, A is the mass of the un-impregnated part in air, B is the mass of the part after impregnating with the oil and C is the mass of the impregnated part in water. Either of two procedures may be used for oil impregnation.

1. The specimen is immersed for 4 hours minimum in oil 'Saybolt Universal' at 380°C held at a temperature of 82 ± 4°C and then cooled to room temperature.

2. The pressure over the specimen immersed in oil at room temperature is reduced to a maximum of 50 mm Hg pressure for 30 minutes. The specimen then remains immersed in oil at atmospheric pressure for 10 minutes.

The amount of interconnected porosity (P in %) can be calculated as follows:

$$P = \frac{B-A}{(B-C)S} \times 100\tag{11.2}$$

where, A is the mass of the lubricant-free sample, B is the mass of the oil-impregnated sample, C is the mass of the oil-impregnated sample immersed in water and S is the specific gravity of the impregnant at test temperature. All the masses are determined to the nearest 0.1%.

11.3 SINTERED POROSITY AND PORE DISTRIBUTION

Pores in sintered compacts are of two types: interconnected and closed or isolated. In the first case the pores are connected with each other along the particle junctions. The pores are consequently irregular, unless the particles are initially spherical. Such pores can remain as low as 5% of the total porosity. The latter type of pores, i.e., closed pores, are pronounced when total porosity is low (< 5%). They are often, but not necessarily, spherical.

Porosity has no units. It is determined from the density and the theoretical density, both expressed as mass per unit volume. It is an important parameter in sintered parts, but does not convey the complete picture of pore sizes, pore shapes and other microstructural features like connectivity. Table 11.1 summarises various experimental methods for obtaining quantitative information on changes in pore geometry during sintering. The buoyancy method for measuring density has been highlighted in the previous section. In this section, the mercury porosimetry method is described. The method is based on the Washburn equation:

$$P = -\frac{4\gamma \cos\theta}{d} \tag{11.3}$$

where, P is the capillary pressure, θ is the contact angle, γ is the solid–liquid surface energy and d is the pore size. In case the liquid is of wetting-type, $\theta = 0°$, while in the case of non-wetting mercury, it is 130°. It is worth mentioning that this method measures only the open sintered porosity.

Table 11.1 Experimental method for deriving quantitative information on changes of pore geometry during sintering

Method	Quantity measured	Remarks
Dilatometry	Length change	Shrinkage may vary in different directions; relative precision approximately 10^{-6} of sample length
Buoyancy	Density	Impregnation or pore sealing necessary; relative precision approximately 10^{-3}
Gas adsorption	Solid–pore interface	Only for high specific interfaces (> 0.1 m^2 cm^{-3}); closed pores not included; relative precision approximately 10^{-2} for total area >0.5 m^2
Mercury porosimetry	Accessible pore volume	For open and fine pore systems; interpretation of pressure–volume diagrams difficult
Indirect methods	Physical properties	Exact relationships between pore geometry and properties usually not known
Quantitative microscopy	Direct geometric parameters	Tedious but most effective method for complete characterisation of pore geometry

Example 11.1: In reality the pores are not cylinders—an assumption made for mercury porosimetry. Describe how this will affect the measurement.

If the intrusion path is observed at a pressure of 50 kPa, what is the approximate pore size? The surface energy γ for mercury is 0.48 J m^{-2}.

Solution: In such cases, the mercury intrusion and extrusion paths will not be identical, and hence there will be a difference in the volume pressure curve.

According to the Washburn relation:

$$d = -\frac{4\gamma\cos\theta}{P}$$
$$= -\frac{4\times0.48\times0.643}{50000}$$
$$= 0.247 \text{ μm}$$

Since mercury is not a wetting liquid, higher pressures are required to force mercury into smaller open pores. If the amount of mercury found in a sintered sample is recorded as a function of pressure applied, not only the open pore volume but also the size distribution of the open pores can be determined. The calculation assumes that the pores are cylindrical in shape. At the start of a mercury porosimetry experiment, the open pores are evacuated and the sample is surrounded with mercury. Interpretation of mercury porosimetry data is complicated by pore shape and pore connectivity.

11.4 STRUCTURE OF PM MATERIALS

Structure is a general term which is used to cover a wide range of structural features—from those visible to the naked eye down to those corresponding to the inter-atomic distances in the crystal lattice. Structure is classified into four categories: macrostructure, mesostructure, microstructure and nanostructure. Table 11.2 summarises these different scales of material structure in terms of the magnification required in order to observe the features concerned. In old literature, both mesostructure and microstructure were clubbed as mere microstructure.

Unlike wrought metals and alloys, PM materials have heterogeneous microstructure, because of the presence of porosity and non-ideal homogenisation of powder premixes during sintering. Microstructural analysis covers two aspects: qualitative and quantitative. The qualitative part includes the examination of various phases and their distribution. In quantitative metallography, the measurements of grain sizes and amounts of various phases are established. However, the heterogenous sintered microstructure is entirely absent in case of fully dense hot-consolidated PM materials. Here, the uniformity and fineness of microstructure is even better than in ingot metallurgy alloys.

Table 11.2 The scale of structural features, the magnification required to reveal the feature, and some common techniques for studying the microstructure

Scale	Macrostructure	Mesostructure	Microstructure	Nanostructure
Typical magnification	$\times 1$	$\times 10^2$	$\times 10^4$	$\times 10^6$
Common technique	Visual inspection	Optical microscopy	Scanning and transmission electron microscopy	X-ray diffraction
	X-ray radiography	Scanning electron microscopy	Atomic force microscopy	Scanning tunneling microscopy
	Ultrasonic inspection			High-resolution transmission electron microscopy
Characteristic features	Production defects	Grain and particle size	Dislocation substructure	Crystal and interface structure
	Porosity, cracks and inclusions	Phase morphology and anisotropy	Grain and phase boundaries	Point defects and point-defect clusters
			Precipitation phenomena	

11.4.1 Qualitative metallography and ceramography

Metallographic study of sintered products is essential to study the type and morphology of pores which affect various properties. Metallographic preparation of such materials can lead to changes in the specimen surface which can cause erroneous interpretation of the microstructure. Some examples of such changes are:

- Partial closing of pores by plastic deformation during grinding
- Break out of material around pore
- Closing of pores with grinding debris
- Rounding of pore edges.

The preparation sequence of metallographic specimens of porous materials is as follows.

Sample preparation: An abrasive cut-off wheel with water as coolant is used for sectioning purposes. Thorough rinsing with water must be carried out in order to remove any cutting debris. Specimen mounting can be done as is the usual practice.

Grinding: This can be done using SiC paper of 220 grit size using water as a coolant on an automatic grinding wheel. A speed of 300 rpm with a load of 90, 60 and 30 N is used for three grinding steps. After the grinding operation, the specimens are ultrasonically cleaned in an alcohol bath.

Impregnation: The process is necessary to seal the open porosity of the specimen so that abrasives, water and etchants are not entrapped later on. If the specimen is not moisture-free, 'bleeding out' during etching may occur, which causes staining of the surface. Vacuum impregnation is carried out with epoxy resin.

Regrinding: After impregnation, regrinding is carried out on 500 and 1000 grit silicon carbide paper.

Polishing: This operation is generally carried out with 6, 3 and 1 μm diamond polishing spray on an automatic wheel using a load of 90, 60 and 30 N for one minute each. Polishing using a cloth with a suspension of alumina can also be carried out. In the unetched condition, total porosity, pore size and shape, non-metallic inclusions, additions like manganese sulphide for improved machinability, undissolved alloying elements, etc., may be observed.

The common polishing defects are surface relief, rounding of edges, and scratches or plastic deformation. Scratches and plastic deformation can be prevented by selecting a more compliant support for the polishing media, in order to reduce the forces applied to the individual particles and increase the number of particle contacts per unit area of the samples. Conversely, surface relief and edge rounding can be reduced by selecting a less compliant support.

For polishing ceramic specimens, the selection of the polishing medium is important. Its hardness should exceed the hardness of the sample. Diamond is invariably the medium of preference.

Etching: Etching is generally performed by immersion. This facilitates the study of homogeneity of alloying, grain size and the presence of different phases.

If the different phases differentially reflect and absorb incident light, then etching may be unnecessary. Most non-metallic inclusions in alloys are visible without etching, since the metallic matrix reflects incident light, while the inclusion absorbs it and appears dark.

The common etching reagents are listed in Table 11.3. The specimen surface should be examined from time to time, and washed afterwards.

After mounting the specimen, it is examined under the metallurgical microscope. In the beginning the magnification should be kept low at about X100 and then increased as per the requirement of the system. The light microscope has a limit of resolution that is governed by the wavelength of visible light. Figure 11.2 illustrates the schematic of qualitative optical metallography.

Optical microscopes can be arranged to view an object by either reflected or transmitted light. If the object is fairly thick or opaque, for example, metallic systems, the reflective mode is used. In case of transparent and very thin objects, for example, some ceramics, the transmission mode is preferred. All lenses in an optical microscope are susceptible to serious damage from even relatively minor mishandling and therefore greater precautions are needed.

Microscopic preparation of ceramics

For polishing ceramics, two methods are used: fixed abrasives in a variety of solid lap materials, or cloth laps with various loose abrasives. Both methods have particular advantages in terms

Table 11.3 Common etching reagents for some metals and their alloys

Material	Etching reagent	Remarks and composition
Iron and steel	Picral: picric acid 5 g, ethyl alcohol 100 cm^3	For general structure of iron and steel
	2% Nital: conc. nitric acid 2 cm^3, ethyl alcohol 98 cm^3	Heat-treated steels: martensitic or tempered
	5% Nital: conc. nitric acid 5 cm^3, ethyl alcohol 95 cm^3	General structure of iron and steel
		Colours: austenite yellow, martensite white, and tempered structure brown; etching time 5 min
	Fry's reagent: cupric chloride 90 g, hydrochloric acid 120 cm^3, water 100 cm^3	Used to reveal strain markings in mild steel
Stainless steel	Marble's reagent: copper sulphate 4 g, hydrochloric acid 20 cm^3, water 20 cm^3	
	Aqua regia: hydrochloric acid 75 cm^3, nitric acid 25 cm^3	Requires careful handling
	Aqueous ferric chloride: ferric chloride 10 g, hydrochloric acid 30 cm^3, water 120 cm^3	
Copper and its alloys	Aqueous ferric chloride: ferric chloride 10 g, hydrochloric acid 30 cm^3, water 120 cm^3	Suitable for brasses, bronzes, copper–aluminium alloys, phosphorous bronze and German silver
	Alcoholic ferric chloride: ferric chloride 10 g, hydrochloric acid 5 cm^3, ethyl alcohol 200 cm^3	In some cases produces better contrast than aqueous solution
	Heyn's reagent: copper ammonium chloride 5 g, water 120 cm^3. Add ammonium hydroxide until the preci-cipitate which forms redissolves giving a clear blue colour	
Nickel and its alloys	Merica's reagent: nitric acid (70%) 50 cm^3, acetic acid (50%) 50 cm^3	
Aluminium and its alloys	Conc. hydrofluoric acid 10 cm^3, conc. hydrochloric acid 15 cm^3, water 90 cm^3	
	Keller's reagent: water 90%, hydrochloric acid 1.5%, nitric acid 2.5%, hydrofluoric acid 1.0%	

of capability, and may be used together in a sequence to obtain optimum speed of polishing and quality of finish.

Cloth laps: These are suitable for refractories, porcelains, whiteware and some aluminas. If long polishing times is required, grain rounding and pull-out are enhanced, and better results are obtained using a tin or similar lap, using cloth laps only for the final stages. The advantages of using a tin lap with an unmounted sample are increased rate of material removal, better specimen flatness and decreased grain pull-out. However, it is difficult to remove the final

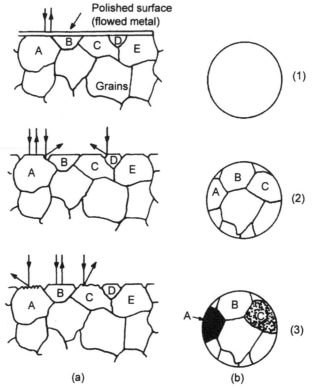

Fig. 11.2 Schematic of a specimen under microscopic study: (a) Side view of a section with the plane surface on top and (b) Top view as seen through the eyepiece: (1) after grinding and polishing, (2) after a light etch, (3) after a deep etch

polishing scratches, and polishing with a soft cloth with γ-alumina or colloidal silica suspensions is resorted to.

Polymer-based laps: These are used as an intermediate grinding stage with an automatic polishing machine to reduce the amount of grain pull-out and give a better surface finish at the 3 μm grinding stage; care must be taken when selecting the load to apply to the specimen. A high load can cause grinding or polishing debris to get lodged in pores, causing scratches and pull-out. The debris is difficult to remove by ultrasonic cleaning.

Grain pull-out and porosity in some ceramics are difficult to distinguish as they often have similar appearances. Sub-surface damage from the initial grinding can be the cause of grain pull-out. The depth of damage can be considerable, and must be completely removed by adequate grinding or polishing on fine grits before final polishing is attempted.

Thermal etching is one of the most appropriate etches for oxides. This involves refiring the sample to a sufficiently high temperature to cause surface relief—typically, temperatures up to 1470°C for periods up to 30 min are suitable. Chemical etching is appropriate for many types of silicates or non-oxide ceramics for which thermal etching is inappropriate. Some care in selecting conditions is required, because the appearance of the microstructure revealed is dependent on the etching conditions employed. Structures can readily be destroyed or greatly modified by over-etching. Some ceramics pose considerable problems, and notable among them are sintered silicon nitride and sialons. Ion etching is said to be more useful, but it can be difficult to interpret the result.

Thin-section techniques are seldom used for ceramics but extra information can be obtained compared with that from reflected light microscopy, particularly when applied to sections of about 10–15 μm thickness. Textured grain orientations produced by the fabrication method can be observed even in very fine-grained ceramics.

In ceramic systems the common etchants are various mixtures of hydrofluoric, nitric, hydrochloric, sulphuric and phosphoric acids. Often, after etching, neutralising and washing operations are carried out. For most acid etchants, a concentrated ammonium hydroxide soak followed by several water washings is usually sufficient.

Scanning electron microscopy

The scanning electron microscope (SEM) became the most important analytical tool in PM industry. It uses X-rays or electrons scattered back from the surface to generate an image with remarkable three-dimensional qualities. Since the beam need not be transmitted, the specimen does not have to be extremely thin and replication procedures are not necessary. SEM has a number of analytical features. For example, X-rays generated as the electrons impact the sample provide a characteristic fluorescence pattern related to the elements present. The interpretation of these X-rays yields a semi-quantitative chemical analysis of the specimen at the point where the beam is focussed.

Many specialised objective–lens assemblies are available. One of the most useful, for both reflection and transmission work, is the dark-field objective which illuminates the specimen with a cone of light surrounding the lens aperture. The light scattered by the specimen into the lens aperture is then used to form a dark-field image in which the intensity is the inverse of that observed in normal illumination.

The most general class of specimens examined in SEM comprises rough, irregular objects which have changes in elevation along the optic axis (z-axis) of the microscope. The z-axis information is lost as the image is simultaneously generated on the cathode ray tube (CRT), because the image is effectively constructed by projecting the three-dimensional surface onto a two-dimensional plane. Stereomicroscopy provides a means of reconstructing the lost third dimension. The technique involves the use of SEM images (the so-called stereo-pair) of the same field of view prepared at slightly different angles and an optical stereo-viewer to trick an observer into seeing a 'three-dimensional stereo-image'. To study stereo-microscopy, we require two different views of the specimen relative to the incident beam. These views are commonly obtained by recording the first image at a low angle of tilt (for example 0°) and then tilting the specimen to a higher angle of tilt. The difference in the tilt angles is typically chosen from a minimum of 4° to a maximum of 10°. For qualitative stereo-microscopy, the choice of the tilt angle depends on the nature of the specimen topography. If the specimen has extremely rough topography, the difference in tilt angle between the members of the stereo-pair should be chosen to be a low value.

Stereo-microscopy is not restricted to scanning electron microscopy, but is equally valid for optical microscopy. Optical stereo-microscopy with conventional optics is limited to magnifications of about 100X.

Example 11.2: How will you distinguish between poorly-sintered and well-sintered PM materials under the microscope?

Solution: Poorly-sintered materials will have unsintered particle boundaries, angular pore edges and small pores located at prior particle boundaries. A well-sintered material on the other hand, will show the disappearance of particle boundaries and small pores, in addition to smoothing of the individual pores.

The depth-of-focus, D, can be expressed as

$$D \text{ (in mm)} = \frac{0.2}{\alpha M} \tag{11.4}$$

where M is the magnification and α is the beam convergence. 0.2 mm corresponds approximately to the minimum diameter of an object which the unaided human eye can resolve in the final SEM image.

11.4.2 Quantitative metallography

Quantitative metallography involves a large number of measurements, especially if good precision is required. Measurements are made on random slices through a three-dimensional object, which may or may not be uniform throughout. Rapid determination of grain size is possible by using a calibrated eyepiece which eliminates the necessity of either projecting the image on a screen or taking a photomicrograph. Currently, many automatic image analysers are available in the market. For determining the relative volume fraction of constituent phases, the analysis methods are areal analysis, lineal analysis or point counting. There is always an equivalence between the volume fraction of the constituent phases and the intercepted area, line or point fractions. Figure 11.3 illustrates the above features. The volume fraction of a second phase can be determined from the areal fraction of the phase, seen on a random planar section, or from the fractional length of the random test line which intercepts the second phase particles in the section, or from the fraction of points on a test array which falls within the regions of the second phase.

In sintered particulate composites—for example, cemented carbide and heavy alloys—another microstructural parameter known as 'contiguity' is measured. This is defined as the fraction of the grain perimeter that is in contact with a similar phase. A simple way to measure contiguity from a two-dimensional image is via contact counting. Test lines are randomly overlaid on the microstructure and the number of the same grain N_A and different grain N_B contacts are counted. It is expressed as

$$C_g = \frac{2N_A}{2N_A + N_B} \tag{11.5}$$

For a successful composite material, the contiguity should be as small as possible.

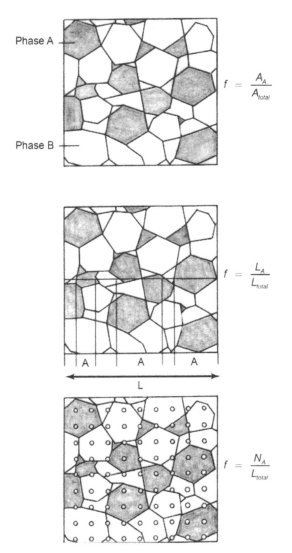

Phase A

Phase B

$$f = \frac{A_A}{A_{total}}$$

$$f = \frac{L_A}{L_{total}}$$

$$f = \frac{N_A}{L_{total}}$$

A A A

L

Fig. 11.3 The volume fraction of the second phase can be estimated from the areal fraction of the phase, seen as a random planar section, or from the fractional length of a random test line which intercepts the second phase particles in the section, or from the fraction of points in a test array which falls within the regions of the second phase

11.4.3 X-ray analysis

When a beam of X-rays strikes a crystalline solid, an interference pattern is produced. Because the exact diffraction (interference) pattern is unique for every crystalline substance, X-ray diffraction has become a standard method of identifying crystalline materials. The X-rays used in diffraction are usually all of one wavelength. The specimen is generally in the form of a fine powder held in a rotating holder. The angles at which diffraction occurs are determined with a moving detector.

The condition under which the diffraction takes place is given by the Bragg equation:

$$d = \frac{\lambda}{2\sin\theta} \qquad (11.6)$$

Since each crystalline material has a unique set of interplanar spacings (d), which can readily be found in tabulations such as the Powder Diffraction File (PDF) issued by the Joint Committee of Powder Diffraction Standards, the details of the diffraction pattern allows identification of the material. If the specimen is a mixture of two phases, identification is more complicated.

Before studying the polycrystalline material, proper selection of target material is called for. In general, absorption increases rapidly with increasing atomic number and with increasing wavelength. Each element exhibits a number of sharp absorption 'edges', where the absorption suddenly falls and then begins to rise again as the wavelength is increased. These edges occur when the incident radiation can excite K, L or M spectrum of the absorption material. The special characteristics of the absorption edge are applied in selecting the filter material. The filter consists of a thin foil of an absorbing element where the K absorption edge lies between the K_α and K_β wavelengths emitted. The K_β rays, which lie on the short wavelength side of the absorption edge, are heavily absorbed in the filter, while the longer K_a rays are transmitted relatively easily. An example is nickel β-filter for K_a copper radiation. For a copper target the excitation parameters are 30 kV and 15 mA, while for a molybdenum target they are 45 kV and 19 mA.

A second analytical method employing X-rays permits the determination of the chemical composition of a material, regardless of whether it is crystalline or glassy. When an intense beam of X-rays strikes a material, it often excites the material itself to emit X-rays. This is known as X-ray fluorescence. This should not be confused with diffraction, because the emitted X-rays come out at all angles to the entering beam. X-rays are characteristic of the kind of chemical element in the material specimen, and the intensity of emission at each characteristic wavelength is a measure of the amount of each element present.

The major advantage of fluorescent X-ray study is its non-destructive character and the fact that very little specimen preparation is required—it being necessary only to present a flat surface, about 2.5 cm to the incident beam. It is immaterial in which form the elements are combined, since only atoms and not molecules are involved. The method is best suited for the analysis of elements with atomic numbers between 20 and 51, using their K_a radiation rays, while elements with atomic numbers higher than 51 can be dealt with using weaker L lines.

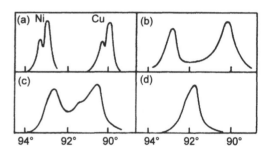

Fig. 11.4 X-ray diffraction peaks of (311) plane of copper, nickel and their (1:1) premix (particle size ~ 100 μm): (a) Prior to sintering, (b) After sintering at 850°C, 1 h, (c) After sintering at 1050°C, 1 h and (d) After sintering at 1050°C, 54 h.

X-ray diffraction analysis is a convenient tool to study the homogenisation process during sintering of green compacts of powder premixes, where alloying takes place. Figure 11.4 shows a typical example of alloying behaviour of sintered compacts made of elemental copper and nickel powders premixed in equal quantities.

Another important application of X-ray diffraction in powder metallurgy is the reliable and simple means of estimating the true crystallite size within the powder. The breadth of a diffraction line, which is effectively the angle it subtends at the specimen, is a convenient method to measure this parameter. Sherrer expressed the breadth of a line in radians at a point where the intensity is half its maximum value (Fig. 11.5). Left-hand side on of the figure shows the scheme for crystals of 'infinite' size, while the right-hand side of the picture is representative of very small crystallites. It must be remembered that it is the observed width.

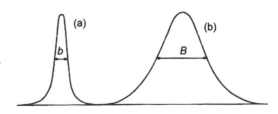

Fig. 11.5 Scherrer half-peak breadths of diffraction lines: (a) Observed breadth b for crystals of effectively 'infinite' size and (b) Observed breadth B for very small crystallites

Thus,

$$B = \beta + b \qquad (11.7)$$

where, β is the width due to diffraction effect and b is the contribution due to extraneous factors such as the divergence of the X-ray beam controlled by camera and specimen geometry, absorption within the specimens, the Bragg angle and the natural wavelength spread of the characteristic radiation employed.

The main job now is to extract the intrinsic breadth β (caused purely by diffraction effects), from the observed breadth B, and to then compute from it the effective crystallite size, t. The detailed theoretical treatment of these aspects is beyond the scope of this book.

The Scherrer expression is given as

$$\beta = \frac{k\lambda}{t\cos\theta} \qquad (11.8)$$

where, k is the Scherrer constant and t is the crystallite size defined as the cube root of the volume. The value of k depends on the way in which the line broadening and the crystalline size are defined and it is not very different from unity. For practical purposes we can assume k to be unity.

Example 11.3: Figure Ex. 11.3 shows the X-ray diffraction pattern of a medium-carbon sintered steel (0.4% C) when CuK_α radiation of 0.155 nm wavelength was used. Index the planes and identify the phases. Can this analysis be misleading in full characterisation of the steel?

Solution: The phases and the characteristic lines, from the PDF file are as follows:

(a) α-Fe (011)
(b) α-Fe (022)
(c) α-Fe (121)
(d) α-Fe (022)

The steel also contains cementite (Fe_3C) phase; since the amount present is small, the intensity of the line corresponding to Fe_3C is very weak. Overlapping of some carbide reflections with those of iron may occur: for example, Fe_3C (031) with Fe (011). A very careful examination of the diffraction spectrum is called for, otherwise one may infer this material is pure iron.

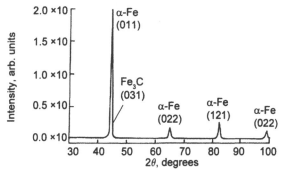

Fig. Ex. 11.3

Example 11.4: You are given a PM nickel—alumina cermet to determine the content of alumina in it by X-ray diffraction analysis. Explain the steps involved.

Solution: The steps involved are:

1. Prepare the diffractogram of the sample and identify various lines.

2. Measure the peak intensities.

3. Determine the ratio of the integrated peak intensities.

4. Plot a calibration curve of the system using different amounts of alumina, or refer an available one in literature (Fig. Ex. 11.4)

5. Fit in the intensity ratio data for the cermet and read the composition in Fig. Ex. 11.4.

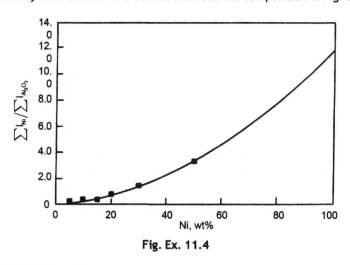

Fig. Ex. 11.4

11.5 DIFFERENTIAL THERMAL ANALYSIS

Any raw material that changes composition or structure on heating may be identified by comparing the temperature differences that develop when the specimen and a stable standard material (often aluminium oxide) are slowly heated side by side at the same controlled rate. The specimen absorbs heat (endothermic reaction) or liberates heat (exothermic reaction) when it undergoes structural changes, decomposes or melts. Since the standard material undergoes no such changes on heating, an endothermic reaction will cause the specimen to remain cooler than the standard material until the reaction is completed. The reverse is true in the case of the exothermic reaction. Typically, the temperature difference between the specimen and the standard material is determined by thermocouples and is plotted against the temperature of the standard material. The reactions appear as peaks or 'inverse' peaks—the endothermic peak being opposite in direction to the exothermic peaks. Analysis of such data is known as differential thermal analysis (DTA) and the plot is called a *thermogram*. This method is commonly used in refractories and ceramic industries. It is also useful in sintering studies. DTA data are frequently obtained simultaneously and correlated with gravimetric changes occurring

Example 11.5: How would you determine the amount of monoclinic and tetragonal zirconia on the surface of a zirconia wear plate?

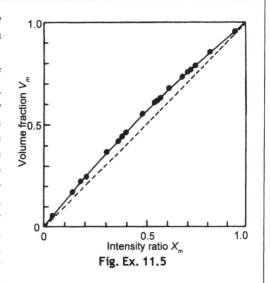

Solution: The quantitative analysis of ZrO_2 systems with monoclinic tetragonal polymorphs is significant, as it describes the martensitic transformation details. Such a study is required to examine the transformation toughening of zirconia. The X-ray diffraction method is very handy as the intensity ratio of the lines pertaining to two phases can be used in conjunction with the calibration curves obtained by plotting the percentage in powder mixtures. Many researchers have studied this aspect, where the linearity of the formula was assumed. Toraya et al report the non-linear calibration curve for volume fraction versus intensity ratio in the monoclinic/tetragonal ZrO_2 system (Fig. Ex 11.5).

Fig. Ex. 11.5

X-ray examination of the polished specimen is carried out in the range $20° \le 2\theta \le 90°$ with CuK_α radiation. A step-scan technique can be used at $0.05°$ intervals in 2θ with a fixed time of 20 s. A divergence slip of $1°$, receiving slip of $0.15°$ and the scintillation counter can be used. Toraya proposed the empirical formula using least squares fit as:

$$V_m = \frac{1.311\,X_m}{1+0.311\,X_m}$$

where,

$$X_m = \frac{I_m\,(111)+I_m\,(11\bar{1})}{I_m\,(111)+I_m\,(11\bar{1})+I_t\,(101)}$$

The reason for selecting the empirical formula over the theoretical one is due to the fact that more intensity data were utilised in determining the volume fractions by powder pattern fitting than in the theoretical derivations using the calculated intensity of only three reflections.

in the specimen during the heating cycle. This time/temperature weight loss procedure is called *thermogravimetric analysis* (TGA). It is useful for oxidation or sulphidation studies on materials.

11.6 THERMAL EXPANSION

The determination of linear thermal expansion coefficient involves the measurement of small length changes as a function of temperature. The most common method is to study the use of dilatometers. The dilatometer may employ either a dial gauge or a linear variable differential transformer (LVDT) to measure total expansion. This property measurement is very useful in studying the sintering kinetics of a system. In structural ceramics, the property measurement is necessary as the dimensional changes generate high stress levels, which may lead to probable structural failure. The magnitude of the dimensional change on heating is directly proportional to the original length, the change in temperature and the mean coefficient of linear thermal expansion of the material. The volume thermal expansion coefficient of a material is approximately three times its linear thermal expansion coefficient, provided the material is isotropic.

The advantages of a dilatometry study in PM systems are:

- The sintering behaviour of various powders can be determined rapidly and effectively for the purpose of comparison and production control.
- Dilatometric curves provide a good indication of the most suitable sintering temperature in the production floor.
- Dilatometer tests are of significance in fundamental studies, e.g., in respect of the dimensional behaviour during the sintering process, with or without liquid phases.

11.7 THERMAL SHOCK RESISTANCE

This property is of direct relevance to PM cutting tool materials and structural ceramics. Thermal shock failure is due to the large temporary differences in temperature between the surface of the specimen and its interior, when the specimen is subjected to extremely rapid heating or cooling. The thermal shock resistance in general improves with an increase in strength and thermal conductivity and with a decrease in thermal expansion coefficient and modulus of elasticity. The microstructure of the material also plays a role in thermal shock resistance.

Either the modulus of elasticity or modulus of rupture of the specimen is measured after the thermal shock. The fall in these properties due to increased cracking propensity after each thermal shock can be used as a measure of shock resistance. Another convenient method of measurement is to determine the maximum temperature difference that can be tolerated during a quench without causing any cracking. Some of the values for ceramics are 950°C for Sialon, 500°C for partially stabilised zirconia, 350°C for SiC and 200°C for alumina.

11.8 THERMAL CONDUCTIVITY

It is measured by determining the rate of heat flow through and the temperature drop across a carefully insulated piece of material heated from only one side. The thermal conductivity is expressed as,

$$k = \frac{Qx}{A(T_2 - T_1)} \qquad (11.9)$$

where, Q is the rate of heat flow through a specimen of cross-sectional area A and thickness x when there is a temperature difference $(T_2 - T_1)$ across the thickness. Temperature T_1 is kept at a fixed 'reference' temperature. As k varies with temperature, the above equation is used over a number of temperature intervals, so that an accurate plot of k versus average temperature can be made. Sintered parts with a variety of pore sizes and shapes have distinctive conductivities. Thermal diffusivity is the ratio of thermal conductivity to heat capacity of the material. The laser source is commonly used to measure thermal diffusivity, from which one can calculate the thermal conductivity. The unit of thermal conductivity is $W(m°C)^{-1}$.

11.9 OPTICAL PROPERTIES

Optical properties, like thermal properties, are related to the interaction of a material with radiation in the form of waves or particles of energy. The wavelengths of the visible spectrum range from 0.4 μm (violet) to 0.7 μm (red). When light strikes a material, it can undergo one of the three transitions. For a given wavelength, an incident radiation may be reflected, absorbed or transmitted, depending on the characteristic properties of the medium. No solid reflects or transmits all the incident radiation, but a certain amount of it is also absorbed. A transparent material becomes translucent and then opaque if light is scattered at internal imperfections such as grain boundaries and dispersed crystallites in the amorphous matrix. The presence of impurities may lower the energy gap of the crystal, imparting a colour to the transparent crystal, for example, Cr^{3+} in Al_2O_3 imparts a pink colour. The colour arises because of partial absorption of white light and selective transmission of the other colours of the spectrum. Another example is zirconia, which is doped to produce a variety of colours. TiN physical vapour deposition on cemented carbide substrate gives rise to a pleasing golden colour, apart from the tailored surface hardness of the tool.

When a single wavelength or narrow band of wavelengths is absorbed by a material, our eyes will see the colour of the remaining wavelengths that pass through it or are reflected back. For example, if the orange colour of the spectrum is absorbed, the material will appear blue. If green is absorbed, the material will appear purple. The colour we see is referred to as the complimentary colour. An example of this is the addition of cobalt in ceramic powders, which produces 'cobalt blue', which is significant in Chinese porcelain decorations. Addition of iron filters out indigo and green to give colours varying from yellowish to rust. Sometimes the ceramic is crushed into powder form and added to other materials as a colourant or pigment, especially when the material or object is created or used at a high enough temperature at which other types of pigments are destroyed.

The surface finish of a material is directly governed by the material's optical properties. We will see more of this in the next section.

11.9.1 Roughness

The roughness of sintered parts is governed by the successive operations the material is subjected to. In the powder pressing stage, it is influenced by the wear of the dies. In the sintering stage the governing factors are the lubricant burn-off and the reduction of any oxide on the powder surface. During the sizing operation the surface condition of tools is important. The roughness would be different if measured on the surfaces normal to the pressing direction or on surfaces parallel to it.

Conventional profilometer readings give an erroneous impression of surface finish for sintered parts, because a different surface condition exists from that found on the machined or ground surfaces of wrought materials. Conventional readings (RMS) take into account the peaks and valleys of machined surfaces, while PM parts have a series of very smooth surfaces which are interrupted by varying size pores. A chisel stylus is preferred for PM parts because it bridges the negative gaps caused by the pores and will still measure any protrusions on the surface of the part.

11.10 HARDNESS

The hardness of a material is related to the resistance to indentation. As it implies a resistance to deformation, this property is useful for studying the strength and heat treatments carried out on materials. There are three major types of hardness measurements, depending on the manner in which the test is conducted—scratch hardness, indentation hardness and rebound or dynamic hardness. *Scratch hardness* is of importance to a mineralogist and is measured on Moh's scale. This consists of 10 standard minerals arranged in the order of their ability to be scratched. Moh's scale is not suited for metals since the intervals are not widely spaced in the high hardness range. *Indentation hardness* is most common for metals and ceramics. In *dynamic hardness* measurements the indenter is usually dropped on the material surface and the hardness is expressed as the energy of impact. The most common method in this group is *Shore Scleroscope*, where the hardness is measured in terms of the heights of rebound of the indenter.

11.10.1 Brinell hardness test

In the Brinell test, a hardened steel ball is pressed for 10–15 s on the surface of the material by standard load. After the load and ball have been removed, the diameter of the indentation is measured.

Brinell hardness number (BHN) is given as

BHN = applied load/surface area of indentation

$$= \frac{P}{(\pi D / 2)(D - \sqrt{D^2 - d^2})} \tag{11.10}$$

where, P is the applied load in kg, D is the diameter of the ball in mm, and d is the diameter of indentation in mm.

11.10.2 Vickers hardness test

This test involves a diamond indenter, in the form of a square pyramid with an apex angle of 136°, being pressed under load for 10–15 s into the surface of the material under test. The result is a square-shaped impression. Like the Brinell test, the Vickers hardness number (HV) is obtained by dividing the applied load by the surface area (mm^2) of the indentation.

$$\text{Area of the indentation} = \frac{d^2}{2\sin\theta/2} = \frac{d^2}{1.854}$$

where, d is the mean diagonal length and the apex angle 2θ is equal to $136°$. Therefore,

$$HV = \frac{1.854P}{d^2} \qquad (11.11)$$

The biggest advantage with the Vickers test is that the hardness value is independent of the magnitude of the load used. Typically a load of 30 kg is used for steels, 10 kg for copper alloys, 5 kg for pure copper and aluminium alloys, and 2.5 kg for pure aluminium.

11.10.3 Rockwell hardness test

This test utilises the depth of indentation under constant load as a measure of hardness. This depth is directly indicated by a pointer on a calibrated scale. The test uses either a diamond cone or a hardened steel ball as the indenter. There are a number of Rockwell scales (Table 11.4), the scale being determined by he indenter and the additional force used. In any reference to the results of a Rockwell test, the scale letter must be quoted. For metals the B and C scales are most common. In case of hard materials like cemented carbides, the A scale is used. A variation of the Rockwell test is used for thin sheets, and is referred to as the Rockwell superficial hardness test. Smaller loads are used and the depth of indentation which is correspondingly smaller is measured with a more sensitive device.

Like the density, the hardness value has great importance and provides indications of the mechanical behaviour of sintered products. The indentation hardness of a sintered material is also strongly affected by its density, because, voids in the structure of a material do not contribute to the support of indenter. The indentation of a porous material should be considered in apparent hardness. The ISO standard on apparent hardness recommends Vickers as the reference method but allows Brinell and Rockwell methods as alternatives. It forbids direct conversion from one hardness scale to another. Though it is possible to compare one hardness scale with another—provided both tests have been done on the products concerned—there are practical shortcomings in the tests currently specified. In Vickers and Brinell tests, surface preparation is critical, the tests are slow and require visual judgement. Furthermore, automation is not practical.

Table 11.4 Rockwell hardness scales

Scale	Indenter	Load kg	Typical applications
A	Diamond	60	Extremely hard materials
B	Ball 1.588 mm dia.	100	Softer materials, e.g., Cu alloys, Al alloys, mild steel
C	Diamond	150	Hard materials, e.g., steels, high-alloy steels
D	Diamond	100	Medium case-hardened materials
E	Ball 3.175 mm dia.	100	Soft materials, e.g., Al alloys, Mg alloys, bearing metals

Note: The diameter of the balls arise from standard sizes in inches, 1.588 mm being 1/16 in, 3.175 mm being 1/8 in.

The Rockwell test, on the other hand, uses heavy loads of 60–150 kg which are less responsive to the metallurgical structure of the material than density variations. In addition, the test is insensitive in the most common hardness ranges used in sintered materials.

11.10.4 Microhardness test

Apart from testing procedures, there are problems associated in measuring true hardness of a multicomponent system. For example, the porosity may be filled by a softer or harder material than the skeleton material, and at times the true hardness may not be an indication of functional parameters, such as abrasion wear. In order to avoid such complexities, the best thing is to express true hardness for a 'specific constituent'. This is determined by microhardness testers utilising Knoop or diamond pyramid hardness indenters. It measures the true hardness of the structure by eliminating the effect of porosity. In case the indentor strikes a pore, the diamond mark will exhibit curved edges and the reading must be discarded. Since the data tend to be scattered, compared to fully dense material, it is recommended that 5 to 10 indentations be made, anomalous readings discarded and an average of the remainder taken.

The Knoop indenter is a modification of the Vickers, as it has a diamond-shaped indentation with the long and short diagonals in the approximate ratio of 7:1, resulting in a state of plane strain in the deformed region. The Knoop hardness number (KHN) is expressed as:

$$\text{KHN} = \frac{P}{A_p} = \frac{P}{L^2 C} \qquad (11.12)$$

where, P is the applied load in kg, A_p is unrecovered projected area of indentation in mm^2, L is the length of the long diagonal in mm, and C is the constant for each indenter supplied by each manufacturer. The majar advantages for such a test method are efficient testing of a thin layer and testing of brittle solids, since here the tendency for fracture is proportional to the volume of the stressed material. The surface of the specimen in case of microhardness testing must be carefully prepared, similar to metallographic polishing. Work hardening of the surface during polishing can influence the results. Hot microhardness tests are a convenient method for studying creep deformation.

Example 11.6: Why does micro-indentation hardness evaluation for a PM part require far more care as compared to macro-indentation test?

Solution: In PM parts a well polished and prepared surface is essential. It is necessary to remove any smearing of the metal that may cover the pores as a reading at that spot would be misleading. PM parts must be well ground, polished, etched to remove smears, repolished, and adequately etched to show the microstructure. This sequence may have to be repeated more than once.

11.11 STRENGTH

Powder metallurgy materials are tested for strength in the loading of (i) tension, (ii) compression or (iii) bending, depending on the nature of the material. For example, tensile test is carried out in ductile metallic materials, while compression and bending strengths are carried out for brittle materials such as ceramics or cermets.

The Metal Powder Industries Federation (MPIF) of USA has adopted the concept of minimum strength values for PM materials for use in a structural application. The PM process offers equivalent minimum tensile strength values over a wide range of materials. It is seen as an advantage of the process that equivalent strengths can be developed by varying chemical composition, particle character and density/processing techniques.

The material may be specified on the basis of properties obtained in test samples made under similar condition, but it is understood that the properties of sintered parts may not be identical to the test pieces because of shape effects. The test methods and instrumentation used are similar to those used for wrought products. The various national and international standards have been limited to standardising the types of testpieces.

11.11.1 Tensile strength

The tensile properties of sintered products are directly influenced by porosity. Due to the presence of porosity, the tensile properties are somewhat lower than those of wrought materials of the same composition and structure. For sintered materials, machined test pieces are almost never used and test pieces are invariably obtained by pressing and sintering. Figure 11.6 illustrates a typical MPIF test piece for tensile properties evaluation. With sintered materials, care should be taken when storing them before testing, because their interconnected porosity may give rise to internal corrosion. In the case of fully dense materials like PM-forged, machined test pieces may be used.

Porosity has a more pronounced effect on ductility than on strength. A pore content of a few per cent can be rather detrimental to ductility. However, production variables, particularly in sintering, also have a significant effect on ductility, such that the ductility of similar materials of the same porosity but of different origins, may differ widely.

11.11.2 Compressive strength

The compressive strength of a material is a measure of its ability to bear crushing or pressing loads. The general test fixture is shown in Fig. 11.7. The compressive strength of the material is given by:

$$S = P/A \qquad (11.13)$$

where, P is the load at fracture and A is the cross-sectional area of the test piece.

In order to obtain a reliable strength value, a lot of care is required in specimen preparation and alignment. The specimen faces bearing the load must be absolutely flat and parallel. If these criteria are not met, the load will be carried unevenly by the specimen, causing premature failure rate. The measured compressive strength is influenced by the specimen size, configuration and loading.

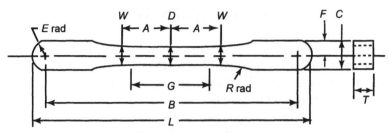

Pressing area = 1.00 in²

Note: Dimensions specified except G and T are those of the die

Dimensions

	in	mm
A Half-length of reduced section	5/8	15.88
B Grip length	3.187 ± 0.001	80.95 ± 0.03
C Width of grip section	0.343 ± 0.001	8.71 ± 0.03
D Width at centre	0.225 ± 0.001	5.72 ± 0.03
E End radius	0.171 ± 0.001	4.34 ± 0.03
F Half-width of grip section	0.171 ± 0.001	4.34 ± 0.03
G Gauge length	1.000 ± 0.003	25.40 ± 0.08
L Overall length	3.529 ± 0.001	89.64 ± 0.03
R Radius of fillet	1	25.4
T Compact to this thickness	0.140 to 0250	3.56 to 6.35
W Width at end of reduced section	0.235 ± 0.001	5.97 ± 0.03

Fig. 11.6 Standard flat tensile test specimen for PM materials

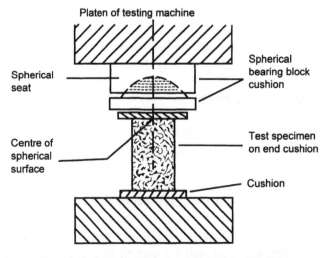

Fig. 11.7 Compressive strength test configuration

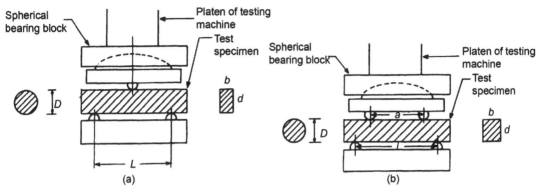

Fig. 11.8 Modulus of rupture (MOR) or transverse rupture strength (TRS) testing configuration: (a) Three-point loading and (b) Four-point loading

11.11.3 Transverse rupture strength (Bending strength)

This test is applicable only to materials of negligible ductility. In the ceramic industry, the property is commonly named as *modulus of rupture* (MOR). The test is carried out on a long rectangular bar, supported near its ends with a load applied to the central portion of the supported span. There are two methods for applying the load: three-point loading and four-point loading (Fig. 11.8). In order to give correct results, a bar in three-point loading must fracture at the exact centre, but a bar in four-point loading can fracture at any point between the inner two loading points and still provide a valid measurement. Because of this less restrictive fracture requirement, four-point loading is preferred.

For three-point loading of a rectangular specimen the *transverse rupture strength* (TRS) is given by

$$\text{TRS} = \frac{3PL}{2bd^2} \qquad (11.14)$$

where, P is the breaking load, L is the span between the outer support, b is the width of the specimen and d is the depth of the specimens.

For four-point loading the TRS is given by

$$\text{TRS} = \frac{3P(L-a)}{2bd^2} \qquad (11.15)$$

where, a is the distance between the inner load application points.

The important factors that control the TRS measurement are the loading rate, the ratio of span to specimen thickness (L/d) and the specimen alignment.

11.11.4 Modulus of elasticity

For ductile materials, e.g., metallic systems, the modulus of elasticity is measured from the stress–strain plot in tensile test. In the case of ceramics (brittle materials), the sonic test is a convenient method. It is based on the fact that the modulus of elasticity of a material is proportional to the square of a particular natural vibrational frequency, i.e., resonant frequency of the specimen. The resonant frequency of ceramic materials can be measured electronically. This sonic test has become a convenient quality control tool for ceramic parts, where standards are set up for resonant frequency, and parts that have frequencies below a set limit are either re-sintered or rejected.

The modulus of elasticity of a particulate composite (E_c) can be determined from the following relation for the upper bound:

$$E_c = E_m V_m + E_p V_p \tag{11.16}$$

where, E and V are elastic modulus and volume fraction respectively, and m and p designate matrix and particle, respectively.

11.11.5 Effect of grain size and dispersed hard particles

In polycrystalline materials the role of grain size is significant: smaller the grain size, higher the tensile yield strength (σ_y). These parameters are related to the Hall–Petch equation:

$$\sigma_y = \sigma_o + k_y d^{-1/2} \tag{11.17}$$

where, k is a material constant, σ_o is the yield stress of a single crystal of similar composition and dislocation density and d is the average grain diameter. However, it may be noted that the Hall–Petch relation is not universally true for any grain size.

It is a well-established fact that by producing a fine and stable particle (10–100 nm) dispersion in an alloy, the yield stress of the composite material is increased. This is known as dispersion strengthening. The degree of strengthening depends on the distribution of particles

---

Example 11.7: The yield stress of a PM material drops from 2 × 10⁷ N m⁻² to 10⁷ N m⁻² when grain size is increases from 100 μm to 1000 μm. Determine the yield stress for the material when the grain size changes to 500 μm.

Solution: The Hall–Petch relation is expressed as

$$\sigma_y = \sigma_1 + k d^{-0.5}$$

So,

$$2 \times 10^7 = \sigma_1 + k(100 \times 10^{-6})^{-0.5}$$
$$1 \times 10^7 = \sigma_1 + k(1000 \times 10^{-6})^{-0.5}$$

Solving we get

$$k = 146.2 \times 10^3$$
$$\sigma_1 = 0.538 \times 10^7 \text{ N m}^{-2}$$

Substituting the above values,

$$\sigma_{y500} = 0.538 \times 10^7 + 146.2 \times 10^3 (500 \times 10^{-6})^{-0.5}$$
$$= 1.192 \times 10^7 \text{ N m}^{-2}.$$

in the matrix, volume fraction, average particle diameter and mean inter-particle spacing. The increase in stress due to the presence of second phase hard (non-deformation) particle is inversely proportional to the size of the particle. As many dispersion-strengthened alloys are prepared through the powder metallurgy route, this relation is of significance to control the ultimate mechanical properties of such alloys.

11.12 IMPACT TEST

Impact toughness is a measure of the ability of a material to absorb energy up to fracture. Both Izod and Charpy tests involve the same type of measurement but differ in the form of the test pieces. Both involve a pendulum swinging down for specified height to hit the test piece and fracture it. Figure 11.9 shows the test piece for Charpy test. This test is very common for PM materials which have some porosity. In the Izod test, one end of the specimen is firmly held between vices which might promote fracture of the test piece (porous) during fastening.

Test pieces for the impact test are generally obtained by pressing and sintering. The die cavity has a rectangular section of dimension 55 × 10 mm and is preferably made of cemented carbide. The thickness of the test piece must be 10 mm with a tolerance of ± 0.2 mm on both the dimensions of the cross section. If the sintered materials are particularly tough it is possible to make standard U- or V-notches as with wrought materials. The pressing direction must be noted on the test pieces as the direction of impact must be perpendicular to it.

In the whiteware ceramic industry, the impact test is a common test for quality control.

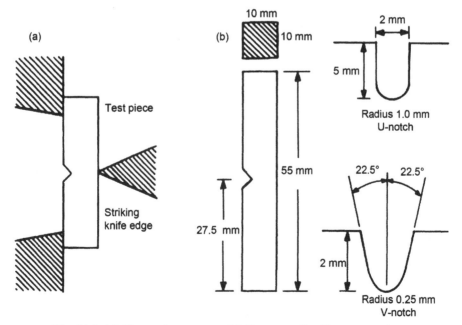

Fig. 11.9 (a) Charpy impact test; (b) Geometry for Charpy test pieces

11.13 FRACTURE TOUGHNESS

For the evaluation of fracture toughness, the resistance to crack propagation under non-dynamic conditions of stress is measured. The impact toughness test described in the previous section has major drawbacks.

1. The result obtained is purely comparative and cannot be used as a design parameter.

2. The values obtained when testing a sintered product are very low, inaccurate and hard to reproduce.

On the other hand, the fracture toughness test offers many advantages. The value obtained is a quantitative assessment of the toughness of the material which can be used as a design parameter. In most of the cases, for the range of materials covered by powder metallurgy (excepting fully dense sinter forgings), it is not necessary to induce an initial crack by fatigue stresses, and a suitable machined notch is adequate

The stress intensity factor K characterises the severity of the crack and depends on crack size, stress and geometry. So far, the material is assumed to behave in a linear-elastic manner. Fracture toughness signifies the resistance of brittle fracture, as long as K is below a critical value K_C. The values of K_C for a material depends on temperature, loading rate and thickness of the material under test. The lower limiting value of the critical stress intensity factor is called the *plane strain fracture toughness* and is denoted by K_{IC}: lower the value of K_{IC}, the less tough the material is. The SI unit of K_{IC} is MPa m$^{1/2}$. Fracture toughness can be as low as 0.5–1 MPa m^{-2} for glass and as high as 150 MPa m^{-2} for alloy steel.

The validity of applying fracture toughness testing in a given sintered product is established if it meets the following criteria:

- The shape and size of the test pieces must ensure plane strain during the three-point bending.

- An accurate method of measuring load and crack opening displacement (COD) employed.

- It must be established by testing that the measured stress intensity factor (K_C) is in fact equivalent to the critical or minimum fracture toughness (K_{IC}) for each type or range of materials considered.

The test consists of a three-point slow bend test similar to that used for measuring transverse rupture strength. The toughness K_C is calculated from:

$$K_C = \frac{P}{BW^{1/2}} \times Y \tag{11.18}$$

where, P is the fracture load, Y is the geometrical factor also known as compliance factor, B is the specimen thickness, and W is the specimen width.

Provided plane–strain conditions apply (largely determined by correct specimen geometry),

$$K_C = K_{IC} \tag{11.19}$$

In the laboratory, fracture toughness testing involves test pieces with sharp notches being strained until the crack propagates and the test piece fails. The problem in obtaining test pieces is producing the sharp notches. Figure 11.10 shows the dimensions of typical test pieces for three-point bending and tensile loading, according to British Standards BS7448. For the test a strictly increasing load is applied to a test piece, which causes the notch to open. The maximum value of the load (P) is recorded before breaking occurs as a result of crack propagation.

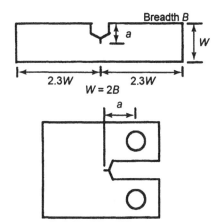

Fig. 11.10 Fracture toughness test pieces for three-point bending (top) and for tensile loading (bottom)

The Palmquist method is another very handy method to measure fracture toughness of brittle materials, particularly cemented carbides. It involves measurement of crack lengths emanating from the corners of indentation diagonals in a Vickers hardness testing machine. This is also known as 'short crack' method. Considering the problems involved in machining brittle materials into standard test specimens required for long crack materials, the Vickers indentation method is extensively used in industry.

In case of brittle solids there is a statistical variation of strength. The distribution of crack lengths in such solids would have effects. A large sample will fail at a lower stress than a small one, because it is more likely that it will contain one of the larger defects. There is, therefore, a volume dependence of the strength. Weibull defined the survival probability $P_s V_o$ as the fraction of identical samples, each of volume V_o, which survive to a tensile stress σ. According to him

$$P_s V_o = \exp\left\{ -\left(\frac{\sigma}{\sigma_o} \right)^m \right\} \tag{11.20}$$

where, σ_o and m are constants. When $\sigma = 0$ all the samples survive and $P_s V_o = 1$. As σ increases, more and more samples fail and $P_s V_o$ decreases. Large stresses cause virtually all the samples to break, so $P_s V_o \to 0$ as $\sigma \to \infty$. The constant m is the indicator showing how rapidly the strength falls as one reaches σ_o. This is called the *Weibull modulus*: lower the value of m, greater the variability of strength.

11.14 FATIGUE BEHAVIOUR

The influence of porosity is more important in fatigue tests than in other mechanical tests. The kind of test piece used for the tensile test after pressing and sintering, as shown in Fig. 11.6, can also be used for reverse bend or tension compression fatigue tests.

Fatigue tests are important after surface treatment, hardening and nitriding of sintered steels. All such treatments raise the fatigue limit, as in pore-free materials. Above the fatigue limit microcracks initiate at the pores and inclusions and link together to form the final crack, which generally gives in a mixed transgranular–intergranular manner.

Sintered materials, if not of full density, tend to provide lower fatigue properties than the corresponding wrought compositions. Common fatigue test piece geometries and tests are the bending beam, rotating beam and axial tensile ones. Flat sintered specimens are more convenient to use. The cylindrical specimens used in rotating beam method, which provides full stress reversal on each cycle, are to be machined. The *S–N* curve provides the stress value (*S*) as a function of the number of cycles (*N*) to failure. Higher the probability of survival desired for an application, lower the allowed stress.

Compared with wrought alloys, PM materials are relatively sensitive to changes in the mean stress. Their response is more comparable with that of cast alloys. PM materials are therefore particularly suited to fatigue loading regimes with negative values of stress ratio *R* (S_{min}/S_{max}), i.e., with loading being predominantly compressive.

Typically, rolling contact fatigue is studied on gears. The controlling factor in such fatigue is Hertzian contact stress. The Hertzian contact stress *S*, between, for example, two parallel cylindrical rollers, is given by:

$$S = \frac{0.35F\left(1/r_1 + 1/r_2\right)}{\left(1/E_1 + 1/E_2\right)} \qquad (11.21)$$

where, *F* is the applied load, r_1 and r_2 are the radii of two rollers, respectively, and E_1 and E_2 are the moduli of elasticity for the two roller materials.

PM materials below full density have an elastic modulus lower than that of conventional alloys. So, for a given load, these materials operate at a lower Hertzian stress. It is therefore important to use the correct elastic modulus when carrying out design calculations on PM materials.

11.15 CREEP BEHAVIOUR

Creep is important in engineering applications involving high temperature, such as steam turbine plants, jet and rocket engines, nuclear reactors and furnace refractory linings. Creep data is generally generated either in tension, bending or compression stress mode, but it may not be forgotten that in actual engineering components, multi-axial stresses may occur.

Comparative creep tests between different materials are routinely carried out in the ceramic industries producing structural parts. The deformation under constant load at a given temperature is measured as a function of time. This can be measured optically, but the instrumented device based on linear variable differential transformer (LVDT) is now universally common.

After a rapid initial deformation, the creep rate gradually decreases to a constant rate, which continues for long periods of time. This is the rate of general intent. The time taken to reach the constant creep rate is very short at relatively high temperature, and increases greatly as the temperature decreases. Figure 11.11 schematically presents the effect of creep rate and temperature on the nominal stress during secondary stage creep.

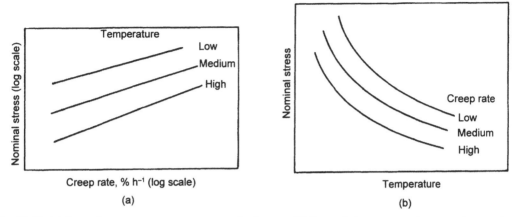

Fig. 11.11 (a) Second stage creep rate versus nominal stress; (b) Second stage creep rate versus temperature

11.16 FRACTURE BEHAVIOUR

Fracture in sintered materials may involve similar mechanisms known in wrought materials. As inclusions and cavities are important in fracture, the porosity in the sintered materials controls fracture and is responsible for its early onset. As sintered materials generally contain a relatively large volume of coarse pores, localised internal necking is able to start at relatively low plastic strains. The large pores lead to high stress concentrations, thus accelerating the spread of fracture. The real sintered materials are therefore more complex in the stress strain response; their behaviour depends on the work hardening characteristics of the matrix and on the pore size distribution and shape. Intergranular and cleavage fractures are less common than ductile failures, although intergranular fracture may occur in sintered parts at inter-particle necks which may be imperfectly bonded, e.g., due to segregation of impurities at the interface. The pulling out of particles from a fracture surface may be an indication of intergranular fracture.

11.17 WEAR RESISTANCE

Wear is defined as deterioration of the surface due to use. No machine element is immune from wear. Scuffing of piston in internal combustion engines, pitting in power transmission gears, frettage in press-fitted assemblies and cavitation corrosion in cylinder liners are some of the examples. Wear is affected by a variety of conditions: the type and mode of loading, speed, quantity and type of lubricant, temperature, hardness, surface finish, presence of foreign materials and the chemical nature of the environment. Sometimes, wear may be due to a combination of one or more elementary forms. The most common types of wear are abrasive wear and adhesive wear. In the former, hard particles in one of the sliding elements (or in the form of hard third particles) cut grooves in the softer materials in a manner similar to that involved in fine grinding. In the case of adhesive wear, the high points (asperities) on material surfaces come close enough to establish strong bonds. If a weak point in the bulk material happens to lie in the vicinity of a contacting asperity, a wear particle is generated.

It is found empirically, that for both abrasive and adhesive wear, the wear volume (V) is the function of applied load (F), the sliding distance (L), and the hardness of the softer of the two sliding elements (H). This is expressed as

$$V = K \frac{LF}{H} \tag{11.22}$$

where, K is a constant dependent on the materials in sliding contact and the sliding conditions. This non-dimensional quantity is called the wear coefficient. Some of the examples of adhesion wear are galling, scuffing, scoring and seizing. Depending on severity, abrasive wear may be of the gouging and scratching form.

Erosion and impact wears are other forms of wear. In the former, the loose abrasive particles abrade a surface, while in the latter, the reason is the impact over the surface. Deburring by vibratory finishing and tumbling is an example of impact wear.

Testing for abrasive wear in a laboratory is usually done using hard mineral particles, which produce no significant adhesive phenomenon. However, a great variety of shapes and mechanical properties of abrasive particles and diverse loading conditions give rise to variable stresses at the contact. The easiest method to determine abrasive wear is by rubbing the material against emery paper, which provides a high repeatability and accuracy of results under strictly defined conditions—low pressures and sliding speeds, abrasive grits of high strength and hardness, protection against temperature and environmental effects, and prevention of abrasive grits from movement and fracture. However, these test conditions are often inadequate to simulate diverse operating conditions found in engineering practice.

Example 11.8: Figure Ex. 11.8 (left-hand side) illustrates the typical physical phenomena occurring during a hot forging operation. List the various properties for which data is needed to understand the basics of these processes.

Solution: Data regarding various properties is given on the right-hand side of Fig. Ex. 11.8.

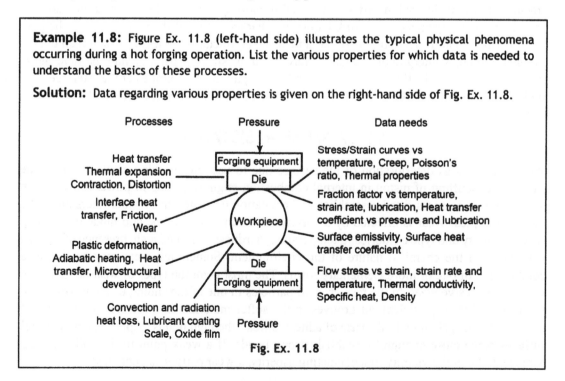

Fig. Ex. 11.8

Another method for studying abrasion wear is the rotating rubber wheel, which presses against the test piece. The testing is performed by rubbing abrasive sand particles—fed through a hopper—between the rotating rubber cylinder and the sample. Wear resistance is measured by the volume or mass lost from the sample after a preset number of revolutions. The preset number is generally 1000 or 10,000.

11.18 MACHINABILTY

The machining of PM material is complex, and standard machinability test methods for all cutting processes in such materials are lacking. Often, one type of machinability test may produce an acceptable result, while the same sample will fail in another type of test. Two tests—drilling test and turning test—are common.

Drilling test
For this test, the following features must be taken into account:

- Number of holes drilled before drill failure occurs
- Time to drill a hole
- Drill life
- Material removal rate
- Productivity measurements.

The most frequently investigated machining parameters are drilling speed, feed rate, axial-thrust force and torque.

Other variables such as workpiece material, lubrication, tool material and tool geometry must be kept constant throughout testing of one type of material.

Table 11.5 gives the average machinability values for PM steel samples. The materials are listed in the ascending order for the number of holes drilled. Once 192 holes—the limit for the test—is reached, the list continues by decreasing drilling force values. The MPIF specification for ferrous PM materials is described in Chapter 12 (Section 12.15). In the MPIF Standards 35, only corresponding rating values for machinability are listed (Table 11.5). They are complemented by the corresponding number of holes drilled, cutting edge wear and axial drilling force values. For the purpose of this standard, the machinability is determined by measuring the number of holes that can be drilled using a set of test controls, and normalising the results to a well-established baseline. For this measure of machinability the PM steels were compared with wrought AISI 1045 steel in the normalised condition. A value of 100 was established for 1045 steel. More machinable PM steels (greater the number of holes drilled) have a rating greater than 100. The test conditions used to develop the machinability rating is given in Table 11.5. The listed machinability ratings do not include the effect of density, since above the density range of 6.7–7.3 g cm^{-3} no meaningful difference was found. In general, a carbon content greater than 0.5% reduces machinability; adding 0.5% MnS to the steel or resin impregnation following sintering improves machinability by drilling. These results may or may not be applicable to other machining methods.

Table 11.5 Standard machinability rating of PM steels Drilling test: 9.5 mm drill, drill speed 1250 rpm, v = 37.2 m per minute, feed 0.23 mm per revolution, blind hole depth 25.4 mm, failure criteria: drill breakage or cutting edge wear > 0.28 mm, no coolant

Material	Number of holes drilled	Ax. drill (thrust) force		Cutting edge wear, mm	Rating
		lbf	N		
FD-0208	6.5	664	2948		5
FN-0208	22	648	2977		18
FC-0208	27	624	2771		22
FL-605	28.5	667	2961		24
FL-4205	42.5	637	2828		35
FD-0405	53.5	535	2373		44
FD-0208 + 0.5MnS	67.5	394	1439		55
FD-0405 + 0.5MnS	81	364	1616		66
AISI 1045(wrought)	122	479	2127	0.21	100
FL-4405	128	496	2202	0.21	105
FLN-4205	134.5	548	2433		110
FLN-4205 + 0.5 MnS	139	292	1296	0.41	114
FN-0205	178	494	2193	0.41	146
F-0008	192	488	2167	0.28	153
FC-0205	192	469	2082	0.36	157
FD-0205	192	416	1847	0.18	160
FC-0208 + 0.5MnS	192	391	1736	0.30	168
FL-4205 + 0.5MnS	192	365	1718	0.33	171
F-0005	192	377	1674	0.41	177
FN-0208 + 0.5MnS	192	336	1638	0.33	181
FL-4605 + 0.5MnS	192	305	1621	0.33	184
FC-0205 + 0.5MnS	192	336	1492	0.20	201
FN-0205 + 0.5MnS	192	305	1354	0.38	220
FD-0205 + 0.5MnS	192	304	1351	0.38	220
F-0008 + 0.5MnS	192	300	1332	0.33	222
F-0005 + 0.5MnS	192	231	1023	0.28	263
FC-0208 + IMP	192	193	857	0.20	286
FC-0208 + 0.5MnS + IMP	192	161	715	0.18	305
FN-0208 + IMP	192	153	679	0.13	310
F-0008 + IMP	192	140	622	0.08	317
FC-0205 + IMP	192	123	546	0.20	328

IMP - resin impregnated: 1 1bf = 4.44 N

Turning test

This test involves straight or face turning specimens of cylindrical shape of various outer and inner diameters and lengths. In the short-time tool life test, the turning occurs from the surface of the central hole to the circumference of the ring-shaped specimen at constant feed and depth of cut. The test is finished at the cutting speed at which the tool totally fails. For testing interrupted machining, rectangular specimens are used. Cutting speed and depth of cut in combination with cutting tool (material, geometry) and the workpiece are the major variables in the turning test.

11.19 ELECTRICAL RESISTIVITY

Electrical conductivity is an important property for evaluating the performance of electric contact materials, which are often prepared by powder metallurgy route. The contact area and, thus, electrical resistivity is affected by several factors, including oxide films, lubricants, compacting pressure and sintering parameters, as well as powder characteristics. The physical properties of the metal concerned will also have to be considered. For measurement of electrical resistivity a Kelvin double bridge set up is useful. Spring-loaded knife edge contacts are required with side-feeding of the current. The potential difference is then measured across two points within the length of the compact. The width, thickness and resistance of the specimen are measured. The apparent resistivity (ρ) is found from the equation:

$$\rho = \frac{RA}{L} \tag{11.23}$$

Another method is based on eddy current principles and the instrument is designated primarily to measure the electrical conductivity of non-ferromagnetic specimens with a flat surface larger than 2 mm diameter, and thickness greater than 0.3 cm. A single probe or test coil carrying 50,000 Hz AC frequency is placed on the specimen and the electrical conductivity is found by adjusting a calibrated dial to bring the deflection of the galvanometer to zero. This operation compares the impedance of the probe with that of a standard coil inside the instrument case. Before use, the electrical circuits must be adjusted by using two specimens of known conductivities. The conductivity reading obtained is not affected by surface roughness or surface films, provided the probe is no more than 0.1 mm from the specimen surface. Since the area covered by the probe is only about 3 cm^2, the instrument is useful for detecting inhomogeneities, segregation and small surface cracks in flat strips obtained after powder roll compacting.

11.20 MAGNETIC PROPERTIES

The non-destructive quality that is widely used to test ferrous alloys and cemented carbides is the magnetic property. Magnetic properties are affected by the presence of pores because of:

- a lower quantity of magnetically active material per unit of volume,
- an alteration in the distribution of magnetic flux.

In addition, the pore size, shape and their distribution along the flux path must be taken into account.

Devices for the measurement of magnetic properties of soft magnetic materials permit two types of test pieces: ring or bar. In the former, the closed-ring test pieces, square in section, are obtained directly by pressing and sintering, to have uniform section in all directions with minimum variation in density. In such tests, a direct magnetising current is used, with a stabilised generator and a ballistic galvanometer to measure the induction. Induction and density are correlated in linear mode, while there is a less evident correlation between the maximum permeability or coercive force and the sintering conditions.

In the latter case, i.e., bar test pieces, the method is used less frequently, because of the lack of accuracy of the method of measuring the circuit. The bars are inserted in a core or yoke, which is a part of the measuring circuit. They must be straight with a constant cross section along their whole length. Both types of test pieces—ring or bar—must be deburred before winding, and their dimensions should be checked.

In case of hard magnetic materials, unlike soft magnets, magnetic tests are carried out directly on the sintered parts, as long as their dimensions are compatible with the instrument, the shape is cylindrical or parallelepiped and the flux is in the direction of the axis. In case the part is of complex shape, it is necessary to make it constant cross section in the direction of the flux. The main features of such a magnet are obtained from the demagnetisation curve traced on the second quadrant of the graph of the hysteresis cycle. The test piece is placed between the poles of an electromagnet in which a direct amount of variable intensity and sign is fed. The field is measured with a Hall probe and gaussmeter, while induction is measured with a coil wound around the sample and an integrating fluxmeter. Such measurements are helpful in relating composition and any heat treatment on the permanent magnets. In the case of cemented carbides, because the factors influencing coercive force are complex and interactive, the test is employed to indicate departure from a norm, rather than to determine absolute values of cobalt content, grain size or carbon deficiency. When both cobalt content and grain size for a particular grade are controlled within reasonable limits, the test can give a clear indication of carbon deficiency and the consequential presence of the embrittling η-phase.

11.21 CORROSION RESISTANCE

All corrosion consists essentially of reactions between a solid phase and its environment. One of the most popular test methods for corrosion study is the weight loss method, where the weight of the specimen is semi-continuously recorded while corroding in a test solution. However, this method is not applicable when testing passive alloys, such as stainless steels, as the degree of weight loss is near the vanishing point.

Another method consists of measuring and interpreting polarisation curves (Fig. 11.12). The specimen to be investigated (2) is connected as an electrode into an electrolytic circuit together with a platinum electrode (1). The solution usually contains NH_2SO_4 and has to be entirely free of oxygen or reducible ions. The idea is to change the potential of the specimen electrode from negative to positive values, and in the same time record both the electrode potential and the current density value. These operations are done with a potentiostat and from the measurements the polarisation curve is constructed (Fig. 11.13).

The polarisation curve and the coordinates represent a kind of plot which describes the electrochemical activity, that is, the corrosion behaviour of the material as a function of its potential in solution. All current values above the zero line represent anodic current densities, or corrosion currents. Values below the zero line represent corresponding cathodic current densities, i.e., mainly the rate of hydrogen evolution on the material surface.

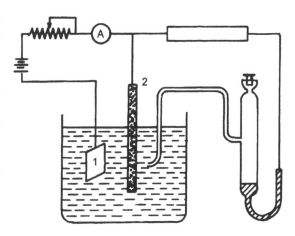

Fig. 11.12 Schematic diagram of apparatus for measuring polarisation

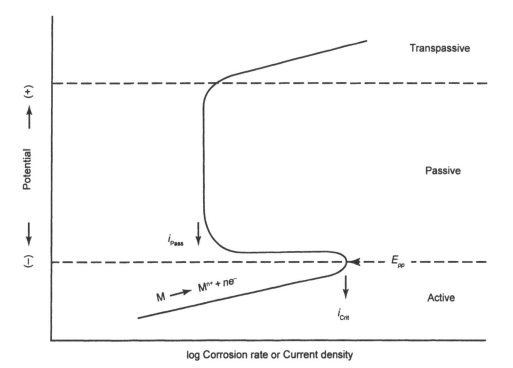

Fig. 11.13 A typical polarisation curve

The active–passive transitions exhibited by metals is caused by the formation of a surface film of hydrated oxide that interferes with the anodic reactions and drastically lowers the corrosion rate. Figure 11.13 illustrates the presence of some critical E_{pp}, such that, below it the metal corrodes at a relatively high rate. Depending on the oxidising power of the

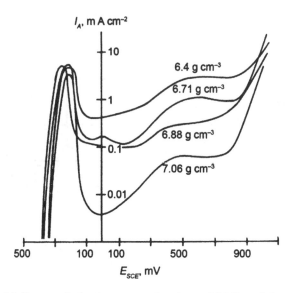

Fig. 11.14 Some polarisation curves for sintered 316L stainless steel

solution, an alloy may exist in the passive state above E_{pp}, or in the active state below it. Passivity must not be considered as the ultimate solution for corrosion resistance. In case the passive film is brittle, its failure may result in localised form of corrosion, including pitting, crevice corrosion or stress corrosion cracking. A more severe condition of higher acidity and temperature generally decreases the passive potential range and increases current densities and corrosion rates at all potentials.

Figure 11.14 illustrates some polarisation curves for 316L stainless steel sintered to different densities. It is clear that higher the density, lower the passive current density, but at the same time there are only small differences in the passivation current densities.

11.22 QUALITY CONTROL OF PM PARTS

The quality and reliability of a PM part, like any other engineering component, must be defined as a basic part of the specification of the product. Many aspects of quality and reliability can be expressed in objective and measurable terms; but certain aspects will always be subjective and based on the user's opinion. Quality can be achieved at three stages in the life of a product: design stage, manufacturing stage and service stage. Variation in the quality of manufactured parts is inevitable and decisions must therefore be made about the standards of quality and reliability which will be achieved, and the cost which will be incurred in meeting these standards.

Some of the specific quality aspects of manufacturing planning include:

- Choice of machines, processes and tools capable of holding the tolerances
- Choice of instruments of adequate accuracy to control the processes
- Planning the flow of manufacturing information and inspection criteria
- Planning process quality controls.

Quality control can be carried out at three stages.

1. Raw materials are checked to ensure that the standards specified are present.
2. Components are checked as they are being made within the company's own manufacturing unit.
3. Completed products are tested in the operation environment to ensure that they are able to perform to specification.

Sintered products are often used to replace traditional materials for cost saving or technical improvement, or ideally both. In either case it is necessary to quantify the service parameters in terms of physical, mechanical and chemical properties. Moreover, while dealing with sintered products, one must emphasise the need for function awareness when specifying any material. Define the function of the component and develop functional tests which will determine whether the part will do the job. This brings out the importance of differentiating between material tests and product tests. The former provides information about the properties of the material of a test piece in relation to its intended applications. However, the product tests specifically check the immediate fitness for service of the product and the batch from which it has been drawn.

Some factors which may handicap the quality of sintered products are:

- Testing and inspection procedures which do not realistically reflect actual use situation, e.g., particle size yield in floor screening by vibroscreens versus routine laboratory screening.
- Arbitrary material substitution by purchasing or manufacturing departments, without adequate engineering evaluation. This is very important in case of tool material selection for complex PM parts.
- Crash design revisions to incorporate new features in existing designs with minimum tooling changes. Here a prototype development can help considerably in building up confidence.
- Failure to apply the same evaluation methods to purchased components or powders as are applied to internally manufactured ones.
- Failure to anticipate misapplication of the product by the user: for example, in selecting the proper grade of cemented carbide for different cutting purposes.
- Too little consideration given to the wide variations in the physical and intellectual abilities of customers.
- Interpretation of the statistical quality control function as absolute quality assurance rather than as a basis for action.
- Inadequate advice to the user of safety procedures related to the product. For example, iron base metal powder sintered bearings are more suitable for higher loads and slower speeds than copper base bearings.
- The PM process allows considerable cost variation if specific part requirements are not clear. Such a situation would consequently bring forth a considerable quality variation too. For widely different costs there may be various reasons such as revision of tolerance, difference in manufacturing practices, introduction of supplementary processes such as repressing, etc., and lack of any specified minimum density.

In brief, gross differences in quality may lead to a serious misunderstanding of requirements. A tendency for reduced manufacturing cost and increased short time profit is another handicap in PM part quality.

The inspection of any cracks in green PM parts is important in order to efficiently utilise the sintering capability of the plant. Moreover, the recycling of green parts is much more efficient than the finished off-quality rejected sintered parts. For such inspection, a multiprobe electric equipment is developed, where the relatively high electric resistivity of PM green parts permits the creation of substantial variations in electric fields inside and on the surface of the samples in response to an applied current. The resistivity voltage patterns established on the surface provide sufficient information to detect both surface cracks and sub-surface defects. The process can be computer controlled.

In the following sections, the testing and quality control of some typical sintered parts are described.

11.22.1 Filters

For filters, the following functional performances are evaluated:
- Geometric characteristics
- Fluid permeability
- Filtration threshold.

While most tests of different porous PM parts are similar, no standard test fixture has so far been established in the industry. It is important for each test procedure to be worked out jointly between supplier and customer. Fluid permeability tests may be carried out with a simple instrument. The type shown schematically in Fig. 11.15 conforms to the MPIF standard 39–68. The filtration threshold is measured by the bubble test method, following the procedures of ISO standard 4003. Chemical composition of the filter material is an additional check, when there is onset of corrosion during operation.

11.22.2 Porous bearings

Porous bearings or self-lubricating bearings are those which retain a considerable amount of lubricant in their pores. This lubricant comes out of the pores and provides a film between shaft and bearing when the shaft begins to turn. The major material properties to be specified in the case of porous bearings are adequate porosity, pore size distribution and adequate strength. The quality of oil is also an important factor, since it affects the performance of the bearing more than the material of which it is made. Porosity in terms of interconnected void space should not be less than 18% in case of such bearings. In addition, radial crushing strength should not be less than the value calculated by:

$$P = \frac{KLT^2}{D-T} \qquad (11.24)$$

where, P is the radial crushing load in kg, K is the strength constant for the grade and type specified, L is the length of the bearing in cm, T is the wall thickness of the bearing in cm and D is the outside diameter of the bearing in cm.

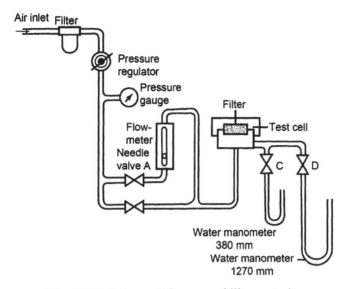

Fig. 11.15 Instrument for permeability control

Radial crushing strength is determined by compressing the test specimens between two flat surfaces at a 'no load' speed of 2.5 mm min^{-1}, the direction of the load being normal to the longitudinal axis of the specimen. The point at which the load drops, due to the first cracks, is considered the crushing strength. This test is applied to plain cylindrical bearings; flanged bearings are tested by cutting off the flange and compressing the two sections separately.

Bearing performance test is generally carried out on oiled bronze bearings using a bearing tester at nominal PV values of 115 MPa m min^{-1}. The torque, peak and steady-state temperatures, bearing wear and oil loss are measured. Maximum speed and load are applied from cold start. Performance tests are completed in about five to six hours.

Hour glassing and out-of-round (see Section 8.1) are important properties in the ability to size the bearings to closer tolerances. The hour glassing and dimensional changes gradient on the premix bronze bearings can be attributed to the phase changes occurring in the bearing during the alloying of tin, and the liquid phase sintering.

Air permeability is a good indicator of the general pore size and oil flow through the bearings. Finer pore size typically results in lower permeability on Darcy number.

11.22.3 Structural parts

In powder metallurgy structural parts to begin with design aspect must be carefully looked into. The shape of the parts is to be optimised and the material they are made of must be selected so that they perform without deformation and without loss of integrity. Following design, problems can arise in service, such as crack initiation, which require calculations of the stress and strain distributions. Finally, failure assessment of the part must be done. To reach these objectives, more sophisticated design tools are available for engineers, provided they are suitably supplied with reliable data concerning the mechanical behaviour of materials.

In any part testing, it is important to conform rigorously to the standards, in order to avoid any dispute between client and supplier, and also to achieve the most meaningful results. The provisions of standards result partly from compromises based on considerations that are barely scientific, but also on sound theoretical considerations and on round-robins. It is not always easy to understand the reason for which a particular requirement is imposed. Sometimes, its importance is underestimated, particularly when difficulties are found in the literal application of procedures.

As an illustration, component testings of some of the structural PM parts are given below.

Baulk ring in automotive gear box: The test is done to determine bursting load. A typical burst load for a baulk ring is 1.5 tonnes, at which there is an adequate safety margin for actual service.

Timing-chain sprocket: Such sprockets are used to drive the chain which conveys motion from an engine crank shaft to the camshaft. The test comprises loading an individual tooth and measuring the torque to cause failure.

Bevel gear: The gear is placed in a mesh with a solid fixed gear in the jig and 2.7 kg (6 lb) weight is dropped from a height of 230 mm on a collar located in the root of one of the teeth of the bevel gear. The teeth around the gear, which can be checked individually, must withstand this impact of 6 Nm without breaking.

Car seat belt latch and cog: In this test, the latch and cog are placed in the jig and subjected to a torque load of 135.6 Nm.

Differential pinion: In the impact fatigue testing, a 90 N hammer is dropped initially on the striker from a height of 15 cm. For each successive blow, the drop height is increased by 2.54 cm until failure occurs. While the data from the type of test was somewhat difficult to treat statistically, the advantage of this test is that only a few blows (less than 50%) are required to fail the pinion. Differential fatigue testing differs from impact fatigue testing in that it is primarily a running test and during each revolution of the pinion member, all possible loading positions are encountered. Consequently, by maintaining constant applied torque to the differential gear set, the maximum bending stress in the root fillet of a pinion tooth is always obtained and the test becomes insensitive to tooth load height position. However, a major disadvantage of the method is that two pinion members and side gear members are required to make up the differential assembly for one test.

Figures 11.16a–11.16d illustrate some examples for the structural part testing fixtures.

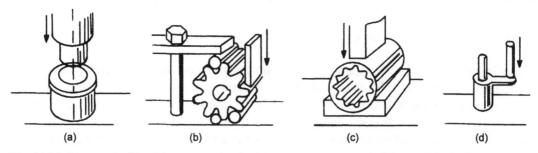

| (a) | (b) | (c) | (d) |

Fig. 11.16 Some typical breaking test modes for (a) cup-type component, (b) gear, (c) tubular component and (d) lever-type component

Example 11.9: For the successful performance of an engineering component, the evaluation of its mechanical properties becomes more important than that of the material itself. Show schematically what mechanical tests on a sintered synchronising ring for an automobile transmission system will be recommended.

Solution: See Fig. Ex. 11.9. Such testing of the component in a laboratory is not sufficient and actual performance in the field is to be studied.

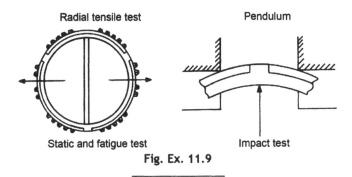

Fig. Ex. 11.9

Example 11.10: For high performance PM gears, surface densification is carried out. Describe the significance of the same and also the actual product testing. What is a typical heat treatment for such gears?

Solution: PM gears are typical examples of structural parts which undergo arduous service conditions. Depending on the strength requirements of the gear, different sintered (core) densities can be provided. Low strength requirements can be fulfilled by part densities ~7.1 g cm⁻³, while more highly loaded applications may require core densities greater than 7.3 g cm⁻³ and full density in the critical stress regions at the tooth flank and tooth root region. For the heaviest-duty applications, hot forging after sintering can be used to supply gear performs with densities greater than 7.7 g cm⁻³.

After preform production, the surface layer can be densified by one of the techniques, namely precision rolling or surface densification. The main purpose of the former is to improve the dimensional precision of a gear. Depending on the process design, a minor amount of densification of the surface layer, typically below 0.2 mm, may occur. In the latter case, i.e., surface densification, the primary aim is to create a fully dense layer within the stressed area along the tooth profile. Both the tooth root and tooth flank can be rolled and densified in order to considerably increase the tooth root and tooth flank performance of sintered gears. Here, the depths of densification up to approximately 1 mm or more can be achieved. The most critical design part in this case is the profile modification of the rolling tool. By applying the computational method for preform and rolling tool design, it is possible to modify the initial rolling tool precisely and deliberately, based on the outcome of the first rolling trial.

For testing the service behaviour of PM gears, one of the common test rigs (Fig. Ex. 11.10) is known as FZG (Forschungsstelle fur Zahnräder und Getriebebau), as it originated in Germany. This test rig is used to determine the scuffing load capacity, pitting resistance, tooth load carrying ability, and generate the S-N fatigue data. (Continued on next page)

The test rig contains a set of test gears and a set of drive gears of the same tooth geometry/tooth count. The test gears are loaded on the open end of the machine, and there is a closed loop power cycle. The two parallel shafts are connected to each other by the gears on both ends of the shaft. The torque is applied via a split shaft coupling (load clutch). The coupling allows torque to be applied to the shafts, which in turn apply torque to both sets of gears. Once torque is applied, the split shaft coupling is tightened and the torque remains in the system. The motor then turns the shaft, and the gears are run at the preloaded torque. This method of testing reduces the need for a brake or other friction device to apply torque to the gears which eliminates power losses due to braking. This allows for a smaller amount of input power to compensate for losses in the system.

The torque is applied using predetermined load intervals, and the test is run for a predetermined number of cycles. These test conditions are detailed in ASTM D5182-92, and are summarised in Table Ex. 11.10A.

Each load stage runs for a total of 21,700 revolutions of the pinion. After load stage 4, the gears are inspected for tooth wear, spalling or tooth breakage. When the sum of the total damage on all the teeth is equal to or greater than one gear tooth width, the specimen is considered to have failed. In some cases, after load stage 8 (Table Ex. 11.10A), if the gears do not show signs of failure, they are further subjected to stage 8 testing until failure occurs. This is then reported as the failure stage. For example, if a gear was to run through load 8 three times prior to failure, the load stage at failure would be labelled as an 8-3 failure.

Table Ex. 11.10B specifies typical test gear geometry.

Tooth-surface fatigue is caused when the loading force on one tooth is transferred to another tooth by intimate contact. The load area moves up and down the gear tooth profile as the gear and the pinion proceed through the mesh.

The most common heat treatment is carburising, which comprises a single boost/single diffusion cycle at 925°C for 90 min in a carburising gas followed by holding at the same temperature for 90 min. (Continued on next page)

[For full details refer: PM Gear Technology Focus issue, *Int. J. of Powder Metallurgy*, 42(1), 2006].

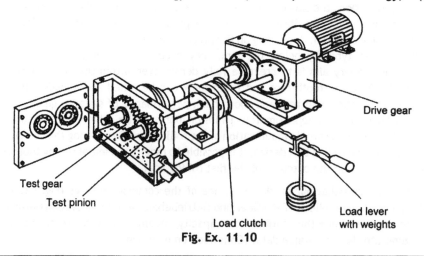

Drive gear

Test gear

Test pinion

Load clutch

Load lever with weights

Fig. Ex. 11.10

Table Ex. 11.10A FZG test conditions

Load stage	Torque on pinion, N-m	Tooth normal force, N	Hertzian contact pressure, N mm^2
1	3.3	101	124
2	13.7	421	254
3	35.3	1,084	407
4	60.8	1,867	535
5	94.1	2,890	665
6	135.5	4,161	798
7	183.4	5,632	928
8-X	239.3	7,349	1,061

Table Ex. 11.10B Test gear data

Diametral pitch (in^{-1*})	8	
Nominal pressure angle, deg	20	
Module, mm	3.175	
Base pitch, mm	9.373	
Profile contact ratio	1.319	
	Pinion	**Driven**
Number of teeth	20	24
Outside diameter, mm	71.920	84.620
Effective outside diameter, mm	70.690	84.133
Reference pitch diameter, mm	63.500	76.200
Normal tooth thickness at Ref PD, mm	5.461	5.728
Start of active profile, mm	61.109	74.199
Maximum specific sliding ratio	1.793	1.338

* 3–9 per 100 mm

Example 11.11: Describe the factors affecting the fatigue testing of a PM connecting rod.

Solution: The connecting rod for an automobile offers a case of classical fatigue application. Fatigue testing of such parts can be dealt with as follows:

Specimen shape: The test piece specimens in sintered materials have to be looked at from additional considerations. Almost all sintered parts have rectangular cross section, edges and unmachined surfaces. Therefore, data obtained with machined, unnotched, round specimens are difficult to be applied for fatigue design—especially when they are determined under fully reversed ($R = -1$) rotational bending.

Notched vs unnotched specimen: Notched specimens deliver data which are the closest to reality, because parts without notches do not exist. Round bars can be notched, but machining of PM part influences the properties significantly. Central notched specimens are suitable for the assessment of bores (e.g., oil bore of connecting rods) only under axial loading. On the other hand, the double-edged notched specimens are suited for the assessment of fillet notches. Therefore, the specimens may be specified with different radii and consequently different loading mode dependent notch factors.

Fracture mechanics role: The connecting rods manufactured through powder metallurgy have shear crack, due to the interaction between local powder flow and shear stress during the compaction stage (Fig. Ex. 11.11). In the preliminary prototype test in fired engines, it is essential to establish whether this defect will lead to a premature failure of the rod. In case there is a premature failure, the fatigue behaviour analysis of other critical areas of the component is unnecessary. The fracture mechanics role of crack propagation during fatigue may be applied in case of connecting rods. For the engine test to succeed, the ΔK value must be lower than the threshold value.

Based on experience, and assuming a logarithmic Gaussian distribution, it is possible to calculate the failure probability by coupling the various scatters. The safety factor is a function of the standard deviation of the loading, the standard deviation of the strength values around a mean value, and the standard deviation of the mean value fluctuations. The deviation of allowable stresses with a safety factor presupposes a certain production reliability and knowledge of the maximum loading which occurs in actual service. If the above-mentioned scatters are partly unknown, the safety factor must be increased, as a result of which, the design curves must be lowered to smaller allowable stresses.

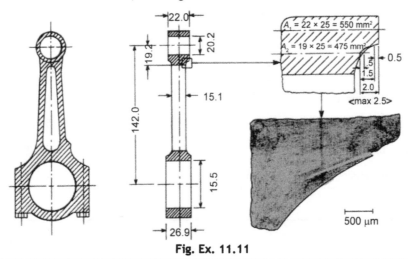

Fig. Ex. 11.11

11.22.4 Cemented carbide cutting tools

There is a large extent of ignorance on the part of customers of cemented carbides as to what in fact they really need. The customers of cemented carbides are interested in 'service'. Most of the properties measured are those which assist in controlling the quality of hard metal during manufacture, for example, hardness, density, transverse rupture strength, coercive force, etc. However, the direct relationship of these properties with application of these hard metals is still far distant. Basically, there are three major difficulties in the evaluation of cemented carbides. The causes of such difficulties are:

- The material is basically brittle, which introduces a psychological barrier in the mind of designers.
- The materials are applied in extreme conditions of temperature, pressure or time. These are difficult to describe and investigate.
- In the case of light machining applications, wear resistance could roughly be related to hardness. Wear in the case of hard metals is a collective term which signifies such categories like erosion, frittering, oxidation, corrosion, and most important of all, diffusion wear. In other words, currently it is impossible to measure wear resistance in hard metals as a 'single' property.

11.23 NON-DESTRUCTIVE TESTING

Since flaws and their morphology are critical to successful application of PM products, the development and application of non-destructive test (NDT) techniques are essential. If flaws are detected in the green body, the expense of processing a part that is no good is avoided.

Radiography: Conventional radiography uses energetic γ-ray or X-ray sources and very fine grain film to detect flaws in solid bodies. The size of the defect that can be detected in this manner is dependent on the thickness of the part, its X-ray absorption characteristics, the size of the flaw, the orientation of the flaw, and the X-ray opacity of the flaw relative to the flaw. The X-ray image has to be taken in more than one orientation to observe all the defects, since thin defects, such as closed cracks, will not be observed when the X-rays are parallel to the cracks.

Ultrasonic NDT: Ultrasonic NDT can be used to detect sub-surface flaws in PM components. A schematic showing the principle of this method is given in Fig. 11.17. A piezoelectric transducer near the part emits ultrasonic waves that pass through the part. When the part emits ultrasonic waves that pass through it and when these waves hit a discontinuity in the material, secondary waves are generated due to scattering or reflection. A receiver detects these secondary waves and converts them to an image. As in the case of radiography, the inspection has to be made in more than one orientation. Ultrasonic inspection is best for flat-sided parts while complex shapes present patterns that are difficult to interpret.

Dye penetrant: Penetrant fluids are employed to detect surface flaws. In this method, a fluorescent dye is painted on the part and then dried. Wherever there is a surface crack or pore,

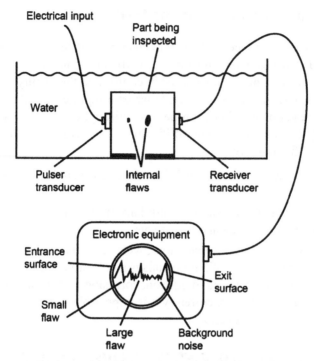

Fig. 11.17 Schematic illustrating the basic principles of ultrasonic NDT

the penetrant is trapped and observed by means of ultraviolet light. However, if the part has interconnecting porosity, the method will not work, since the entire part will fluoresce under UV inspection.

11.24 STATISTICAL QUALITY CONTROL

Measurements of the set of properties required for quality control bring forth two important parameters: accuracy and precision. *Accuracy* is a measure of the closeness of a particular measurement to the true value of a property. *Precision* refers to the reproducibility of multiple measurements of the same property. A poorly calibrated tool may give reproducible measurement values which are far from the true value; the results from using such a tool would be precise but inaccurate. A tool yielding reproducible values very near to the true value would be both precise and accurate.

Whenever a large number of measurements are made during quality control testing in a production plant, there is always some variation in the results of these measurements. It is true even when all items measured are from the same production run and are all tested in the same way. Part of this variation is a result of actual differences in the items being measured, and part is a result of errors made during the measurement.

The usual way to answer the variations is to first reduce the large volume of test data down to two or three representative quantities called statistics. If the results of a large number of

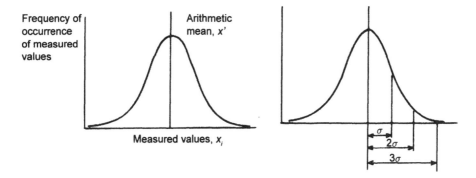

Fig. 11.18 Normal distribution curves for variable data exhibiting: (a) Arithmetic mean and (b) Standard deviation

measurements of the same property are graphed as frequency of occurrence of a specified value versus the magnitude of the value, the result will usually resemble Fig. 11.18. The curve is called a *normal* or *Gaussian* distribution. Sometimes, the curve may be slightly distorted or skewed, but usually this does not much affect the statistics calculated for the distribution. It is possible to express both the clustering and dispersing tendency of the data by calculating two simple statistics from the group of measured values. These are *arithmetic mean* and *standard deviation.* The arithmetic mean is calculated according to the formula:

$$X' = \frac{\sum\limits_{i=1}^{n} x_i}{n} = \frac{x_1 + x_2 + x_3 + \ldots + x_n}{n} \tag{11.25}$$

where, n is the total number of individual measurements. The value of X' represents the average of all the individual measurements x_i and normally corresponds to the value at the peak in the frequency distribution. The units on the mean are the same as those on the individual measurements.

The dispersion tendency of the data can be expressed by several different statistics, the most common of which is the standard deviation. This statistic is computed as:

$$\sigma = \left[\frac{\sum\limits_{i=1}^{n}(x_i - X')^2}{n} \right]^{1/2} = \left[\frac{(x_1 - X')^2 + (x_2 - X')^2 + (x_3 - X')^2 + \ldots + (x_n - X')^2}{n} \right]^{1/2} \tag{11.26}$$

The difference between each individual measurement (x_i) and the arithmetic mean (X') is squared; these quantities are averaged, and the square root of this average yields the standard deviation. The units of the standard deviation are the same as the units of the individual measurements. [Note: Some treatments of statistics utilise $(n - 1)$ rather than n in the denominator of the equation].

SUMMARY

- In the absence of interaction between different testing procedures, more than one type of test may be carried out on the same test piece.
- Unlike wrought metals and alloys, PM materials have heterogeneous microstructure.
- Lineal or point counting is used for determining the relative volume fraction of constituent phases in PM material.
- X-ray diffraction analysis is a convenient tool to study the homogenisation process during sintering of green compacts of powder mixes, where alloying takes place.
- Differential thermal analysis data are frequently obtained simultaneously and correlated with gravimetric changes occurring in the specimen during the heating cycle.
- A laser source is commonly used to measure the thermal diffusivity from which one can calculate the thermal conductivity.
- The influence of porosity is more important in fatigue tests than in other mechanical properties.
- The polarisation curve describes the electrochemical activity, i.e., the corrosion response of the metal as a function of its potential in solution.

Further Reading

Anonymous, *Common Cracks*, Metal Powder Industries Federation, Princeton, 2004.

Beiss P, Dalal K and Peters R, *International Atlas of Powder Metallurgy Micro*structures, Metal Powder Industries Federation, Princeton, 2003.

Bowdens FP and Tabor D, *The Friction and Lubrication of Solids*, Oxford University Press, Oxford, 1950.

Brandon D and Kaplan WD, *Microstructural Characterization of Materials*, John Wiley and Sons, Chichester, 1999.

Dieter GE, *Mechanical Metallurgy*, McGraw Hill, New York, 1988.

Dowling NF, *Mechanical Behaviour of Materials*, Prentice Hall, New Jersey, 1993.

Fontana MG and Green ND, *Corrosion Engineering*, 2nd edn, McGraw Hill Co., New York, 1978.

Haynes R, *The Mechanical Behaviour of Sintered Metals*, Freund Publishing House, London, 1981.

Hertzberg RW, *Deformation and Fracture Mechanics of Engineering Materials*, John Wiley and Sons, NY, 1989.

Lall C, *Soft Magnetism: Fundamentals for Powder Metallurgy and Metal Injection Molding*, Metal Powder Industries Federation, Princeton, NJ, 1992.

Mosca E, *Powder Metallurgy Criteria for Design and Inspection*, Mechanical, Metallurgical and Allied Manufacturer's Association, Turin, Italy, 1984.

Rice RW, *Porosity of Ceramics*, Marcel Dekker, New York, 1998.

Robinowicz E, *Friction and Wear of Materials*, 2nd edn., John Wiley and Sons, New York, 1995.

Standard Test Methods 2007, Metal Powder Industries Federation, Princeton, 2007.

Toraya H, Yoshimura M and Somiya S, Calibration curve for quantitative analysis of the monoclinic-tetragonal ZrO$_2$ system by X-ray diffraction, *J. Am. Ceram. Soc.*, 67:C119–21, 1984

Upadhyaya GS and Anish Upadhyaya, *Materials Science and Engineering,* Viva Books, New Delhi, 2006.

Wachtman JB, *Characterization of Materials*, Butterworth-Heinemann, Oxford, 1973.

EXERCISES

11.1   Give three examples each of structure-sensitive and structure-insensitive properties of engineering materials.

11.2   Which optical property of a material is important for determining the preparation of a sample for optical microscopy?

11.3   Give one example where the dark-field illumination is far more advantageous than the bright-field illumination.

11.4   Why is a ceramic specimen far easier to polish for microscopic examination than a metallic specimen?

11.5   Give reasons why γ-Al$_2$O$_3$, and not α-Al$_2$O$_3$ powder is used for metallographic sample polishing.

11.6   Many include porosity as the microstructural feature of a sintered part, while others do not. What is your opinion?

11.7   Mention some of the quantitative microstructural parameters, which are not routinely determined in PM industry, but have much significance in studying basic densification mechanisms.

11.8   The number of particles of a second phase per unit volume of the material cannot be estimated from a planar section without making an assumption about the particle shape. Comment on the statement.

11.9   Why is the resolution attainable in the electron microscope so much better than that obtained from an optical microscope?

11.10   Why is it not useful to analyse regions much less than 1 μm in size in the scanning electron microscope?

11.11   Explain why the lattice parameter determination of sintered nickel is far simpler than that for α-alumina.

11.12   What is the significance of knowing the chemical composition of a phase observed in a microstructure?

11.13   What technique would you use to analyse the concentration of impurities in grain boundaries of sintered alumina of 10 μm grain size?

11.14   The mean intercept length determined for a polished and etched section of a sintered material exhibiting regular grain growth is 1.3 μm. What is the mean grain size?

11.15   A multicomponent oxide refractory possesses a heterogeneous microstructure—crystalline as well as amorphous (glassy). Identify the controlling phase responsible for better thermal insulation behaviour of the refractory.

11.16   Grain size can be measured either by optical or X-ray diffraction methods. What is the additional advantage in the case of X-rays?

11.17   What contributes to the background in X-ray recording spectra? Can it be eliminated completely?

11.18   Retained austenite is a common feature in high-speed steels. What is the easiest method to detect it in the sintered steel compact, when the amount is very small?

11.19 What special feature must be incorporated to analyse elements with atomic number less than 20 by X-ray fluorescence technique?

11.20 For X-ray diffraction studies different target elements are used. After a search from the *Handbook*, write down the K_α wavelengths for the following targets and suggest the filter elements used to absorb the β-radiation: (a) Cr, (b) Mn, (c) Fe, (d) Co, (e) Ni, (f) Cu.

Is there a pattern in the change of the wavelengths and excitation potential of the above targets?

11.21 Figure P. 11.21 shows the X-ray diffraction pattern of a solid solution TiCN powder carried out in copper K_α radiation. Index the same. It is not known what is the ratio of carbon and nitrogen in this compound. Describe how you will establish the composition to the best possible accuracy. Given the lattice parameters of TiC and TiN are 0.4328 nm and 0.424 nm, respectively, what assumption do you have to make?

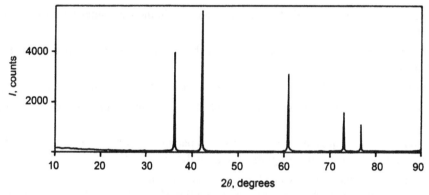

Fig. P. 11.21

11.22 Figure P. 11.22 shows the X-ray diffraction (copper K_α) pattern of two phase W–Ni–Fe PM composite prepared through liquid phase sintering at 1500°C. The phases are cubic tungsten and face-centered cubic Fe–Ni solid solution binder. Using the PDF, designate and index all the ten lines marked in the figure.

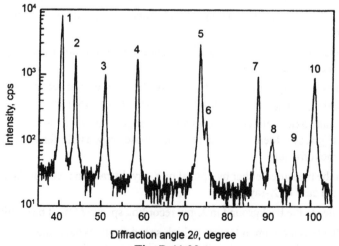

Fig. P. 11.22

11.23 Chemical analysis of a complex multiphase ceramic raw material gave two components 1% SiO_2 and 5%–10% dicalcium silicate. X-ray diffraction revealed strong SiO_2 peaks, but not so in case of silicate. What is the reason for this? What special technique will you apply to determine the silicate quantity?

11.24 X-ray diffraction pictures for clays yield extremely complex powder patterns with heavy backgrounds. What are the reasons?

11.25 A full density PM-rolled aluminium alloy sheet was subjected to cold reduction followed by heat treatment. The material was later subjected to X-ray diffraction studies. Describe the mode for studying the surface and interior details.

11.26 In case a thin PM sheet material is highly absorptive to X-rays, what variable would you think of for carrying out transmission X-ray studies?

11.27 A vendor has supplied three gold medals of different carats prepared by PM route. How will you identify them using different techniques? What effect does the change in caratage have on the pseudo-binary phase diagram?

11.28 With the Vickers hardness test a 30-kg load gave a sample of PM steel an indentation with diagonals with a mean length of 0.53 mm. What is the hardness?

11.29 Why are ceramics usually much stronger on compression followed by bending and then by tension?

11.30 Sintering temperature and time are major sintering variables. In case you obtain similar sintered density values for sponge iron powder compacts prepared from the same green density level by manipulating sintering period and temperature, justify which variable will be more helpful for attaining better strength values? Why?

11.31 Calculate the diametral tensile strength of a sintered compact if the diametral load causing failure is 5000 N and the compact is 50 mm thick and 30 mm in diameter.

11.32 How does the elastic modulus of a porous sintered material vary with applied stress?

11.33 What are the precautions to be taken before measuring the hardness of a sintered component?

11.34 For a brittle sintered material the 4-point bend test is more reliable than the 3-point test. Why?

11.35 What are the distinctive structural parameters which control the strength of sintered materials with (a) high porosity and (b) very low porosity?

11.36 Many publications on PM materials mention mechanical properties like tensile strength and ductility, but often ignore to report a property which is vital for PM composites as a structural material. What is that mechanical property? Describe its significance and method of quick evaluation.

11.37 In particulate composites, the Young's modulus is much less than predicted by the rule of mixture. Why?

11.38 TD nickel contains 2 vol% ThO_2 particles dispersed in nickel. What effect would there be for splitting the particles in half to double the number of particles, while maintaining the same volume fraction?

11.39 A sintered low-alloy structural steel is tested for fatigue strength in as-sintered as well as in heat-treated (quenched and tempered) states. In both cases a number of tests were carried out . After heat treatment, the fatigue limit of the steel increased; but at the same time the data obtained had a wider range than as sintered samples. Comment on this feature.

11.40 A TiCN-based cermet was vacuum sintered followed by HIP. The density and hardness data are given in the table below.

Operation	Density, g cm^{-3}	Hardness, HRA
Vacuum sintering	6.39	92.2
HIP (30 min)	6.51	93.0
HIP (90 min)	6.45	92.5

What other mechanical test would you like to perform in order to optimise your HIP operation?

11.41 The mechanical properties of HIP-consolidated material have smaller sample-to-sample variation than the wrought or cast counterparts. Give reasons.

11.42 Explain why inclusion shape control in any PM forging increases the transverse ductility and toughness, but has little effect on the longitudinal properties.

11.43 An extruded PM aluminium alloy with a crack has a yield strength of 480 MPa and a plane strain fracture toughness K_{IC} of 25 MPa m$^{1/2}$. Using a safety factor of 2, estimate the critical half length.

11.44 Suggest a simple and quick method for qualitatively assessing the density distribution within a green PM part.

11.45 A cemented carbide production plant due to some unavoidable reasons does not have any mechanical testing equipment. However, it has a unit to measure magnetic saturation and coercivity. Explain why you can or cannot rely on these measurements for the quality control of the following grades:

(a) Straight WC-Co alloys

(b) WC-Co containing some cubic carbides

(c) WC-Co containing Cr_3C_2 as grain-growth inhibitor.

Give reasons.

11.46 In a jig saw there is a gear drive and a counterweight. The gear is made of infiltrated PM steel, while the counterweight is made of heat-treated nickel steel. During service an excessive wear problem with the gear cam was observed. Suggest an economical solution to solve this problem.

11.47 Of the following defects in an engineering ceramic, which one has the widest and the shortest size range: (a) grain boundaries, (b) agglomerates, (c) machining flaws?

11.48 Wear and friction are tribosystem properties and not material properties. Comment on the statement.

11.49 Which type of corrosion is more prevalent in press and sintered PM parts? Why?

11.50 What is the most important property which must be measured for a MIM feedstock?

11.51 You are asked to perform an ultrasonic test on a PM metallic engine part prepared by two methods:

(a) Sintered and then welded

(b) Sintered, welded and hot isostatically pressed

In which case is the capability of the inspection improved? Why?

11.52 Which of the following is least likely to increase the structural reliability of a ceramic?

(a) Increased Weibull modules

(b) Increased toughness

(c) Increased strength

(d) Increased grain size

11.53 Which material would you expect to have the greatest resistance to sliding contact stress:

(a) Reaction-bonded Si_3N_4

(b) Transformation-toughened zirconia

(c) Silicon carbide

11.54 An Al-Si alloy proposed by casting and HIP routes was fatigue tested. Elaborate on the origin of fatigue failure in each of these.

11.55 NDT of powder components formed by HIP presents a particular challenge. Comment on the statement.

11.56 A transformation-toughened ZrO_2 sample fractures at a stress of 950 MPa at a flaw estimated to be about 50 μm by observation of the fractured surface. Assuming that the geometrical constant Y is 1.3, what is the approximate value of the fracture toughness of the sample?

11.57 In which type of loading—bending or axial—is the notch sensitivity during fatigue testing of a PM material low?

11.58 In sintered steel based on pre-alloyed powder, why is it difficult to estimate the combined carbon content metallographically?

11.59 In rock drilling using cemented carbide bits, the following wear mechanisms may be involved.

(a) Surface impact spalling

(b) Surface impact fatigue spalling

(c) Thermal fatigue

(d) Abrasion

Suggest which mechanisms are operative while drilling the following rocks: (i) quartzite, (ii) granite, (iii) calcite, (iv) magnesite and (v) sandstones. [Hint: Refer the properties of rocks from any mineralogy book]

11.60 Observe the fracture surfaces on cemented carbide drill bit inserts and highlight the root causes for the fracture.

11.61 In the case of micro-indentation hardness measurement, gauge qualitatively the repeatability and reproducibility of the values for a hardened PM steel part as compared to the straight sintered counterparts.

12

Metallic and Ceramic PM Materials

LEARNING OBJECTIVES

- Low- and high-alloy PM steels
- Major non-ferrous alloys related to Cu, Al, Ag, Ni and Ti refractory metals
- Intermetallics
- Ceramic systems (oxide and non-oxide), cermets, ceramic–ceramic composites
- Sintered nanocrytalline metals and ceramics, including functionally graded PM materials
- Material codes of the Metal Powder Industries Federation

Powder metallurgy engineering parts are made of varied materials such as steels, light and heavy non-ferrous alloys, intermetallics and ceramics. Apart from this, metal–ceramic and ceramic–ceramic composites are also useful additions. Late developments in the field of nanocrystalline materials have opened an entirely new avenue in which powder metallurgy processing plays an important role. In addition, functionally graded materials have given rise to properties that are superior to those of homogeneous or multilayered materials. This chapter covers the material aspects of powder metallurgy parts.

12.1 LOW-ALLOY FERROUS MATERIALS

Low-alloy PM steels constitute the largest share of engineering products produced through the powder metallurgy processing route. Apart from carbon, other alloying additions have dramatic hardening effect, particularly P, Si, Mn, etc. In Chapters 2 and 4, pre-alloyed and premix routes for alloying were discussed. Master alloy additives in iron powder prepared by pre-alloying are very common. During sintering they become liquid, wetted and diffused into the ferrous powder to produce a homogeneous sintered alloy steel. Hybrid low-alloy steel powders cover PM materials manufactured from pre-alloyed low-alloy steel powders using nickel, molybdenum and manganese as the major alloying elements, to which varying amounts of elemental metal powder(s) have been admixed. Hybrid low-alloy powders are normally used in medium- to high-density PM applications. These metals provide greater hardenability than is possible with admixed copper or nickel steels. When the final density is to be 7.0 g cm^{-3} or more, these materials may be produced by pressing, pre-sintering, repressing and sintering. Table 12.1 summarises the historical development of various types of PM steel alloys and their strength and ductility values.

Table 12.1 Characteristics of sintered steels

Period	Material types	Production cycle	Characteristics		Competitive materials
			UTS, MPa	Elongation, %	
From 1940	Fe	$P_1 + S_1$	200–300	5–25	Common cast iron
	Fe-C	$P_1 + S_1 + P_2 + S_2$	250–400	0–5	
From 1945	Fe-Cu Fe-Ni Fe-Cu-Ni	$P_1 + S_1$	350–450	1–5	Special cast iron, unalloyed steels
About 1950	Fe-C Fe-Cu-C Fe-Ni-C	$P_1 + S_1$	400–500	0–3	Unalloyed steels
From 1950	High-carbon steels	$P_1 + S_1$	450–600	0–2	Low-alloy steels
From 1965	Ternary diffusion-bonded steels	$P_1 + S_1$	450–600	2–4	Medium-alloyed steels
From 1970	Ternary diffusion steels with very high-compressibility powders	$P_1 + S_1$ $P_1 + S_1 + P_2 + S_2$	500–800 600–1000	3–5 3–6	Quality construction steels
From 1994	Diffusion-bonded steels but processed through Ancordense patented warm compacting	$P_1 + S_1$ $P_1 + S_1 + P_2 + S_2$	550–700 650–800	3–5 3–6	Quality construction steels

P_1 = Pressing; S_1 = Sintering (or first sintering); P_2 = Repressing; S_2 = Second sintering.

12.1.1 Fe-C alloys

The Fe-C PM alloys are useful for high tensile strength products, where ductility is of low importance. In order to understand various microstructures resulting in plain carbon steels after cooling from elevated temperatures, it is important to know the binary Fe-C phase diagram (Fig. 12.1a). This system undergoes three types of reactions: (i) peritectic at 1495°C, (ii) eutectic at 1147°C and (iii) eutectoid at 727°C. The composition near 0.8% C (exactly 0.77% C) is known as *eutectoid composition*; the austenite transforms into a eutectoid mixture of ferrite and cementite known as pearlite. Steels with carbon less than the eutectoid composition are known as hypo-eutectoid steels and those with carbon greater than eutectoid composition as hyper-eutectoid steels. Alloying additions like Ni, Mn, Cu and Co in Fe-C alloys are known as *austenite stabilisers*; while elements like Cr, W, Mo, V, Al, Si and P are *ferrite stabilisers*. All alloying elements lower the eutectoid composition to values less than 0.8% C. Manganese and nickel lower the eutectoid temperature while Cr, W, Si, Mo and Ti raise the eutectoid temperature. As an example, addition of 2.5% Mn to a steel containing 0.65% C will produce a completely eutectoid, i.e., pearlitic structure.

The time–temperature transformation (TTT) curve is very useful in understanding the basics of steel heat treatment. The TTT diagram of eutectoid (0.77% C) steel is shown in Fig. 12.1b. The diagram is also called the *isothermal transformation diagram*, since transformation kinetics is studied at constant temperature, which may be varied. However, care must be taken to cool the steel from austenite range to the new lower temperature. The rate of cooling should be controlled such that metastable austenite is present prior to isothermal treatment. In case the steel is water quenched from austenite, a metastable martensite results.

If we switch to hypo- or hyper-eutectoid steels, the nature of the TTT diagram is similar, except that the formation of pro-eutectoid ferrite or cementite must also be shown in the isothermal transformation plots.

The TTT diagram of eutectoid steels can be divided into three regions: (i) pearlitic region, (ii) bainitic region and (iii) martensitic region. In Fig. 12.1b, cooling curve 1 gives rises to pearlite, while cooling curve 2 gives rise to upper bainite. The pearlite field can again be divided into coarse pearlite and fine pearlite, depending on the temperature of isothermal transformation. It must be understood that the amount of martensite formed is a function of the temperature to which the austenite is cooled and not a function of time. It is clear from Fig. 12.1b that martensite formation takes place in a range of temperatures from M_s (start of martensite) to M_f (finish of martensite). In between these two temperatures, one gets martensite and austenite, both in metastable conditions. It is worth noting that the hardness of martensite increases, while M_s and M_f temperatures drop with the carbon level in the steel.

In practical industrial heat treatment, it is not possible to cary out the operations isothermally and the steels are cooled continuously from the stable austenite field. The cooling rates can be varied, i.e., furnace cooling, air cooling, water cooling and brine solution cooling. In this order the cooling rate increases. For studying the phase evolution under such circumstances, we take recourse to continuous cooling transformation (CCT) plots. The CCT plots could be approximated by shifting the TTT plots to lower temperature but longer period.

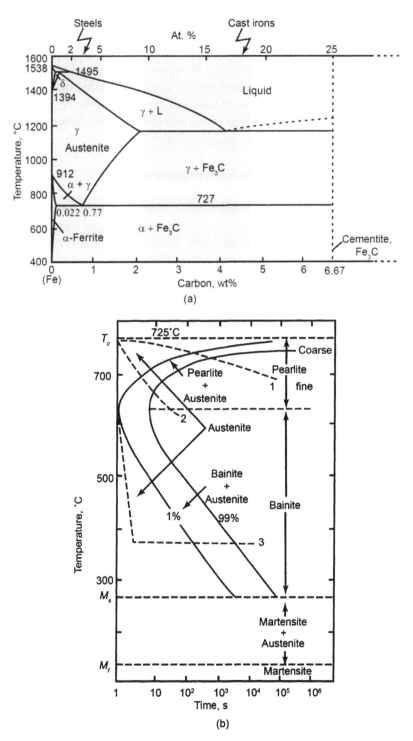

Fig. 12.1 (a) Fe-C phase diagram; (b) TTT diagram for a eutectoid steel

One of the most convenient methods for preparing PM steel is to use iron–graphite powder premixes. Oxides in the iron powder would be detrimental as they will consume some graphite during reduction, before the graphite actually diffuses into iron. The graphite powders to be used should be finer. The use of amorphous carbon in preparing sintered steels is not advisable due to carbon's very strong affinity with oxygen. Of late, it is noticed that primary synthetic graphite is as good as natural graphite. Purity has a considerable influence on the properties of sintered plain carbon steel. Minute amounts of mineral dust prevent carbon pick up, while, as little as 0.03 SiO_2 causes reduction in the solubility rate of carbon.

12.1.2 Copper- and nickel-containing steels

In the previous section the low ductility of sintered Fe-C alloys has been mentioned. To minimise this in practically all low-alloy steels, some copper is invariably added. This is in contrast to cast and wrought steels. Copper facilitates liquid phase sintering. The dimensions of the sintered parts grow after copper addition, reaching a maximum of 8%–10% Cu, which corresponds to the solubility limits of copper in iron. To compensate for the shrinkage of plain iron compacts during sintering, addition of minor amounts of copper (1%–2.5%) is common. Copper growth can be decreased by changing the sintering temperature, sintering time or heating rate within practical limits. Another method involves carbon or phosphorus addition and a decrease in the particle size of starting iron and copper powders. The cooling rate from sintering to room temperature has a significant effect on final properties. Rapid cooling gives high tensile strength and hardness, but lower elongation. Infiltration of an iron PM compact by copper at a temperature higher than the melting point of the latter is common in densifying the porous skeleton.

The addition of both copper and carbon to iron powder offers greater strength and more hardness than the addition of either one of them. Some of the difficulties in sintering iron–copper (copper growth) and iron–carbon (decarburisation) are more or less eliminated. There is also a range of possibilities for a trade-off between good machinability with the higher cost of iron–copper and poor machinability with the lower cost of iron–carbon. The sintering atmosphere is of non-decarburising type and the temperature is generally above the melting point of copper, unless double sintering is used. The tensile strength of Fe-Cu-C steels increases with increasing compressibility of iron powder.

Addition of nickel in Fe-Cu-C steels is advantageous, as, apart from increasing the hardenability, it impedes the copper growth. The property is useful in maintaining dimensions of Fe-Cu-Ni alloys. The nickel particles are encapsulated by iron, since iron diffuses more rapidly over nickel than nickel over iron. Statistically, nickel concentration in the contact areas between the iron particles is higher, resulting in a nickel martensite structure. In carbon-containing nickel steels, the amount of martensite is significantly higher after furnace cooling. The ferrite–pearlite microstructure disappears almost entirely after oil quenching. Nickel steels are used purposefully in the heat-treated state in order to derive full benefit from the addition. Figure 12.2 shows typical microstructures of Fe-1.5Cu-4 Ni-0.5 Mo-0.45C steel (Distaloy AE) in as-sintered and after quenching and tempering treatment. The inhomogeneities in sintered steels produced from powder mixtures are unavoidable, but it should not be regarded as a disadvantage.

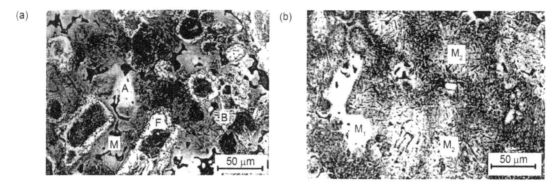

Fig. 12.2 Typical microstructures of PM alloy steel (Distaloy AE: Fe-1.5Cu-4Ni-0.5Mo-0.45C, sintered density 7.13 g cm^{-3}): (a) As sintered: A = austenite, F = ferrite, P = pearlite, B = upper bainite, M = Ni-rich martensite; (b) Quenched and tempered: M_1 = Ni-rich martensite with low-carbon content, M_2 = low-Ni martensite with high-carbon content.

12.1.3 Manganese- and chromium-containing steels

Both manganese and chromium are cheap alloying additions in steels. Because of high oxygen affinity, the small size particles of these alloying elements get rapidly covered with oxides, when exposed to air. These oxide layers form diffusion barriers during sintering and thereby prevent uniform homogenisation of the steels. The oxidation can be reduced or even prevented if these alloying elements are pre-alloyed with other elements which are less sensitive to oxidation, or if compounds that are more resistant to oxidation are used. Ferromanganese and ferrochromium have been used as additives and have good results. One of the disadvantages of these alloying additions is the use of relatively high sintering temperature, which makes the use of walking beam furnace a necessity for mass production sintering. Figure 12.3 illustrates the tensile strength

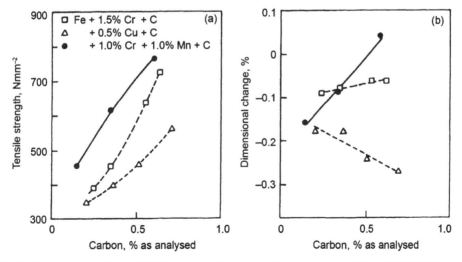

Fig. 12.3 (a) Tensile strength and (b) dimensional change of steels prepared from SC 100.26 + graphite and other alloying additions; sintered at 1250°C for 30 min in N_2-5H_2; sintered density 7.0 g cm^{-3}

and dimensional change variation of Fe-Cr-C, Fe-Cu-C and Fe-Cr-Mn-C steels with respect to carbon content. It is noticed that at combined carbon content of about 0.6% the steels containing Cr or Cr-Mn are superior to those without these elements. However, it may be mentioned that from the dimensional tolerance point of view, the steels containing only chromium are least dependent on the combined carbon content. This is important when sintering precision parts, where a scatter of the combined carbon content in the parts of about ± 0.05% is not unusual.

Oil atomisation has been found attractive for producing low-oxygen high-quality Mn-Cr steels (AISI 4100 grades). By adjusting the particle size distribution in the powder charge and high temperature sintering at 1250°C in nitrogen for 30 min, followed by oil quenching and tempering, a tensile strength of the order of 1300 MPa for 0.6 C steel has been reported.

Example 12.1: Figure Ex. 12.1 shows the pseudo-binary phase diagram of (Fe-75%Mn)-C system, which indicates a eutectic reaction at 1095°C with a eutectic alloy at 3.8% C. With further increase in the carbon content, the liquid phase gets further stabilised. After literature survey, describe what has been done in developing low–melting point master alloys for ferrous sintering. Give some important references.

Solution: After the introduction of a patent by Stadles at GFE (Gessellschaft für Electrometallurgie, Nütnberg, Germany), Sintermetallwerk Krebsöge joined in developing a master alloy based on complex carbides and called them MCM and MVM. This was done in order to lower the oxygen affinity of the alloying elements. Since the desired alloying elements Cr and Mn are known as carbide formers, and since carbon was also needed in the steel as an alloying element, alloys which contained the alloying elements in the carbide form were selected. These master alloys are produced by fusing commercial high-carbon ferroalloys in vacuum induction furnaces, followed by cleaning, preliminary jaw crushing, fine grinding (up to 0.15 mm) and attritor milling. During coarse grinding the oxygen pick-up is negligible, while during fine and ultrafine grinding (average particle size 5 μm) special measures are taken. It was also noted that the addition of lubricants during attritor milling had noticeable influence on the oxygen content of the powders, although they improved storage life. On the other hand, certain lubricants caused undesirable agglomeration during transportation and handling. The grinding fluids were selected while ensuring that the final powder did not contain more than 0.15% O_2. (Continued on next page)

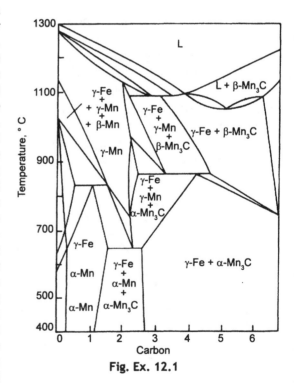

Fig. Ex. 12.1

In summary, such master alloys fulfilled the following conditions:

- They contain carbon in the combined form
- They contain at least two of the elements Mn, Cr and V
- They are stable during the heating-up time to sintering temperature
- They dissociate under sintering conditions
- They minimise production cost.

They have additional advantages such as:

- They contain additional molybdenum, which is a well known alloying element due to its positive effect on hardenability.
- As they include carbon, the amount of added free carbon required for the desired composition is reduced.
- They allow the development of sintered steels with higher mechanical strength by using relatively lower amounts of alloying elements.
- Their hardness is lower than that of the individual carbides, thus causing less tool wear.
- The oxidation stability is sufficient.

The later development was a master alloy containing Mn, Mo, Fe and C (MM), where Cr(V) is further replaced by manganese. Such a MM master alloy is obtained by melting high-carbon ferromanganese, ferro-molybdenum, graphite and iron in an induction furnace using a susceptor. The resulting base alloy is broken into chips, which are subsequently ground in a planetary mill under cyclohexane. It was noted that gross melting occurs at about 1100°C, while in the case of MCM, partial melting was below 1200°C.

The chemical and phase compositions of the abovementioned master alloys are presented in Table Ex. 12.1a. MCM and MVM mainly consist of complex carbides, but their phase compositions and structures are very different. The amount of metallic phases and the oxygen content also deviate from each other (Table Ex.12.1a). Since only 1%–2% of the master alloy is used for the manufacture of sintered steel, the amount of oxygen introduced in this way is less than 200 ppm at any rate.

From a close study of ternary phase diagrams on the various iron–metal–carbon systems, the formation of a liquid phase during sintering of MCM, MVM and MM steels is expected. Some of these systems showing low-melting liquids are summarised in Table Ex.12.1b. The formation of a liquid phase can be identified from microstructural observations, including fracture surface and hot-stage microscopy. The liquid phase is transient in nature because the constituents of the master alloy particles diffuse into the iron matrix. (Continued on next page)

Table Ex. 12.1a Chemical and phase compositions of master alloys

MCM, wt%						MVM, wt%						MM, wt%				
Mn	Cr	Mo	Fe	C	O	Mn	V	Mo	Fe	C	O	Mn	Mo	Fe	C	O
25	23	22	22	7	−0.18	25.5	23	25.5	20	5	0.2–1.0	40	20	32	7	−0.3
80% (Cr, Mn, Fe, Mo)$_7$C$_3$ 10% (Fe, Mo, Cr, Mn)$_6$C 5% (Fe, Mn, Mo)$_3$C 5% α-(Fe, Mn, Cr, Mo)						25%(V,Mo)C 25%(V,Mo)$_2$C 20%-25% α-(Mn,Fe,Mo) 20%-25% γ-(Fe,Mn,Mo) MnO (percentage depending on oxygen content)						Three carbide phases with M$_7$C$_3$ as a main constituent, little metallic α-phase				

Table Ex. 12.1b   Liquid phases occurring in some iron base alloys

System	Reaction components	Melting temperature, °C
Fe-Cr-C	$L + Cr_7C_3 = \gamma + Fe_3C$ (peritectic)	1184
Fe-Mo-C	$Fe_3C + \gamma = Mo_2C + L$	1080
Fe-Mo-C	$Fe_3C + C + C + Mo_2C + L$	1120
Fe-Mo-C	$\gamma + Mo_2C + \eta + L$	1150
	$\alpha + \gamma + \eta + L$ where $\eta = Fe_3Mo_3C$	1210
Fe-Mn-C	$L + \gamma + Fe_3Mn_3C$ (20 wt% Mn, 4.2 wt%)	1080
Fe-Cr-Mo-C	5-6 at% Mo + Cr	1160
	8 at% Mo + Cr 20 at% Mo + Cr with 15 at% C (−3.6 wt% C)	1120

Reprinted with permission from the Metal Powder Industries Federation, Princeton, USA.

German Patent, No. 2204886, 1973.

Retelsdorf HJ, Fichte RN, Hoffmann G and Dalal K, *Metall.*, 29(10):1002, 1975.

Zapf G and Dalal K, In Hausner HH and Paubenblat PW, (eds), *Modern Developments in Powder Metallurgy*, Vol. 10, Metal Powder Industries Federation, Princeton, NJ, 1977.

Zapf G, Hoffmann G and Dalal K, *Archiv Eisenhuttenwessen*, 46:287 1975.

Banerjee S, Schlieper G, Thummler F and Zapf G, In Hausner HH, Antes HW and Smith GD, (eds), *Modern Developments in Powder Metallurgy*, Vol. 12. Metal Powder Industries Federation, Princeton, NJ, 1981.

Schlieper G and Thummler F, *Powder Metallurgy International*, 11(4):172, 1979.

12.1.4 Phosphorus-containing alloys

Phosphorus is an attractive and cheap alloying addition in sintered steels. It is a ferrite stabiliser and promotes densification since the diffusivity in open body-centred cubic ferrite structure is higher when compared to face-centred cubic austenite phase. In an Fe-P

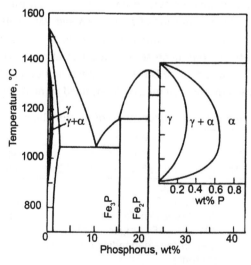

Fig. 12.4 Iron–phosphorus phase diagram

system (Fig. 12.4), at a temperature greater than eutectic (1040 ± 10°C), a melt is formed; it promotes transient liquid phase sintering. The time required for complete solution of the melt is a function of both phosphorus content and the sintering temperature. Completion of this process is reported in only 15 s in the case of a Fe-1.5P premix when sintered at 1200°C. Phosphorus added in small quantities in iron powder in the form of Fe_3P (15.6% P) does not affect compressibility. Figure 12.5 shows the variation of mechanical properties and dimensional changes of various Fe-P alloys.

Investigation on an Fe-0.5P alloy after sintering in the two-phase region (ferrite–austenite) at 1000°C or 1100°C revealed that even after solid-state sintering rounded pores were observed, due to the high diffusivity rates in the ferrite phase. Compacts pressed and sintered several times to obtain a very low pore fraction (0.04–0.05) exhibited extensive grain coarsening.

A major drawback with the use of phosphorus in iron powder metallurgy is the relatively higher shrinkage of parts during sintering. The addition of copper was considered as a possible solution.

Phosphorus-containing sintered iron has many soft magnetic applications. The addition has a favourable effect on magnetic properties like permeability and coercivity and reduces remanence slightly.

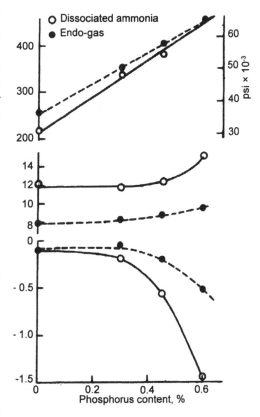

Fig. 12.5 Mechanical properties and dimensional changes of sponge iron powder (NC 100.24) with varying additions of Fe_3P and 0.8% zinc stearate compacted at 590 MPa and sintered at 1120°C in dissociated ammonia for 1 h (full curve) and in endo-gas for 24 min (dashed curve). Sintered density: ~ 6.9 g cm^{-3}

12.2 HIGH-ALLOY STEELS

Two important high-alloy steels—high-speed steels and stainless steels—will be considered in this section.

12.2.1 High-speed steels

High-speed steels (HSS) are basically a group of iron-based alloys containing about 20%–30% of mainly carbide-forming alloying elements and can be hardened to a level of up to 65–70 HRC; no appreciable softening takes place until temperatures in the region of 600°C are reached. The microstructure consists of a tough, tempered martensite matrix with a dispersion of wear resistant, high-hardness alloy carbides. HSS has a sufficient temperature gap between liquidus and solidus temperatures. In conventional ingot steel making, this makes it prone to carbide segregation during solidification, which imparts an embrittling effect. PM processing is attractive in solving these problems where segregation is limited within individual particles (maximum 250 μm dia.). Due to relatively high cooling rates achieved during atomisation processes, segregation within each powder particle is minimised. Additionally, PM HSS powder can contain a large quantity of small size carbide particles distributed homogeneously in the matrix. Other advantages of PM HSS are increased alloying flexibility, improved toughness, superior grindability, less distortion during heat treatment, isotropic mechanical properties, good machinability and improved tool life. The powders used for producing parts are made from either water- or gas(N_2 or Ar)-atomised routes. The major problem in water-atomised powder is the presence of a considerable amount of oxygen in the form of oxide inclusions. These inclusions are almost completely reduced by annealing in either a vacuum or a hydrogen atmosphere. During such treatment, martensite and retained austenite, if any, are transformed into a matrix of ferrite and spheroidal carbide. This facilitates the powder to be softer and ductile and more suitable for green compaction.

The sintering of HSS parts is carried out by super solidus liquid phase sintering method, which has been described in Chapter 7. Vacuum sintering is usually the best choice, as it avoids trapping of gas in the pores. Hydrogen-sintered HSS requires a lower sintering temperature than vacuum-sintered HSS to achieve high density. This is due to the good thermal conductivity and also to the fact that during sintering a fresh oxide surface of steel powder is immediately reduced by hydrogen. The sintering period should not be high since carbide coarsening will take place. Sintering temperatures generally lie between 1150°C and 1350°C. If the sintering temperature is extremely high, incipient melting may occur in local zones and result in eutectic formation during cooling. This may cause slumping of the parts. Since the temperature range required for sintering to full density is extremely narrow, a uniform temperature distribution in the furnace is necessary.

HIP is an important technology for producing quality HSS products. Consolidation is typically carried out at 1100°C at 100 MPa for 1 h. The compacts are further processed to the desired billet and bar sizes by conventional hot rolling and forging. Figure 12.6 illustrates the flowsheets for two methods of PM HSS production—Asea-Stora and Crucible steel methods.

Table 12.2 summarises the fracture toughness values of ingot metallurgy and powder metallurgy processed high-speed steels, including some composites.

12.2.2 Stainless steels

Stainless steels are iron base alloys that contain a minimum of approximately 11% chromium. Few stainless steels contain more than 30% Cr or less than 50% Fe. Their stainless characteristics are due to the formation of an invisible and adherent chromium-rich oxide surface film. This oxide forms and heals itself in the presence of oxygen. Other elements added to improve particular characteristics include Ni, Mo, Cu, N, S and Se. Carbon is usually present in amounts ranging from less than 0.03% to over 1% in certain martensitic grades. Figure 12.7 provides a useful summary of some of the compositions and property linkages in the stainless steel family. With specific restrictions on certain types, stainless steels can be fabricated and shaped by PM techniques.

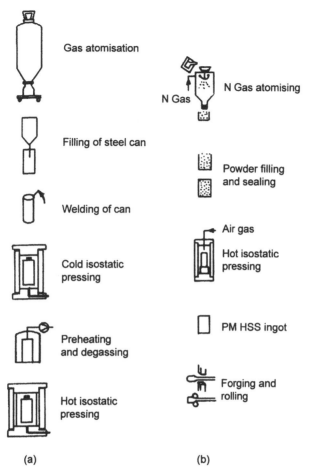

Fig. 12.6 Schematic flowsheet for (a) ASEA- STORA and (b) Crucible steel methods for production of PM high-speed steels

Stainless steels are divided into four main classes:

- *Martensitic*—straight chromium grades that can be hardened by quenching;
- *Ferritic*—straight chromium grades not hardenable by conventional quenching;.
- *Austenitic*—contain sufficient nickel or nickel plus manganese to produce a face-centred cubic structure.
- *Duplex austenitic-ferritic*—find increasing use as they constitute a set of materials with properties that are a compromise between those of purely ferritic steels, which tend to have low toughness, and austenitic steels, over which they exhibit superior stress corrosion resistance. This two-phase alloy is not sensitive to intergranular attack in various corrosive media. However, these steels have considerable difficulties from the point of view of ingot metallurgy manufacture, due to the precision in the chemical composition and temperature control needed. Such difficulties serve to make PM an attractive manufacturing option.

Table 12.2 Fracture toughness of sintered and ingot metallurgy HSS

Grade	Condition	K_{IC}, MN m$^{3/2}$
M2	WL	18, 15–20
	WT	24
	FS	30
	Forged	15–20
T1	WL	21
	WT	22
	FS	25
T6	WL	16
	WT	18
	FS	31
T42	WL	12
	WT	12
	FS	13
M3/2	As sintered, fully heat treated	19
M3/2-TiC-MnS composite	As sintered, fully heat treated	21.4
M3/2-NbC-MnS composite	As sintered, fully heat treated	22.7
M3/2-Cu$_3$P^b	As sintered	14.5
M3/2-TiC-Cu$_3$P^b composite	As sintered	13.9
C23	HIPed heat treated	18.7
ASP 30	HIPed heat treated	14.8
ASP 60	HIPed heat treated	12.9

^{a} W: wrought; L: longitudinal; T: transverse; FS: sintered to full density.

^{b} Sinter assisted by copper–phosphorus.

The production of sintered stainless steel parts requires more attention than other sintered ferrous alloys, from powder selection up to final sintering, mainly in this last stage of the process due to the influences of temperature and time on the sintering operation, and of the furnace atmosphere used during sintering. Sintered stainless steels have a smaller market than PM iron and steel parts. The reasons for this are:

- It is expensive, due to the high alloy content.
- Compressibility of fully pre-alloyed stainless steel powder is far inferior to that of unalloyed iron powder as a result of solid-solution hardening.
- Sintering conditions require more attention than for standard sintered steels.

Powders are generally compacted at a pressure of 600–1000 MPa to achieve green densities in the range of 5.9–6.9 g cm^{-3}. However, for certain grades of stainless steels, such as ferritic stainless steel, it is very difficult to obtain densities higher than 6.5 g cm^{-3}; applying pressure beyond 900 MPa does not improve density. Prior to the actual sintering operation, the lubricant, which is used with the pre-alloyed powder to permit compaction of the powder and

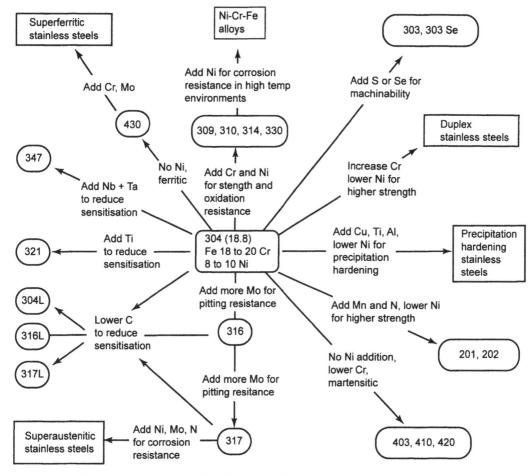

Fig. 12.7 Compositional and property linkages in the stainless steel family

the ejection of the part from the die, must be removed. Residual lubricant carbon deposits have highly adverse effects when stainless steel is sintered, resulting in lower corrosion resistance. Incomplete carbon removal results in an elevated level of carbon in the part. When coupled with relatively slow cooling following sintering, this results in sensitisation. This is due to the precipitation of chromium-rich carbides at the grain boundaries, leaving the surrounding matrix depleted in chromium and subject to corrosion attack. Maximum lubricant burn-off is achieved in the temperature range 425–540°C.

Stainless steel parts are generally sintered at temperatures ranging from 1100°C to 1500°C. Sintering time varies from one to several hours in an appropriate atmosphere. Argon, nitrogen, vacuum and reducing atmospheres are used for such steels. Argon is neutral and is the best sintering atmosphere for certain grades of steels like 316L. Hydrogen is the most strongly reducing sintering atmosphere of all those commercially available. To maintain reducing conditions a dew point of at least –35°C to –40°C is required in the furnace. During vacuum sintering, care has to be maintained that the furnace pressure does

Table 12.3 Typical mechanical properties of various sintered stainless steels

PM material designation AISI	Density g cm^{-3}	Mechanical properties (typical values)			
		YS, MPa	UTS, MPa	Elongation, %	H, Rockwell
303	6.4	230	270	1	62 B
	6.4	275	300	7	60 B
	7.0	120	390	20	45 B
304	6.4	260	290	1	61 B
	6.6	275	400	10	68 B
	7.0	120	390	22	45 B
314	6.4	240	290	1	59 B
	6.6	275	475	10	69 B
	7.0	160	375	16	45 B
410	6.7	440	600	1	26 C
	7.0	100	360	20	53 B
420	6.7	810	850	0.5	30 C
434	7.0	200	350	14	55 B
432	6.9	350	470	2	78 B

not fall below the vapour pressure of chromium. Any chromium depletion would adversely affect corrosion resistance.

Table 12.3 summarises the mechanical properties of various grades of sintered stainless steels.

12.3 COPPER ALLOYS

PM copper and its alloys are used extensively in engineering industries. Figure 12.8 shows three binary phase diagrams of copper-based systems: Cu-Zn, Cu-Al and Cu-Sn. It is evident that these systems contain a number of intermetallic compound phases, β, γ, etc. A knowledge of these diagrams is necessary to identify these phases in sintered alloys. Table 12.4 lists the compositions and mechanical properties of some of the PM copper alloys. The most attractive property of copper is its high electrical conductivity. The effect of sintering and subsequent working of copper powder preforms on sintered density and electrical conductivity of copper is illustrated in Table 12.5. Regardless of sintering conditions, working imparts comparatively dense parts with substantially the same conductivity.

Among the various copper base alloys, tin bronze is produced to the greatest extent. Liquid phase sintering is followed. The water-atomised copper and air-atomised tin powders are subjected to reduction treatment in cracked ammonia at 600°C. Many times tin is introduced in the form of a Cu-Sn master alloy. In Chapter 8, an account of the difference in the dimensional change of the parts during sintering has been given, where it is shown how the premix powder-based bronze parts are subjected to greater dilation (Fig. 8.23).

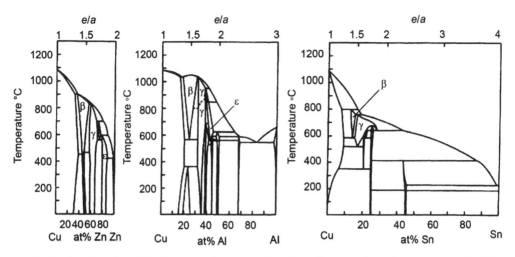

Fig. 12.8 Cu-Zn, Cu-Al and Cu-Sn phase diagrams. The three diagrams have been drawn with different length of the *x*-axis in order to have the same scale for the *e/a* (electron-to-atom) values.

Table 12.4 Typical properties and applications of some copper-based powder metallurgy materials

Material	Density, g cm^{-3}	Tensile strength, MPa	0.2% proof stress, MPa	Elongation, %	Vickers hardness, HV	Remarks and application
Copper	7.0–8.0	80–140	45–75	6–11	30–50	Electrical components
Lead brass (78.5% Cu, 1.5% Pb, remainder Zn)	7.0–8.4	120–200	45–85	7–13	40–55	Lock parts
Nickel brass (18% Ni, 18% Zn, remainder Cu)	7.2–8.4	140–280	-	7–12	45	Corrosion resistance
Tin bronze (90% Cu, 10% Sn)	6.5–7.5	120–190	45–85	5–12 3	5–55	Strength and corrosion resistance
Lead tin bronze (10% Sn, 10% Pb, remainder Cu)	6.7–7.6	-	-	-	-	Abrasion and corrosion resistance
Graphite leaded bronze (8%–12% Sn, 5%–8% Pb, 6%–8% graphite, remainder Cu)	6.0–7.4	30–130	-	0.5–2	25–55	High-temperature bearings
Graphite bronze (9.5%–10.5 % Sn, 1.5% max. graphite, remainder Cu)	6.8–7.2	125	-	3	-	Oil-impregnated bearings

Table 12.5 Comparison of density and conductivity of PM materials with wrought copper

| | PM material | | | | Wrought copper | |
| Condition | Sintered, 5 min | | Sintered, 30 min | | | |
	Density, g cm^{-3}	Conductivity, % IACS	Density, g cm^{-3}	Conductivity, % IACS	Density, g cm^{-3}	Conductivity % IACS
As sintered: Vacuum-dissociated ammonia	7.86–7.95	82.6–84.5	8.03	87.2	-	-
			8.06	86.8	-	-
Hot forged	8.86	98.8[a]	8.89	99.1[a]	8.95	101.5[a]
Hot-rolled and annealed strip	8.8	99.3	8.8	98.7	8.9	101.5
Hot-rolled rod, cold drawn to wire, and annealed	8.8	100.4	8.8	100.4	8.9	101.4

[a] Annealed in steam at 500°C after forging

An improved sintered cupro-nickel coinage alloy (75Cu/25Ni) after double pressing and double sintering (sintering temperature 1050–1120°C) under dissociated ammonia (dew point −40°C to −50°C) has been achieved.

Steel-backed Cu-Pb-Sn sintered materials are used to replace solid bearings. Some of the typical compositions are Cu-25Pb-0.5Sn, Cu-25Pb-3.5Sn, Cu-10Pb-10Sn and Cu-50Pb-1.5Sn.

Oxide dispersion-strengthened (ODS) PM copper alloys—containing normally 0.1%, 0.2%, 0.35% and 0.6% Al$_2$O$_3$—are produced by selective reduction of cuprous oxide in a mixture containing Al$_2$O$_3$, followed by hot extrusion. Alumina particles in the alloy provide excellent thermal stability and permit the retention of hardness, electrical conductivity and thermal properties, even after exposure to relatively high temperatures. Figure 12.9 shows the variation of electrical conductivity and mechanical properties of such alloys.

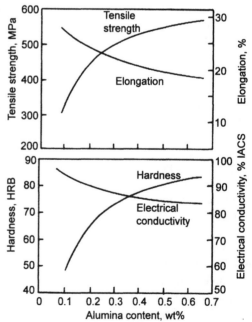

Fig. 12.9 Mechanical and electrical properties of alumina dispersion-strengthened copper

12.4 ALUMINIUM ALLOYS

Aluminium alloys have many attractive properties. Some of them are:

- Lightness
- Resistance to corrosion
- High strength-to-weight ratio
- Excellent workability
- High electrical and thermal conductivity
- Toughness at low temperature—a property in which steel is comparatively poorer.

Aluminium alloys containing a small percentage of other metals are more significant as they provide strength, toughness and other properties. If one looks at the binary phase diagrams based on aluminium, it is evident that other metals have limited solubility in it. This is the reason why, to enhance solubility, rapid solidification is taken recourse to.

Pure aluminium alloyed with different metals is classified in numerical series (Table 12.6).

The production methods for aluminium and its alloys powders have been described in Chapter 2. In green die compaction of aluminium powder, considerable difficulties, like scoring, galling and seizure, are faced, due to the intrinsic softness of this metal. Another problem is its high affinity for oxygen. When exposed to air, an oxide layer of approximately 100 Å coats the aluminium particles. For these reasons aluminium powders are invariably used as premixed alloys. These premixes have good compressibility. Stearic acid or stearates, which are widely used in ferrous sintered parts, cannot be used as lubricants because they react with aluminium. Aluminium alloy premixes are compacted at comparatively low pressures (400–480 MPa). The tenacious oxide film over the aluminium particles is ruptured during compaction, thus resulting in metal-to-metal contact.

Table 12.6 Aluminium alloy designations

Principal alloying element	Alloy series
Aluminium 99% min.	1XXX
Copper	2XXX
Manganese (S + Cu/Mg)	3XXX
Silicon	4XXX
Magnesium	5XXX
Magnesium and silicon	6XXX
Zinc	7XXX
Other elements	8XXX
Unused series	9XXX

For sintering aluminium alloys, liquid phase sintering is the most common method. Sintering of aluminium alloys containing copper, magnesium and silicon as alloying elements, which form low melting eutectics with aluminium, is an example of transient liquid phase sintering. A dry and non-oxidising sintering atmosphere is required to prevent oxidation, thus promoting diffusion and alloying. A correlation of volume change after liquid phase sintering confirms that shrinkage is observed in cases which had low solid solubility of solute in aluminium, and high content in the eutectic alloy (Table 12.7).

Table 12.7 Character of volume changes experienced during sintering of aluminium compacts with various additions

Addition	Addition concentration, wt%	Sintering temperature, °C	Max solid-state solubility in Al, at%	Al content of melt at eutectic temperature, at%	Character of volume changes
Cu	1–5	560–640	2.5	82.7	Shrinkage
Mg	1–30	440–640	19	81.1	Growth
Si	0.5–1	580–640	1.6	88.7	Shrinkage
Sn	5	360–610	Very small	2.2	No change
Zn	20	400–600	66.5	11.3	Growth
Ni	1–5	645–655	Very small	97.3	Shrinkage
Pb	30	620	Very small	0.02	No change
Ag	5	640	5	62.5	Shrinkage

Table 12.8 Nominal composition of several commercial aluminium alloy powder premixes

Premix*	Element, wt%,							Lubricant wt%
	Copper	Silicon	Magnes- ium	Zinc	Chrom- ium	Manganese	Alumin- ium	
Alcoa 601AB	0.25	0.6	1.0	-	-	-	rem	1.5
Alcoa 201AB	4.4	0.8	0.5	-	-	-	rem	1.5
Alcoa 202AB	4.0						rem	1.5
Alcan 24	4.4	0.9	0.5	-	-	0.4	rem	-
Alcan 69	0.25	0.6	1.0	-	-	0.10	rem	-
Alcan 76	1.6	-	2.5	5.6	0.20	-	rem	-

* Manufacturer identification is not intended as an endorsement of the product.

Some of the commercial grades of aluminium powder premixes are given in Table 12.8. Full density PM consolidation of aluminium alloys is more common than press-sinter technology. These alloys have been developed in three major categories:

- **High–room temperature strength alloys**: This mainly covers the 2XXX, 6XXX and 7XXX series alloys.

- **High-modulus and low-density alloys:** This includes lithium-containing aluminium alloys.

- **Alloys for elevated temperature service:** This includes Al-Fe, Al-Fe-Ce alloys for service at 230–345°C.

The alloys cited above are prepared by rapid solidification or more recently, mechanical alloying followed by hot extrusion. Among the various methods of consolidation of aluminium alloy powders, hot pressing, hot extrusion and hot forging are common. Hot isostatic pressing is also practiced for high-performance products. Figure 12.10 shows the schematics of these processes. It is worth mentioning that vacuum degassing, prior to hot consolidation, is an important prerequisite. Otherwise, the rapid evolution of hydrogen in the confined die at high temperatures would create a safety hazard. The detailed description of degassing is given in Chapter 4, relative to powder treatment. Figure 12.11 presents the mechanical properties of hot extruded PM 2014-T6 aluminium alloys.

Dispersion-strengthened aluminium alloys containing Al_2O_3 and Al_4C_3 dispersoids (called DISPAL) have been commercially produced in Germany by the hot extrusion route (temperature 550°C, extrusion ratio 20:1). Table 12.9 gives the typical physical and mechanical properties of DISPAL 2.

Discontinuously-reinforced composites with an aluminium alloy matrix by PM process are becoming more popular, with most of the work done on SiC in either particulate or whisker form. The processing is easier than with continuous fibre-reinforced composites, which is carried out in two stages of preform fabrication of fibres and metal infiltration. Table 12.10 presents the mechanical properties of HIPed and extruded 6061 Al-30 vol% SiC particulate composites.

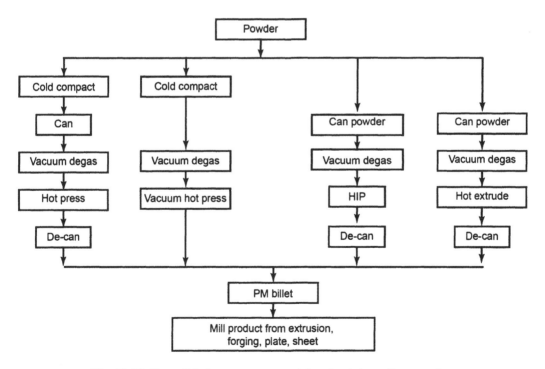

Fig. 12.10 Consolidation processes used for aluminium alloy powders

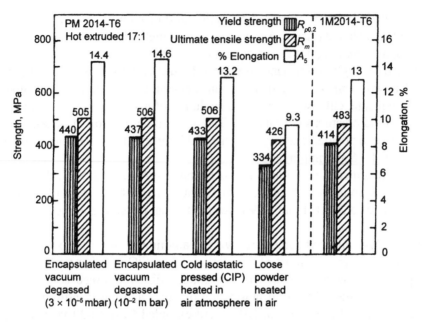

Fig. 12.11 Tensile properties of PM processed 2014-T6 Al alloy compared to those of ingot metallurgy (I/M) alloy

Table 12.9 Properties of DISPAL 2

Dispersoid content[a]	vol%	14
Density	g cm^{-3}	2.7
Hardness	HV$_{30}$	100
Ultimate tensile strength, $R_{p\,0.2}$	MPa	370
Yield strength, R_m	MPa	340
Elongation, A_s	%	10
Young's modulus[b]	GPa	70
Fatigue strength	MPa	115
Impact strength	J cm^{-2}	14
Electrical conductivity	S m mm^{-2}	24
Coefficient of thermal expansion (20–500°C)	10^{-6} K^{-1}	25
Thermal conductivity	W (m K)$^{-1}$	159

[a] 1 wt% C $\cong$ 4.5 vol% Al$_4$C$_3$: 1 wt% O$_2$ $\cong$ 1.6 vol% Al$_2$O$_3$ (combined content of Al$_4$C$_3$ and Al$_2$O$_3$)
[b] Taken from tensile tests

Table 12.10 Mechanical properties of HIPed and extruded 6061 Al-30 vol% SiC$_p$ composites

Process	Density, g cm^{-3}	0.2% YS, MPa	UTS, MPa	E, GPa	Elongation, %
HIP	2.837	395	416	117.5	0.7
Extrusion	2.763	300	328	109.4	2.8

Example: 12.2 Distinguish between short fibre/whisker-reinforced and particulate-reinforced PM aluminium composites under the following heads:

- Reinforcement type and size
- Role of matrix material
- Reinforcement vol% and isotropicity in mechanical properties
- Cost
- Applications

Solution: See Table Ex. 12.2

Table Ex 12.2

Properties	Short fibre/whisker-reinforced	Particulate-reinforced
Reinforcement type, size and application	Short fibres/whiskers of length 0.2 to 250 mm and L/D ratio 2 to 250. Whiskers are single crystal, whereas short fibres are polycrystalline.	Equiaxed particulate (dia 0.1 to 250 µm), polycrystalline in nature.
Role of matrix material	If L/D is less than a critical value, matrix strength plays a dominant role in composite srengthening, else their role is similar to that observed in continuous fibre composite.	Matrix strength plays a critical role in composite strengthening. Reinforcement also helps in matrix strengthening by dispersion or precipitate strengthening mechanisms.
Reinforcement volume % and isotropicity in mechanical properties	Control of reinforcement distribution is difficult and beyond 40% agglomeration becomes severe; properties are relatively less anisotropic than fibre composites.	Control of reinforcement distribution is equally difficult and beyond 40% addition is avoided. Properties are nearly isotropic.
Cost	These are cheaper than continuous fibre composites. Reinforcement cost is still high.	These are the cheapest among all composites.
Applications	Excellent prospect in non-defence applications.	Automobiles and other home appliances. Growth rate of this class of composites is highest.

12.5 SILVER ALLOYS

Commercially pure silver is 99.9% pure. Because of its resistivity to oxidation, silver is suitable for electrical connections and contacts, and hence the sintered silver parts have a premium.

The sintering response of silver is very sensitive to the sintering atmosphere and the green density of the compact. When sintered for 3 h at 870°C, the best densification is obtained after argon atmosphere sintering for low green densities (~ 6 g cm^{-3}), while for vacuum sintering it is for high green densities (~10 g cm^{-3}).

Of all the silver-based sintered alloys, Ag–CdO is the most common and is used for electric contact applications. The dispersed CdO has no solubility in silver and, therefore, does not basically change the properties of the silver matrix. It imparts resistance to arc erosion and seizing, apart from improving the hardness and strength of the material. Figure 12.12 illustrates various PM processes for Ag-CdO production. CdO is in the range 5%–15%, such that the electrical conductivity falls from 85% to 65% IACS. Route I gives less uniform mixing,

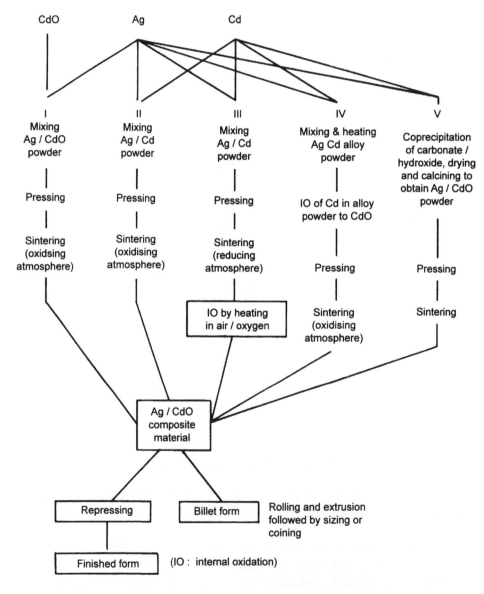

Fig. 12.12 Various PM processes for the production of Ag-CdO contact materials

whereas route V (coprecipitation) causes the most homogeneous mixing of fine silver and CdO particles. The impurity content increases as we proceed from route I to V. Coprecipitated powders are extremely fine. The powder mixture is pressed into blocks and sintered. Sintered blocks can be directly hot rolled, which can be further cold formed into wires, rivet heads, etc. The cold workability of the material is due to the extremely fine and uniform dispersion of oxide and the complete absence of CdO in the grain boundaries. The hardness of this metal is much superior to those produced by other processes.

The route for the internal oxidation process of Ag-Cd alloy powder (< 20.0 μm) requires a much shorter time than the internal oxidation for cast and wrought contact pieces or sheets. It takes place in the temperature range 500–800°C in air or oxygen. Lower the temperature selected for internal oxidation, finer the oxide in the Ag-CdO composite powder; higher the sintering temperature and longer the sintering period selected, greater the coarsening effect. This process can be combined with double-layer pressing for the production of double-layer compact contact parts. The processing method has a distinctive effect on the microstructure, even for identical alloy chemistry.

In an attempt to replace toxic cadmium, Ag-SnO$_2$ is most common and competes well with Ag-CdO contacts. Ag-SnO$_2$ composite exhibits lower erosion rates than Ag-CdO. The major disadvantage is the higher contact resistance after the contacts have been stressed by arcs. The erosion of tin oxide-containing material is only half that of standard Ag-CdO or less. This is attributed to the higher thermal stability of the tin oxide which keeps the shape of the Ag-SnO$_2$ composite and the fine oxide dispersion even at temperatures above the melting point of silver. On the other hand, the thermal stability leads to oxide concentration at the surface after arcing.

12.6 NICKEL ALLOYS

Nickel has good corrosion resistance, high electrical resistivity and good magnetic properties. Ni-based superalloys are extensively used in gas turbines for blades, vanes and discs. For example, Nimonics (Ni/Cr/Al/Ti/Fe) and Inconels (Ni/Cr/Fe), both have small amounts of C, Co, Mo and W. These are all basically Ni/Cr solid-solution alloys, precipitation-hardened with Ni$_3$(Al,Ti) plus various carbides. TD nickel is nickel with a fine dispersion of ThO$_2$ particles in a heavily wrought structure. Permalloy (Ni + Fe, Cr, Mo) has high magnetic permeability and low saturation.

A major application of pure nickel powder is in the production of porous electrodes used in alkaline batteries. The sintering temperature is in the range 900–1100°C. Strength and electrical conductivity are requirements in the porous electrodes. A compromise is to be made between these properties and the sintered porosity. Roll-compacted porous strips are also used for electrodes.

Table 12.11 gives the chemical compositions of some of the nickel-based superalloys. Mechanical alloying has generated a new series of oxide dispersion-strengthened nickel superalloys (MA series). Compared with the pressing/sintering approach in conventional PM, hot consolidated superalloy products are common. Figure 12.13 gives the flowsheet for normal PM, ODS superalloys and rapidly solidified PM processes used in nickel superalloy production.

Table 12.11 Nominal wt% compositions of nickel base PM superalloys

	C	B	Cr	Co	Mo	W	Al	Ti	Zr	Hf	V	Nb	Ta	at% γ' formers Al, Ti, Zr, Hf, V, Nb, Ta	Total at% strengtheners, including Mo and W
Udimt 700	0.06	0.025	15	17	5	-	4	3.5	-	-	-	-	-	11.1	14.4
Low C Astroloy	0.02	0.02	15	17	5	-	4	3.5	0.03	-	-	-	-	12.3	15.2
Rene 95	0.07	0.01	13	8	3.5	3.5	3.5	2.5	0.05	-	-	3.5	-	12.8	16.6
AF2-1DA	0.04	0.015	12	10	2.8	6.5	4.6	2.8	0.10	-	-	-	1.5	13.9	17.7
AF-115	0.05	0.02	10.7	15	2.8	5.9	3.8	3.9	0.05	0.8	-	1.7	-	14.3	17.9
MAR-M 247	0.07	0.015	8.3	10	0.7	10	5.5	1	0.06	1.5	-	-	3	15.0	18.7
IN-100	0.07	0.02	12.4	18.5	3.2	-	5	4.4	0.06	-	0.8	-	-	16.0	18.3
Merl 76	0.02	0.02	12.4	18.5	3.2	-	5	4.3	0.06	0.8	-	1.7	-	16.5	18.6

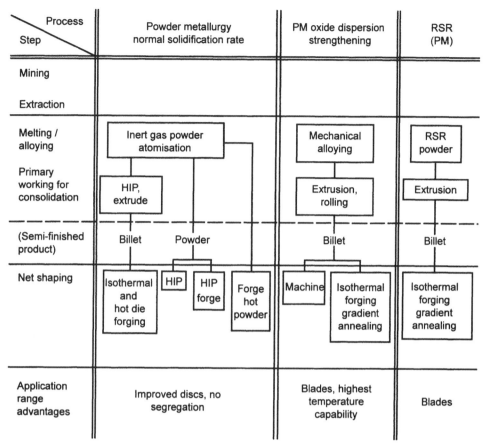

Fig. 12.13 Major processing categories of PM superalloys

12.7 TITANIUM ALLOYS

Titanium exists in two crystalline states: α (HCP structure) at low temperature and β (BCC structure) at temperature greater than 883°C. A range of titanium alloys have been developed. These can be divided into α-alloys, β-alloys and α + β alloys, according to the phase diagrams (Table 12.12). Ti-6Al-4V alloy (α–β alloy) is extensively used and is the work horse of titanium alloys. The alloy is heat treatable.

Powder metallurgy processing for titanium has a scrap loss of 10%–20%, compared to 50% incurred if the part has to be manufactured by the ingot metallurgy route. As titanium has poor machinability, the PM process is attractive.

Sintering of titanium compacts can be carried out in the temperature range 1100–1500°C in vacuum. Unlike many metals, titanium is capable of dissolving its own surface oxides at sintering temperature. Ti-6Al-4V alloy, based on sponge titanium powder (5–50 μm) and Al-V master alloy after vacuum sintering (at 1260°C for 2–3 h) and die forging at 1200°C, has been found to have full density. The alloy is heat treated to produce a stabilised microstructure. The pre-alloyed approach provides cost competitive complex-shaped parts with static and dynamic mechanical

Table 12.12 Classification of titanium alloys

Alpha	Alpha–beta
Ti-0.2Pd	Ti-8Mn
Ti-5Al-2.5Sn	Ti-6Al-4V
Ti-8Al-1Mo-1V	Ti-7Al-4Mo
Ti-6Al-2Nb-1Ta-1Mo	Ti-3Al-2.5V
Ti-6Al-2Sn-4Zr-2Mo-0.08Si	Ti-6Al-6V-2Sn
Ti-5Al-6Sn-2Zr-0.8Mo-0.25Si	Ti-4.5Al-5Mo-1.5Cr
	Ti-6Al-2Sn-4Zr-6Mo
	Ti-5Al-2Sn-2Zr-4Cr-4Mo
	Ti-6Al-2Sn-2Zr-2Cr-2Mo-0.25Si
Beta	
Ti-10V-2Fe-3Al(Ti-10-2-3)	
Ti-13V-11Cr-3Al(Ti-13-11-3)	
Ti-15V-3Cr-3Al-3Sn(Ti-15-333)	
Ti-3Al-8V-5Cr-4Mo-4Zr(Beta-C)	

Table 12.13 Room temperature tensile properties of HIP + worked + heat-treated billets of Ti-6Al-4V

Type	0.2% Proof stress, MPa	UTS, MPa	Elongation, % (5D)	Reduction in area, %
Rotating electrode process (REP)	922	1043	14	41
Centrifugal shot casting (CSC)	952	1072	15	35
Pulverisation sous vide (PSV)	898	1021	15	37
Electron beam rotating disk (EBRD)	910	1026	17	42

properties equal to or exceeding those of cast and wrought ingot metallurgy materials. Table 12.13 gives a comparative view of the room temperature tensile properties of HIPed/hot worked/ heat treated billets of Ti-6Al-4V alloy produced from different sources of pre-alloyed powder.

12.8 REFRACTORY METALS

Refractory metals such as W, Mo, Ta, Nb and V have many applications for high-temperature superconductor, capacitor, aerospace and nuclear applications. Since their melting points are very high, they are generally processed through the PM route. The powder metallurgy of tungsten dates back to the beginning of the twentieth century, when Dr WD Coolidge of General Electric Co., USA, developed the process for producing ductile tungsten incandescent lamp filaments to replace Edison's carbon filaments, which were brittle and short lived.

Presently, molybdenum and tungsten incandescent wires are used in doped form to produce non-sagging creep resistant wires by increasing the recrystallisation temperature. Small quantities of K and Si are introduced in the form of oxides in tungsten powder during the powder preparation stage. After sintering, the amount of dopants decreases; for example, potassium falls from 85 to 55 ppm and silicon from 60 to 55 ppm. After annealing at high temperature, longitudinally distributed bubble rows of potassium are observed.

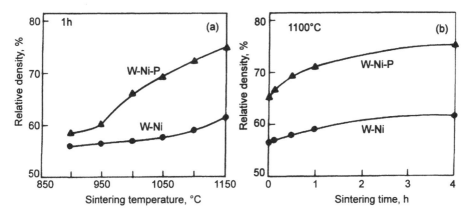

Fig. 12.14 Variation of relative density with respect to (a) sintering temperature and (b) sintering period for W-0.3% Ni and W-0.27% (Ni-P) alloys

Refractory metal powder compacts are activated sintered in solid state with addition of a minor quantity of later transition metals. A full description of the fundamentals of activated sintering is given in Chapter 7. The combination of activated sintering and liquid phase sintering in densifying refractory metals can be attempted. An example is tungsten compacts with the addition of 0.27% Ni and 0.34% P (Fig. 12.14).

12.8.1 Copper-containing alloys

W-Cu and Mo-Cu alloys are important in microelectronic packaging industry. The sintered density close to theoretical is achieved by promoting factors which improve sintering both in the solid state and liquid phase sintering stages, for example, the use of a high sintering temperature, submicron particle size, co-milling and using activators. To reduce copper volatilisation, the sintering temperature is generally limited to 1400°C or less. Another method of preparing PM W-Cu composites is by infiltration of a porous tungsten skeleton by a copper melt in vacuum or in a reducing or inert atmosphere.

12.8.2 Precipitation/Dispersion-strengthened refractory metals

Particle-strengthened refractory metals are vital for increasing the recrystallisation temperature and high temperature strength. Some criteria for selecting the particles are:

- Refractoriness
- Low solubility in refractory metals
- Thermodynamic stability with refractory metals or their dilute solutions under conditions of alloy formation
- Low coarsening of particles.

Figure 12.15 shows the free energy of formation relationship with temperature for some refractory phases. The temperature range suited for sintering refractory metals is indicated at

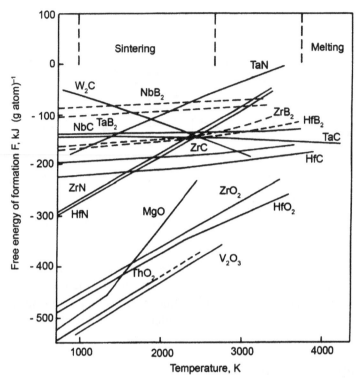

Fig. 12.15 Temperature dependence of free energy formation of refractory metal oxides and non-oxides

---

Example: 12.3 Why is the ductile–brittle transition temperature (DBTT) the most important characteristic to be considered when hot working molybdenum and tungsten? How is it affected by the deformation process?

Solution: The knowledge of DBTT allows one to work in a safe region. As the amount of deformation increases, the DBTT decreases and unrecrystallised wires and thin sheets are ductile at room temperature. As the DBTT is lowered, the temperatures of the metal working operations are reduced accordingly. In addition, high purity and smaller grain sizes also lower the DBTT. A knowledge of the residual porosity of the wrought part as a function of the amount of deformation is important for successful metal working, due to the drastic effect of porosity on the mechanical properties.

---

the top of the figure. It is evident that oxides are most stable, followed by carbides, nitrides and borides. As a matter of fact the best dispersion-strengthened tungsten alloy contains ThO_2 particles, a material used in the filament industry. To develop non-oxide dispersion-strengthened systems by sintering route, care has to be taken to avoid traces of oxygen in the atmosphere to prevent decomposition of less stable nitrides and carbides.

TZM is an interesting group of molybdenum alloys which contain titanium and zirconium. There are other related alloys in which the alloying elements are hafnium and carbon. Table 12.14 gives the composition of TZM and other related alloys. Properties, such as, high

Example: 12.4 Mo-Re products useful for extremely harsh environments, are fabricated via PM route:

(a) The photomicrographs shown below (Fig. Ex. 12.4) correspond to sintered pellets of Mo-41 wt% Re alloy, after 110 h sintering at 1650°C in hydrogen prepared from (i) blended and (ii) composites powders. Indicate the type of original powders. In case the sintering time is reduced to 50 h, what difference in microstructure will be observed?

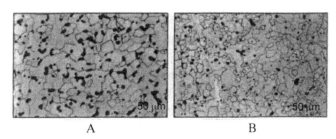

A   B

(b) The table below gives the mechanical properties and density data of Mo-41 wt% Re alloy sintered at 1650°C for 61 h. The initial powders were in blend form and composite powder, respectively. Further, the alloy was cold worked in sheet form. Indicate the type of material/product.

Table Ex 12.4

Condition	UTS, MPa	0.2% YS, MPa	% El	Density, % Theo	Hardness, HRA
A	765	668	13	93.2	58
B	948	754	19	100	65
C	910	834	14	96.3	63

(Reproduced with permission from *Int. J. Powder Metall.*, 9(9), 2003)

(c) Suggest typical applications of this alloy.

Solution

(a) The microstructure (A) exhibits interconnected and triple point porosity and pertains to blended powder compact. The microstructure (B) pertains to composite powder, since the majority of pores are within the grains.

In case the sintering time is lower, say 50 h, the compact based on blended powder would have large-size interconnected porosity and the grains might still be distinct. The composite powder–based compact would show limited interconnected porosity.

(b) (A) is based on blend powder, while (B) must correspond to cold-worked condition, as in this case there is no porosity and the part is fully dense. The condition (C) would be related for the compacts based on composite powder.

(c) Typical applications may be heat sinks, heating elements, thermocouple sheathings, vacuum furnace components and in aerospace.

thermal conductivity, high stiffness, low thermal expansion, high hot hardness and excellent resistance to many molten metals, have found applications in die casting cores, hot working dies, mandrels and heating elements. The alloys are generally produced in PM billets, which are hot forged into pancakes and subsequently machined

Table 12.14 Composition of TZM and other related molybdenum-based alloys

Alloying	Ti, %	Zr, %	Hf, %	C, ppm
TZM	0.4–0.55	0.06–0.12	-	100–400
TZC	1.0–1.4	0.25–0.35	-	700–1300
MHC	-	-	1.0–1.2	600–675

into dies. In general, the physical properties of TZM are approximately the same as those of unalloyed molybdenum. The modulus of elasticity of TZM is significantly higher than that of hot worked steel, and except at intermediate temperatures, also above those of the superalloys. The KI_c value of TZM at room temperature is 16 MPa m$^{1/2}$, which increases to approximately 150% from room temperature to 450°C. The alloy shows little dependence of annealing atmosphere on recrystallisation and grain growth behaviour. TZM has far greater resistance to thermal fatigue (heat checking) than hot worked steel, because of high thermal conductivity, high modulus and low thermal expansion. However, at temperatures over about 550°C,

---

Example: 12.5 Describe the microstructural and preparation features of Non-sag (NS) tungsten.

Solution: Non-sag (NS) tungsten is a dispersion-strengthened microalloy with element K, which is contained as microscopic bubbles in the tungsten lattice. In recent years, another term AKS-doped tungsten has emerged for NS tungsten. These three letters—designates for the German names of these elements: aluminium, kalium (potassium) and silizium (silicon)—are added as doping elements to the tungsten blue oxide. The amount of dopants are 3000–3500 ppm K, 2000–2500 ppm Si and 400–500 ppm Al.

NS doping of tungsten generates submicron, amorphous potassium aluminosilicate particles within reduced tungsten powder. These amorphous potassium aluminosilicates are the precursor to potassium bubbles in sintered ingots. Densification of NS-doped tungsten ingots between 1900 and 2400°C is controlled by grain boundary diffusion.

The potassium bubbles which occur on sintering can be classified into three stages. In the first stage, potassium aluminosilicate particles within the reduced tungsten particles migrate to the necks that are formed during the first stage of sintering. In the second stage, potassium aluminosilicates decompose to suboxides, and aluminium and silicon are removed by volatilisation of their suboxides through surface-connected porosity. The majority of aluminium and silicon evaporates from the ingot in preference to entering solid solution during this stage. During the first and second stages of sintering, potassium is only located in submicron, spherical pores. In the third stage, on prolonged sintering at temperatures at and above 2300°C, oxygen diffuses from the small spherical pores and potassium diffuses out of the submicron pores into the large, irregularly-shaped pores (>2 μm). Generally, potassium is retained in both the small, spherical and the large, irregularly-shaped pores within the tungsten matrix. Once potassium has diffused into the larger pores, its vapor pressure opposes further shrinkage of these pores; this restricts further densification. (Continued on next page)

When the tungsten ingot is deformed by rolling, swaging and drawing, the potassium-containing pores elongate into ellipsoids. There is some indication that the ellipsoids break up along their length during deformation. When a rod or wire containing these ellipsoids is annealed, it either breaks up into a row of spherical bubbles or it contracts to form a single bubble.

Figure Ex. 12.5a shows a filament and micrograph of the cross section of NS-doped tungsten wire, showing the interlocking secondary recrystallised grain structure. Figure Ex.12.5b shows the TEM picture of a tungsten wire exhibiting the fine dispersion of potassium bubbles which generate an interlocked secondary recrystallised grain structure, that can both withstand sagging caused by grain boundary sliding and also increase the high temperature strength.

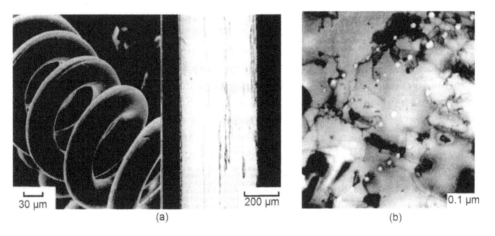

30 μm   200 μm   0.1 μm

(a)   (b)

Fig. Ex. 12.5

The industrial manufacture of doped tungsten metal powder is performed in the following steps:

- Calcination of ammonium paratungstate (APT) to form tungsten blue oxide (TBO)
- NS doping of the TBO with aqueous solution of K, Al and Si
- Subsequent hydrogen reduction of the doped TBO to yield NS tungsten powder with a specific grain size, grain size distribution and dopant incorporation.

TZM oxidises so rapidly in air or oxidising atmospheres, that its use under these conditions is not possible.

12.8.3 Tungsten heavy alloys

Heavy alloys were first discovered by McLennan and Smithells in 1935. Their aim was to prepare a W-Cu-Ni alloy for use as radiation shields. The major attraction of these alloys are their combined high strength, high density and corrosion resistance. The PM processing of this system is a classical example of liquid phase sintering, the basic aspects of which were discussed in Chapter 7. A typical composition of heavy alloys ranges from 80% to 96% tungsten (by weight) with either Cu-Ni or Fe-Ni as the remaining part. Mechanical properties of heavy alloys depend on microstructure, chemistry and processing. Variation of properties has been attributed to alloy composition, impurity segregation on tungsten spheroid/matrix interface, hydrogen embrittlement, intermetallic phase formation (if any), residual porosity and ductile–brittle temperature effects. Rapid cooling from the sintering temperature is done to prevent the formation of WNi_4 intermetallic phase. The sintering temperature is in the range 1420–1450°C and sintering period up to 3 h for 90W-3Cu-7Ni alloy. The size of the tungsten spheroid in the sintered parts is observed to increase with increasing sintering temperature/ nickel content in the binder. Partial substitution of tungsten in W-Cu-Ni heavy alloy by molybdenum has shown better strength, but with a loss in ductility and overall density.

Although W-Cu-Ni heavy alloys were initially developed, currently, Fe-Ni is being considered as a binder due to its better mechanical properties. The sintering temperature for these groups of heavy alloys is higher than for the W-Cu-Ni system, being around 1500°C. Some progress is also reported in the mechanical alloying route, where substantial volume fraction of intermetallics could be tolerated without loss in sintered densities. It is also found that compared to the purity of the powders, a slight deviation from the optimum processing parameters has a significant effect on mechanical properties.

The sintering atmosphere plays an important role in controlling the residual porosity in W-Fe-Ni heavy alloys. Generally, the atmosphere used is hydrogen, but dry hydrogen often causes blistering and swelling during liquid phase sintering. This problem was solved by using wet hydrogen in the sintering cycle (Fig. 12.16). During heating at a temperature of 800°C, a hold period of 1 h in dry hydrogen is given for reducing any oxide over the tungsten powder surface. During actual sintering at a temperature higher than the solidus temperature, the period

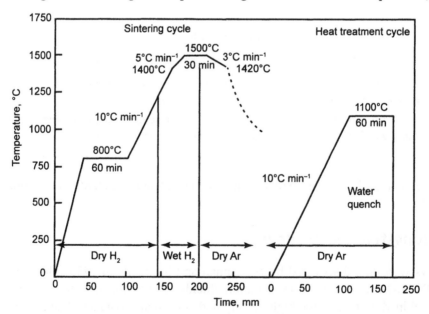

Fig. 12.16 The three-stage atmosphere sintering and heat treatment schedule for a tungsten heavy alloy (Reproduced with permission from TMS, Warrendale, USA, Bose A and German RM, *Metallurgical Transactions*, 19 (A), 1988)

is around 30–45 min, where wet hydrogen is used. During the cooling stage, a slow cooling rate of 3°C min⁻¹ to 1420°C (below the solidus) is used in order to minimise the shrinkage during solidification. A switch to a secondary inert gas atmosphere (dry argon, dew point –60°C) is made near the end of the sintering cycle. This atmosphere switchover is to remove the hydrogen embrittlement problem. Finally, the sintered parts are heat treated by water quenching after 1 h to 1100°C, which effectively suppresses impurity segregation. Table 12.15 summarises mechanical properties of W-Fe-Ni alloys sintered in different conditions.

12.8.4 Tantalum

One of the major applications of PM tantalum is to prepare porous pellets and also wires for use in capacitor devices. The quality and electrical performance of capacitor grade tantalum powder have improved substantially because of improvements made in the stirred reactor reduction of potassium heptafluorotantalite (K_2TaF_7) with sodium and the post-reduction processing of primary powder. Efforts were focussed on modifying the physical properties to improve the pressing characteristics, increasing the surface area to increase specific capacitance, and reducing chemical impurities like oxygen, carbon and sodium. An oxygen content in excess of 3500–4000 ppm has a detrimental effect on the performance of a solid tantalum capacitor. The carbon in the powder ranges from 20 ppm to 70 ppm. Sodium concentration below 2 ppm and potassium concentrations less than the typical detection limit of 10 ppm have been achieved. Unfortunately, the conditions suited to reduce the alkali metal concentration also reduce the surface area and associated specific conductance of tantalum powder.

Figure 12.17 illustrates the effect of alkali metal and sintering temperature on the normal specific conductance of tantalum. Higher powder purity can be attained by lowering the

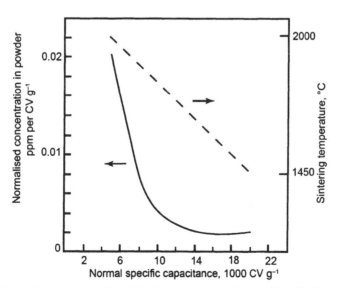

Fig. 12.17 Specific conductance relationship for porous tantalum with alkali metal impurities and sintering temperature

Table 12.15 Ultimate tensile strength and ductility of W-Fe-Ni heavy alloys

Composition wt%	Sintering variables			UTS, MPa	El, %
	Atmosphere	Temp., °C	Period, min		
90W-3Fe-7Ni	Hydrogen	1480 (slow cooling)	60	945	27
90W-3Fe-7Ni	Hydrogen	1460 (rapid cooling)	60	733	8
95W-1.5Fe-3.5Ni	Hydrogen	1460	30	793	4
90W-3 Fe-7Ni	Hydrogen	1445	60	903	29
95W-1.5Fe-3.5Ni	Hydrogen	1450	120	885	29
95W-1.5Fe-3.5Ni	Hydrogen	1520	60	859	18
90W-3Fe-7Ni	Hydrogen	1470	60	850	15
90W-3Fe-7Ni	Nitrogen	1470	60	870	20
90W-3Fe-7Ni	Hydrogen	1460	60	827	30
95W-1.5Fe-3.5Ni	Hydrogen	1470	120 (vacuum annealing)	900	23
90W-3Fe-7Ni	Hydrogen	1470	60	910	36
90W-3.3Fe-6.7Ni	Hydrogen	1475	30	925	23.7
		1490	30	962	24.6
		1510	30	923	19.7
95W-1.5Fe-3.5Ni	Hydrogen	1475	15	1000	21
			30	1000	21
			60	980	15.8
			120	900	6.0
93W-2.1Fe-4.9Ni	Hydrogen + argon	1460	30 H₂	1006	
			10 Ar		22
95W-1.5Fe-3.5Ni	Hydrogen + argon	1460	30 H₂	995	20
			10 Ar		
95W-1.5Fe-3.5Ni	Hydrogen	1470	30	996	18
98W-1Fe-1Ni	Hydrogen	1500	60	749	2.6
90W-8Fe-2Ni	Hydrogen	1500	60	620	4.2
90W-6Fe-4Ni	Hydrogen	1500	60	730	11.0
90W-4Fe-6Ni	Hydrogen	1500	60	810	14.2
90W-2Fe-8Ni	Hydrogen	1500	60	860	15.4

sintering temperature. The green density of pellets vary from 4.0 to 7.0 g cm^{-3}. The sintering of tantalum is carried out in the temperature range 1400–1700°C for 10–30 min.

Capacitor grade tantalum wire is manufactured through PM processing similar to other refractory metals. These are available in a range of diameters from 0.19 mm to 0.76 mm.

The temper ranges from annealed to extra hard. The metal is doped with rare earth oxides, silicon or carbon to provide grain stabilisation.

12.9 INTERMETALLICS

Of late, aluminide intermetallics (Ni_3Al, $NiAl$, Fe_3Al, $TiAl$, Ti_3Al) and silicide intermetallics ($MoSi_2$) have drawn much attention for high-temperature structural applications. Similarly, rare earth intermetallics (RCo_5, R_2Co_{17}, $R_2Fe_{14}B$) have been developed for high-strength permanent magnets.

12.9.1 Silicides

One of the major silicide intemetallics is $MoSi_2$, which possesses a high congruent melting point of 2030°C and excellent high-temperature oxidation resistance. Although brittle at low temperatures, it has a brittle–ductile transition of approximately 1000°C. $MoSi_2$ is thermodynamically stable with potential ceramic reinforcement such as SiC, Si_3N_4, ZrO_2, Al_2O_3, TiB_2 and TiC, and may also be alloyed with other high melting point silicides such as WSi_2. Due to its low electrical resistivity (15 $\mu\Omega$ cm at room temperature) it can be electro-discharge machined, which is a significant benefit for the low-cost fabrication of components. Since the thermal expansion coefficient of $MoSi_2$ is closer to metals, it can be easily joined with metals.

$MoSi_2$ is most stable in the temperature range 1000–1600°C, when a glassy SiO_2 surface layer is formed during oxidation. At low temperatures (600–800°C) $MoSi_2$ exhibits a low oxidation resistance.

Sintered densities upto 99% of theoretical are obtained using a fine-grained powder (2–10 μm). It is isostatically cold pressed, followed by hydrogen sintering at 1700°C. Grain growth during sintering can be decreased by SiC or Al_2O_3 addition, which increases the strength. Hot pressing in graphite dies at 1600–1700°C and a pressure of about 50 MPa with a starting powder size of less than 10 μm yields materials with full density after only 30 min. It is possible to decrease the sintering temperature to 1200°C by chemically activated sintering in a siliconising atmosphere containing fluorides.

12.9.2 Aluminides

Some properties of aluminide intermetallics are summarised in Table 12.16. They possess much lower densities than the existing alloys. They contain a sufficient amount of aluminium to form a thin film of alumina (Al_2O_3) in an oxidising environment, which is compact and protective. The ordered structures exhibit attractive elevated temperature properties. However, the aluminides suffer from brittleness at room temperature and high brittle-to-ductile transition temperatures.

Table 12.16 Important transition metal aluminides and their properties

Alloy	Crystal structure	MP, °C	Density, g cm^{-3}
Ni_3Al	LI_2	1,390	7.50
NiAl	B_2	1,640	5.86
Fe_3Al	DO_3	1,540	6.72
FeAl	B_2	1,250	5.56
TiAl	Al_0	1,460	3.91
Ti_3Al	DO_{19}	1,600	4.20

Aluminides are reactive liquid phase sintered, where the melt reacts with the solid (transition metal) to form the compound. In case the melting point of the compound is higher than the sintering temperature, then the liquid is transient.

The industrial application of PM aluminides is still in the initial stage. Cast Ni_3Al hot pressing dies have replaced 1N-718 dies. The dimensional control during sintering is one of the issues, particularly when post-sintering machining/grinding operations are involved. Of Late, Ni_3Al powder is replacing expensive cobalt binder in same grades of cemented carbides.

12.9.3 Rare earth intermetallics

Among the rare earth intermetallics, RCo_5 is the oldest one, and is widely applied as a permanent magnet. Sm-Co alloy (<81 at%) solidifies as the primary phase $SmCo_5$, while at higher cobalt contents the compound forms by a peritectic reaction between the primary phase Sm_2O_{17} and the melt. The formation of Sm_2O_{17} in $SmCo_5$ should be carefully avoided, as the former greatly reduces the coercive force. Samarium can be partly replaced by Pr, La, Ce and mischmetal.

The most important and recent development in the area of rare earth intermetallic permanent magnets has been $Nd_2Fe_{14}B$. The bonded rare earth cobalt magnets have been almost entirely replaced by bonded Nd-Fe-B magnets. The composition of the alloy is slightly hypo-stoichiometric of the compound $Nd_2Fe_{14}B$ in order to compensate for oxidation during processing. The Nd-rich grain boundary phase is a eutectic. Excess oxidation of the alloy composition during processing results in a shift in the effective composition to the Fe-rich side of the phase diagram. The sintering temperature is 1100°C, the period 20–90 min and the density of the finished magnets is 7.4 g cm^{-3}. They are heat treated for 1–3 h, at temperatures between 600°C and 900°C, in order to maximise the coercivity. Both sintering and annealing is carried out in argon. Prior to cold compaction the powder is prepared by pulverising the cast ingot of Nd-Fe-B alloy (3 μm size). The compaction is done at 200 MPa in a magnetic field (800 kA m^{-1}) to align the powder.

Another method of production is from melt spun ribbons after compacting to a high density, followed by gluing the ribbon fragments together in a bonded-type magnet with dry epoxy resin. A better method is hot pressing.

12.10 CERAMIC SYSTEMS

The composition of ceramic materials varies widely, and both oxide and non-oxide materials are used. Often, for ease of fabrication, some glass forming constituents are added. Classification of ceramic systems can be done based on their functions or composition. Table 12.17 presents the listing according to major functions performed. In subsequent sections, a brief review has been given according to the composition.

12.10.1 Porcelain (Whiteware)

Whitewares includes electrical and chemical technical porcelain and chinawares. These are also categorised as 'traditional ceramics'. They are produced from a mixture of minerals consisting of

Table 12.17 Classification of ceramics by function

Function	Class	Nominal composition*
Electrical	Insulation	α-Al_2O_3, MgO, porcelain
	Ferroelectrics	$BaTiO_3$, $SrTiO_3$
	Piezoelectric	$PbZr_{0.5}Ti_{0.5}O_3$
	Fast ion conduction	β-Al_2O_3, doped ZrO_2
	Superconductors	$Ba_2YCu_3O_{7-x}$
Magnetic	Soft ferrite	$Mn_{0.4}Zn_{0.6}Fe_2O_4$
	Hard ferrite	$BaFe_{12}O_{19}$, $SrFe_{12}O_{19}$
Nuclear	Fuel	UO_2, UO_2-PuO_2
	Cladding/shielding	SiC, B_4C
Optical	Transparent envelope	α-Al_2O_3, $MgAl_2O_4$
	Light memory	doped $PbZr_{0.5}Ti_{0.5}O_3$
	Colours	doped $ZrSiO_4$, doped ZrO_2, doped Al_2O_3
Mechanical	Structural refractory	α-Al_2O_3, MgO, SiC, Si_3N_4, $Al_6Si_2O_{13}$
	Wear resistance	α-Al_2O_3, ZrO_2, SiC, Si_3N_4, Toughened Al_2O_3
	Cutting	α-Al_2O_3, ZrO_2, TiC, Si_3N_4, SIALON
	Abrasive	α-Al_2O_3, SiC, toughened Al_2O_3, SIALON
	Construction	Al_2O_3-SiO_2, CaO-Al_2O_3-SiO_2, Porcelain
Thermal	Insulation	α-Al_2O3, ZrO_2, $Al_6Si_2O_{13}$, SiO_2
	Radiator	ZrO_2, TiO_2
Chemical	Gas sensor	ZnO, ZrO_2, SnO_2, Fe_2O_3
	Catalyst carrier	$Mg_2Al_4Si_5O_{18}$, Al_2O_3
	Electrodes	TiO_2, TiB_2, SnO_2, ZnO
	Filters	SiO_2, α-Al_2O_3
	Coatings	NaO-CaO-Al_2O_3-SiO_2
Biological	Structural prostheses	α-Al_2O_3, porcelain
	Cements	$CaHPO_4$·$2H_2O$
Aesthetic	Pottery, artware	Whiteware, porcelain
	Tile, concrete	Whiteware, CaO-SiO_2-H_2O

Source: Adapted from Kenne GB and Brown HK, High-tech ceramics in Japan: Current and future markets, *Am. Ceram. Soc. Bull.*, 62(5):590–96, 1982.

*Whiteware is a family of porous, dense, fine-grained materials with a glassy matrix, usually containing Al_2O_3, SiO_2, K_2O and Na_2O. Porcelain is a type of whiteware that is non-porous, hard and translucent. SIALON is a solid solution phase with the nominal composition $Si_4Al_2N_6O_2$. Toughened Al_2O_3 is a two-phase material containing a minor amount of doped or undoped ZrO_2.

special clays, fluxing minerals (often feldspar) and finally ground quartz. This mineral mixture is shaped through powder processing methods and then vitrified at high temperatures to produce a dense, hard material. The most abundant true clay mineral is kaolinite, which is a hydrated aluminium silicate with the chemical formula $Al_2Si_2O_5(OH)_4$. Several structural variations of the fixed kaolinite formula exist, depending on differences in internal arrangement of the aluminium, silicon and oxygen atoms in the crystals, but they are less commonly used in porcelain. The feldspars are a group of minerals that form molten glass at moderate temperatures. These are aluminosilicates of sodium, potassium or calcium. The pure feldspars are albite ($NaAlSi_3O_8$), orthoclase ($KAlSi_3O_8$) and anorthite ($CaAl_2Si_2O_8$). The silica used to impart strength and stability to porcelain bodies is a much finer grade form of quartz called Potter's Flint.

Figure 12.18 shows the isothermal section of ternary feldspar–clay–quartz phase diagram at 1300°C, which is the typical sintering temperature for porcelain. The compositions related to porcelains are circled. It is to be noted that the lowest eutectic temperature in this system forms at about 990°C. In ceramic literature, the term 'firing' is often used instead of sintering.

During the vitrification process, the following physical changes occur in whiteware bodies:

- Shrinkage due to loss of open pores
- Development of closed pores
- Development of glassy phase.

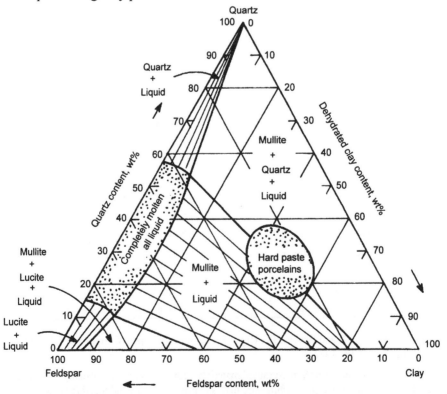

Fig. 12.18 An isothermal section through the phase diagram illustrating the phases present at the firing temperature of hard paste porcelains

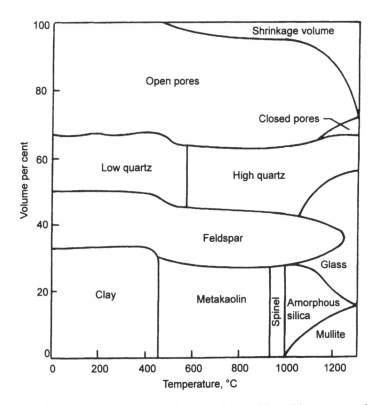

Fig. 12.19 Chart showing presence of various phases and porosities with respect to firing temperature for a porcelain

These physical changes are schematically shown in Fig. 12.19. It is evident from the figure that the open pores are reduced from 33% to almost zero, while the closed pores, which start developing around 1100°C, end up being above 7% of the fired volume. The glassy phase is about 57% of the fired volume.

12.10.2 Alumina, magnesia and zirconia

Among pure oxide ceramics, alumina, magnesia and zirconia find use in a wide range of applications. Table 12.18 includes properties of some major oxide and non-oxide ceramics. Alumina is the most widely used ceramic and it possesses low thermal conductivity, a comparatively high thermal expansion coefficient, high hardness and chemical stability. It can exist in different forms, some amorphous (hydrous and anhydrous) and some as crystalline oxides or hydrates. A third group comprises β-variations with small amounts of alkali or alkaline earth ions lodged into the structure. The α-form is identical with corundum, a natural mineral which is rhombohedral. γ-alumina is generally formed by heating hydrated alumina material such as kaolin $Al_2(Si_2O_5)(OH)_4$ or gibbsite $Al(OH)_3$ at about 900°C. ξ-alumina occurs only in melts which contain lithium.

Table 12.18 Properties of some major ceramics

Material	Crystal structure	Theoretical density, g cm⁻³	Knoop or Vickers hardness, GPa	Transverse rupture strength, MPa	Fracture toughness, MPa m$^{1/2}$	Yound's modulus, GPa	Poisson's ratio	Thermal expansion, 10^{-6} K⁻¹	Thermal conductivity, W m⁻¹ K⁻¹
Al_2O_3	Hexagonal	3.97	18–23	276–1034	2.7–4.2	380	0.26	7.2–8.6	27.2
Mullite	Orthorhombic	2.8	-	185	2.2	145	0.25	5.7	5.2
Partially stabilised ZrO_2	Cubic monoclinic, tetragonal	5.7–5.75	10–11	600–700	-	205	0.23	8.9–10.6	1.8–2.2
Fully stabilised	Cubic	5.56–6.1	10–15	245	2.8	97–207	0.23–0.32	13.5	1.7
TiB_2	Hexagonal	4.5–4.54	15–45	700–1000	6–8	514–574	0.09–0.13	8.1	65–120
TiC	Cubic	4.92	28–35	241–276	-	430	0.19	7.4–8.6	33
Cr_3C_2	Orthorhombic	6.7	10–18	49	-	373	-	9.8	19
SiC	α, Hexagonal	3.21	20–30	-	-	207–483	0.19	4.3–5.6	63–155
Si_3N_4	α, Hexagonal	3.18	8–19	-	-	304	0.24	3.0	9.30
TiN	Cubic	5.43–5.44	16.20	-	-	251	-	8.0	24

Alumina can be easily sintered by addition of liquid-forming sintering oxides. Common examples are Al_2O_3-SiO_2 with a binary eutectic at 1590°C and the more complex Al_2O_3-CaO-SiO_2 and Al_2O_3-MgO-SiO_2, which contain ternary eutectics melting at as low as 1170°C and 1355°C, respectively. These additives form a glassy phase at the grain boundaries, causing loss of high-temperature strength and creep resistance. An addition of MgO to Al_2O_3, has been found to slow down grain boundary migration, thus allowing complete pore elimination by solid-state sintering. Another oxide alloy based on alumina is aluminium titanate which forms at about 1300°C. This is useful for automotive applications where easy machining of formed parts and high wear resistance are needed.

Magnesia is found in only one crystalline form, namely, the cubic(NaCl)-type. It occurs in nature as periclase. Commercially it is produced by the decomposition of magnesite, $MgCo_3$, or the hydroxide, brucite, $Mg(OH)_2$. To be of any use as a refractory, MgO must be heated above 1600°C (dead burned) during which process moisture and carbon dioxide are driven off and complete sintering is achieved.

Zirconia is another important oxide ceramic. There are three polymorphs of zirconia. Monoclinic zirconia is stable upto 1150°C, when it transforms to tetragonal. At temperatures above 2300°C the cubic form exists. Tetragonal–monoclinic phase change is of technological importance, since it is accompanied by a volume expansion of about 4%. This transformation is of the martensitic diffusionless type and if this is allowed to develop in an uncontrolled manner, spontaneous failure of zirconia ceramic can occur. Control of particle size, composition and distribution of the zirconia allows the properties to be engineered.

Partially-stabilised zirconia (PSZ) has high strength and toughness and performs more reliably than zirconia. It is obtained by doping the zirconia with oxides of calcium, yttrium or magnesium. New developments to further improve the properties of PSZ include transformation-toughened zirconia, which has higher toughness than PSZ, because of dispersed tough phases in the ceramic matrix

The rather poor sinterability of cubic zirconia, even at elevated temperatures, is attributed to the low cation diffusion rates that are rate controlling. Maximum densification in ZrO_2-CaO system is noticed for the lowest CaO content needed for stabilisation, and the density decreases linearly with an increasing CaO content in the stabilised zirconia. Densification is more enhanced when stabilisation takes place during sintering at temperatures up to 1500°C than with pre-stabilised material. Oxides obtained by thermal decomposition of compounds sinter better than precalcined oxides.

12.10.3 Barium titanate

Barium titanate ($BaTiO_3$) is used in capacitors, sonar equipments, ultrasonic cleaners, flaw detection equipment and many other applications requiring special dielectric, piezoelectric or ferroelectric properties. It has a very slightly distorted cubic structure at room temperature. The Ti^{4+} and O^{2-} atoms are shifted with respect to each other by almost 0.012 nm, resulting in a permanent dipole. Two important materials in the titanate family are PZT (lead zirconate titanate) and PLZT (lead lanthanum zirconate titanate). Common additives used in sintering

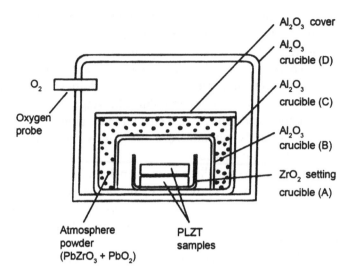

Fig. 12.20 Sintering set up for lead lanthanum zirconium titanate (PLZT) in controlled atmosphere

barium titanate are TiO_2, SiO_2 and Al_2O_3 which exhibit eutectic reactions, thus facilitating liquid phase sintering. The formation of a second phase is observed with the use of excess TiO_2 ($\approx 1\%$) at a sintering temperature above the eutectic melt (1320°C). After cooling, this melt remains for the greater part in a glassy state, predominantly on the triple joint between the crystallites. PZT sintering poses the problem of volatilisation of PbO at the sintering temperature. This can be minimised by reducing the sintering temperature to below 900°C. But, in this case, chemically derived fine powder (<1 μm size) is to be used. Dopants like Nd, La, etc., modify the electrical properties of the sintered components. Hot pressing of PZT powders eliminates optical scattering caused by the presence of porosity, grain boundaries and internal refraction at domain walls. Figure 12.20 shows a sintering facility for PLZT powder sample.

12.10.4 Ferrites

Magnetic ceramics are complex oxides and can be divided into three classes: spinel or cubic ferrite, garnets or rare earth ferrites and magneto plumbites or hexagonal ferrites (Table 12.19). In the last category substitutional solid solution of various divalent cations for Ba^{2+} can occur in the lattice. A variation of Ba^{2+} content is also possible in the $BaO.6Fe_2O_3$ lattice before the equimole $BaO:Fe_2O_3$ phase is formed. These features offer compositional flexibility to tailor the magnetic properties of barium ferrite by substituting a cation, such as Co^{2+} with magnetic moments, for non-magnetic Ba^{2+} ions.

To illustrate the sintering cycle for ferrite, an example of Mn-Zn ferrite may be taken. In this case the spinel phase is stable only over a certain range of atmosphere and temperature conditions. The concentration of iron in the ferrous state is critical to the attainment of the desired properties of low magnetic loss and a maximum in the magnetic permeability. Figure 12.21 illustrates the sintering cycle, where the samples are first sintered at a temperature of 1250–1400°C in an atmosphere of high oxygen partial pressure (0.3–1 atm) to minimise evaporation of zinc.

Table 12.19 Classes of magnetic ceramics

Structure	Composition	Applications
Spinel (cubic ferrites)	1 MeO: 1 Fe_2O_3 (where MeO = transition metal oxide, e.g., Ni, Co, Mn, Zn)	Soft magnets
Garnet (rare earth ferrites)	3 Me_2O_3: 5 Fe_2O_3 (where Me_eO_3 = rare earth Metal oxide, e.g., Y_2O_3, Gd_2O_3)	Microwave devices
Magneto-plumbite (hexagonal ferrites)	1 MeO: 6 Fe_2O_3 (where MeO = divalent metal oxide from group IIA, e.g., BaO, CaO, SrO)	Hard magnets

In this stage both densification and grain growth are completed. In the second stage, the material is annealed (portion A in Fig. 12.21) at a lower temperature (1050–1200°C) in an atmosphere with fairly low oxygen partial pressure (50–100 ppm) to establish the desired concentration of ferrous ion ($Fe^{2+}/Fe^{3+} \approx 0.05$ to 2.00).

12.10.5 Oxide superconductors

$YBa_2Cu_3O_{7-y}$ (123) compound is a typical example of a high-T_c oxide superconductor. Other superconducting oxides are based on bismuth/thallium and copper oxides. Unlike conventional oxide ceramics, high-T_c oxide superconductors are extremely sensitive to their conditions of processing and dopants. Oxygen stoichiometry and associated defects play a key role in influencing the sintering temperature of such materials as well as the rates at which they approach equilibrium during processing.

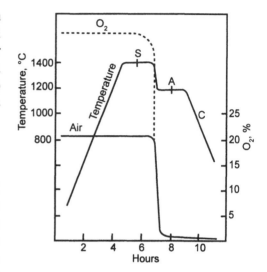

Fig. 12.21 Schematic of heating cycle for manganese zinc ferrites showing controlled atmospheres during sintering, annealing and cooling cycles

For compositions rich in CuO and BaO, a low melting eutectic (M.P. 890°C) occurs; it helps in liquid phase sintering in the temperature range (925–975°C). If a high sintering temperature is applied, much of the oxygen associated with CuO is liberated. Oxygen deficiency is needed for the compound to be a conductor, but too large a deficiency can produce a non-conductor or superconductor with a much lower critical temperature, T_c. Hot pressing in the presence of excess oxygen at 9–11 atm pressure is another process to prepare a high-density product with high T_c. However, the materials are unstable and tend to degrade with exposure to the environment.

The important property of a superconductor, J_c, of YBCO is strongly dependent on processing conditions which affect the homogeneity of the bulk samples, density, oxygen concentration, impurity phase formation and grain boundary chemistry. With an increasing sintered density, samples sintered at 975°C for 8 h give a J_c at 77 K of 20 A cm^{-2} and reveal the formation of $BaCuO_2$ and CuO as secondary phases along the grain boundaries which prevent transport currents.

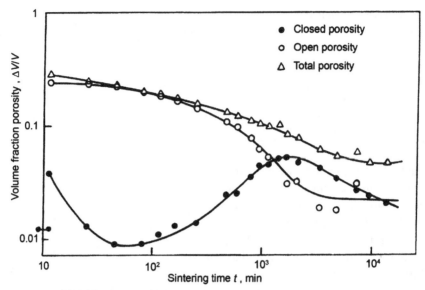

Fig. 12.22 Change in porosity during the sintering of a UO$_2$ powder compact at 1400°C

12.10.6 Oxide nuclear fuels

For power nuclear reactors, the use of oxide fuels is prevalent. Two common oxides are UO$_2$ and PuO$_2$. Uranium oxide is non-stoichiometric UO$_{2-x}$. When reduced in hydrogen, it gets converted to UO$_2$. The starting powder for production of fuel pellets should be very pure, particularly with respect to carbon-, fluorine- and neutron-absorbing elements. Proper powder characteristics of UO$_2$ is very important for satisfactory sintering. The variation of closed porosity with respect to sintering period at 1400°C is rather uneven (Fig. 12.22). For a homogeneously packed powder, the conversion to closed porosity occurs later and more suddenly.

In case of mixed oxide fuels (MOX), UO$_2$ and PuO$_2$ are mechanically milled, pre-compacted at about 75 MPa, followed by granulation and compaction at 300 MPa (green density 52%TD). The pellets are then sintered at 1700°C for 8 h in a reducing atmosphere. The type, amount, size and distribution of pores in ceramic fuels play important roles in the reactor performance.

12.10.7 Silicon carbide

SiC is widely used for structural components. The two major groups of applications are: abrasion- and corrosion-resistant components and heat-resistant components. The bonding in SiC is highly covalent, thus making the solid-state diffusion pretty slow. For this reason, the powder is generally sintered in an inert atmosphere with dopants, typically a mixture of 0.5 wt% carbon and 0.5 wt% boron, at around 2000°C to attain full density. However, the gas is trapped in the closed pores, which can be reduced by using fine SiC powder and maintaining a fine pore/grain size during sintering.

Liquid phase sintering has been extensively applied in order to obtain dense SiC ceramics. It is observed that a 3:5 ratio of Y_2O_3 and Al_2O_3 additives imparts optimum densification in the β-SiC powder. The advantage of using a liquid phase is that the silicon carbide, which is not easily sintered, can be sintered to high densities (> 95% theoretical) at temperatures well below 2000°C. The volume percentage of Y_2O_3 and Al_2O_3 mixture in SiC is generally 10%.

12.10.8 Silicon nitride

In both its structural modifications—α and β—silicon nitride (Si_3N_4) has a unique combination of properties; it is strong, hard, wear-resistant, stable to temperatures above 1800°C, oxidation-resistant, and due to its low coefficient of thermal expansion, it has excellent resistance to thermal shock. It is also half as dense as steel.

In reaction-bonded silicon nitride, the silicon powder mass is shaped, partially nitrided and then machined to product before fully nitriding. The material is porous and has a relatively low strength but finds some applications in metal handling, for example, as thermocouple sheaths. Post-sintered reaction bonding closes the pores and raises the strength to a level approaching that of hot-pressed silicon nitride.

The nitridation reaction commences at approximately 1200°C and the temperature is programmed to 1450°C to achieve a complete reaction. Total reaction time is a function of part cross section and varies from 150–2000 h. Nitrogen–hydrogen or N-H-He gas mixtures are used to give faster and more easily controlled nitridation rates and higher strength of material. The resulting microstructure of reaction-bonded Si_3N_4 is rather complex, consisting of a mixture of fibrous and equiaxed materials of inhomogeneous density.

Since the mechanical properties of silicon nitride are extremely sensitive to residual porosity, full density consolidation processes like gas pressure sintering, hot pressing, sintering followed by hot isostatic pressing, and hot isostatic pressing after encapsulation of powder are used.

The liquid phase sintering of silicon nitride with oxide additives is advantageous in many ways. MgO forms a eutectic melt with SiO_2 at around 1550°C. Normally 3–10 wt% MgO is used and the sintering temperature can be anywhere between 100 and 300°C greater than the eutectic temperature. The use of Y_2O_3 as the additive in place of MgO leads to the eutectic melt formation at a relatively higher temperature (~1660°C). Many silicon nitride ceramics are now sintered with an additive consisting of Y_2O_3 and Al_2O_3. These materials have better high-temperature creep resistance and oxidation resistance compared to the materials containing only Y_2O_3 additive.

12.10.9 Refractory compounds

Carbides, nitrides and borides of early transition metals are commonly called refractory compounds. These are well known for their unique properties such as high melting point, hardness and other interesting physical and mechanical properties. Such non-oxide ceramics mostly have a metallic lustre and electrical and thermal conductivities of the same order as metals. Refractory carbides and nitrides have homogeneity ranges as a result of vacancies in the non-metal sublattice. Such refractory compounds are used in making cermets, which are extensively used in cutting tools, wear and metal forming applications.

The refractory compounds are generally sintered via hot pressing and HIPing methods. The atmosphere is generally vacuum for refractory carbides. Small particle size has much greater sinterability, but care has to be taken for any impurity pick up, particularly oxygen. Very small size (0.3 μm) molybdenum carbide produced by chemical vapour deposition method, when sintered in hydrogen (900–1500°C) and vacuum (1500–1150°C) showed better sinterability in the latter. This is attributed to the removal of free carbon by the reaction of surface oxygen. In the case of refractory nitrides, over pressure sintering in nitrogen is mostly practiced. Investigation on ball-milled powders has shown some activated sintering due to the presence of metal picked up by the refractory compounds. Even the very minor presence of liquid phase during sintering might be the cause for this activation.

As the cost of refractory compounds on weight basis is much higher than the oxide and other non-oxide ceramics, their usage as a monolith is not as spread as cermets, which can be sintered at relatively lower temperatures.

12.11 CERMETS

Cermets based on both oxide and non-oxide ceramics have been developed. Liquid phase sintering is the only established method for producing PM cermets. The wetting of ceramics by metallic melt is the essential pre-requisite for developing cermets. A detailed discussion is given in Chapter 7. The densification is rapid due to a favourable solubility ratio, and good wetting additives may be used to inhibit grain coarsening during sintering. The microstructure and composition control the properties of sintered cermets.

12.11.1 Oxide-based cermets

Alumina is found to be wetted by early transition metals, as compared to late transition metals such as iron, cobalt and nickel. One of the classical oxide-based cermets has the composition Al_2O_3-30Cr. To increase the strength, chromium is alloyed with molybdenum and tungsten, and alumina with chromium oxide and titanium oxide. The sintering of Al_2O_3-Cr cermets is accomplished in high purity hydrogen with some provision being made for a limited oxidation of the chromium powder surface. The idea is to have better compatibility with Al_2O_3 and Cr_2O_3, which form a solid solution. The bending strengths of $34Al_2O_3$-66 (Cr-Mo) cermet at room temperature and 1316°C are reported as 612 MPa and 274 MPa, respectively.

12.11.2 Non-oxide-based cermets including cemented carbides

This group of cermets is extensively used in cutting tools, wear and metal forming applications. It also includes 'hardmetals' or 'cemented carbides'. In literature, the term 'cermet' is used exclusively for titanium carbide–based materials. However, this nomenclature needs to be universalised.

WC-Co system, commonly known as cemented carbides, has a eutectic liquid at 1320°C. The sintering temperature for WC-10Co cemented carbide is 1400°C, where, about 15 mass% compact is a liquid phase. On cooling the liquid phase solidifies with about 20%–25% WC in it; but the solubility decreases with decrease in temperature and, at room temperature the binder phase—cobalt—contains less than 1% WC in solid solution.

Sintering of cemented carbides is carried out either in hydrogen or vacuum, the latter being far superior than the former. A typical sintering cycle of a WC-20Co composition is shown in Fig. 12.23. Carbon control in hardmetals is critical as deviation from the stoichiometric carbon content causes the precipitation of a third phase, either the η-phase or free graphite, and both deteriorate mechanical properties (Fig 12.24).

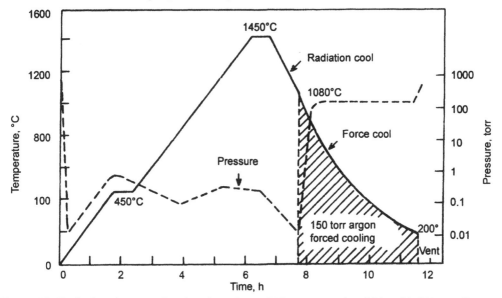

Fig. 12.23 Typical cycle curve for sintering of WC-20Co cemented carbide with 2% paraffin wax

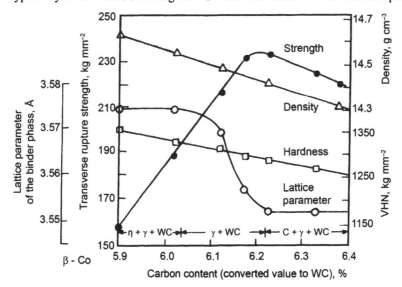

Fig. 12.24 Effect of carbon content on density, lattice parameter, Vickers hardness and transverse rupture strength of WC-10Co cemented carbide

With regard to the role of other refractory carbide additives in WC-Co, TiC is known to reduce the tendency of welding to machined chips, but it has an adverse effect on room temperature transverse rupture strength. The addition of TaC to WC-Co-TiC increases room temperature strength. The effect of a NbC addition is more or less similar to that of TaC as far as the strength of the hard metal is concerned. An increase in the amount of cubic carbides (TiC/TaC) in WC-Co cemented carbide is known to decrease the transverse rupture strength. At elevated temperature ($>800°C$), the difference in TRS values narrows, compared to the lower temperature strengths. Table 12.20 summarises some of the typical compositions and properties of tungsten carbide–based cemented carbides.

The ISO classification divides all grades of cemented carbides into three colour-coded groups:

- Straight tungsten carbide grades (letter K, red)
- Highly-alloyed carbide grades (letter P, blue); used mainly for machining steel

Table 12.20 Properties of selected cemented carbides

Composition					
WC, wt%	94.0	85.3	75.0	78.5	60.0
Other carbides (TiC, TaC, NbC), wt%	-	2.7	-	10.0	31.0
Co, wt%	6.0	12.0	25.0	11.5	9.0
Properties					
Density, g cm^{-3}	14.9	14.2	12.9	13.0	10.6
Hardness, HV$_{30}$	1580	1290	780	1380	1560
Bend strength, MPa	2000	2450	2900	2250	1700
Elastic modulus, GPa	630	580	470	560	520
Fracture toughness, MPa m$^{1/2}$	9.6	12.7	14.5	10.9	8.1
Thermal conductivity, W m^{-1}K^{-1}	80	65	50	60	25
Thermal expansion coefficient, 293–1073 K ($\times 10^{-6}$ K^{-1})	5.5	5.9	7.5	6.4	7.2

Table 12.21 Tungsten carbide–based hard metal groups according to application

Symbol	Application categories	Range of composition, wt%	Hardness, HV	Bend strength, MPa
K	Metals with short chips (preferably cast iron) some non-ferrous alloys, plastics, wood	WC-Co, Co 4%–12%, 0%–3% TiC, 0%–4%(Ta,Nb)C	1300–1800	1200–2200
P	Metals with long chips (steel, steel castings, non-ferrous metals)	WC-TiC-Co, WC-TiC-(Ta,Nb)C-Co, Co 5%–14%, TiC-(Ta,Nb)C up to >50%	1300–1700	800–1900
M	Multipurpose applications	WC-TiC-Co, WC-TiC-(Ta,Nb)C-Co, Co 6%–15%, TiC-(Ta,Nb)C 6%–12%	1300–1700	1350–2100

- Moderately-alloyed grades of carbide (letter M, yellow), which are multipurpose and may be used on steels, complex high temperature alloys, cast iron, etc.

Each grade within a group is given a number between 01 and 50, representing a range from maximum hardness to maximum toughness. Typical applications are given for grades at more or less regular numerical intervals. Table 12.21 illustrates some of the above features.

Cemented carbides are also classified according to the sintered mean grain size of the carbide phase. Table 12.22 summarises these grades.

Table 12.22 Grain size designation convention based on sintered mean grain size

Mean sintered grain size, (µm)	Designation
< 0.2	Nano
0.2–0.5	Ultrafine
0.5–0.8	Submicron
0.8–1.3	Fine
1.3–2.5	Medium
2.5–6.0	Coarse
>6.0	Extra coarse

12.12 CERAMIC–CERAMIC COMPOSITES

The addition of a dispersion of particles or whiskers of a ceramic material which does not react with the ceramic matrix increases toughness. If the particles are irregular in shape or much larger in grain size than the matrix, some bridging can occur. If the particles are significantly different in thermal expansion coefficient than the matrix, some toughening by microcrack formation can occur. Such ceramic–ceramic composites are full density consolidated by hot pressing or HIP methods. Whiskers typically range in size from about 0.5 to 10 µm in diameter and a few micrometres to a few centimetres in length. One of the problems with small whiskers is that they pose health problems. Therefore, they should be handled with proper precautions such as a hood, respirator and careful cleaning of work area. Another type of ceramic–ceramic composite is fibre-reinforced composites. Reinforcing a ceramic with long fibres can increase the distance over which the toughening mechanism acts and lead to enough strain-to-failure fracture that is no longer catastrophic. Continuous fibre ceramic matrix composites are generally formed by

Table 12.23 Examples of particulate reinforced ceramics

Material	Flexural strength, MPa	Fracture toughness, MPa m$^{1/2}$
Baseline hot pressed Si$_3$N$_4$	773	4.6
+ 10 vol% 8 µm SiC	950	5.0
+ 20 vol% 8 µm SiC	763	4.8
+ 30 vol% 8 µm SiC	885	4.9
Baseline Al$_2$O$_3$	420	4.0
Al$_2$O$_3$-5 vol% TiB$_2$	650	6.5
Sintered Al$_2$O$_3$-30 wt% TiC	480	4.4
Hot-pressed Al$_2$O$_3$-30 wt% TiC	583	4.5
Sinter-HIP Al$_2$O$_3$-30 wt% TiC	638	4.4
Baseline SiC	360	~3
Sintered SiC-16 vol% TiB$_2$	478	6.8–8.9

Table 12.24 Critical grain size for physical and structural transitions in nanomaterials

Material, nm	Critical grain size, nm	Nanostructure/ Property	Equilibrium structure/ Property
BaTiO$_3$	120	Cubic	Tetragonal
BaTiO$_3$	120	Drop in T_{curie}*	Constant T_{curie}
TiO$_2$	50	Anatase	Rutile
Y$_2$O$_3$	= 13	Monoclinic	Cubic
ZrO$_2$	8.26	Tetragonal	Monoclinic
W and Ta	>12	A15	BCC

* As particle size decreases, the Curie temperature also decreases (Adapted from Groza JR, *Int. J. Powder Metall.*, 35(7):59, 1999).

chemical vapour infiltration. Table 12.23 lists some particulate ceramic–ceramic composites along with their mechanical properties.

12.13 SINTERED NANOCRYSTALLINE METALS AND CERAMICS

Nanocrystalline materials have given birth to a number of unusual physical and mechanical properties. The reduction in grain size has often resulted in physical properties differing from their coarse grain counterparts: transparency for usually opaque materials, superplasticity in normally brittle materials, very high magneto-resistance, super-paramagnetic properties and controlled band gaps in electronic materials. The main goal of nanopowder consolidation is retention of the initial nanocrystalline structure while achieving full density. The fine grain size may induce effects like alternate structure, extended solubilities or changes in physical properties. The consolidation process, usually performed at high temperatures, will tend to destroy the initial metastable condition of the nanopowders. Thus, during sintering it is important to define the conditions under which metastability is lost. When distinct phase transitions exist, the threshold grain size required to maintain initial property levels in densified parts can be determined (Table 12.24). For nanomaterials, an arbitrary limit ≤100 nm for maximum grain size in the consolidated state is specified. The preparation methods for nanopowders have already been described in Chapter 2.

There are a number of advantages associated with lower sintering temperatures: faster densification, a small grain size, the avoidance of undesirable phase transformations or interfacial reactions, elimination of sintering aids, and less expensive sintering equipment. These low sintering temperatures have extended the use of nanocrystalline powders for applications in which conventional powders are usually not suitable or are not possible.

12.14 FUNCTIONALLY GRADED MATERIALS

Functionally-graded materials (FGM) are materials with deliberately or naturally created gradients in their composition/structure which results in properties that are superior to those of homogeneous or multilayered materials. FGMs need to be competitive with alternative materials in order to become economically and technically feasible. Initially designed as thermal barrier coatings for superior thermal stress relaxation, FGMs have developed into a new field of artificial structures that utilise non-conventional properties resulting from the synergistic integration of different materials in one component. However, it should be recognised that FGMs are largely 'new structures' of 'old' materials and not themselves 'new materials'. PM methods are almost perfectly suited for fabrication of FGMs from the solid phase because of the potential for microstructural control from the outset and the versatility inherent in PM technology.

The methods used to form graded materials by sintering powders are similar to those used for metals and ceramics. A layered geometry is achieved by stacking powders of varying composition prior to consolidation. In case the layer thickness is less than 1 mm in the final dense part, it is rather difficult to achieve layer thickness uniformity. Automated systems have been made to maximise uniformity of layer thickness, allowing very large diameter (up to 300 mm) parts to be fabricated. The biggest concern in these materials is the residual stresses that may develop from different shrinkage behaviour during sintering. This residual stress is different from the one developed due to differential thermal expansion coefficients. The reasons for the mismatch developed during sintering of FGM are due to the difference in initial packing densities, sintering start temperature and sintering rates. Generally, two methods are used to minimise such mismatch. They are:

- Adding different amounts of an organic binder phase to different regions of the FGM so that the initial packing densities are closely matched;
- Changing the constituent particle sizes in different regions to produce different sintering start temperatures, sintering rates and packing densities.

FGM cemented carbide, where the bulk is tough and the surface hard, is a good example for cutting picks used in the mining industry, where subjected to shocks. Scientists have investigated FGM WC-Co cemented carbides with cobalt content varying from 10–30 wt% from one side of the structure to the other, prepared by solid-state or liquid phase sintering routes. In the former case, the graded structure remained after sintering, as there was no risk of homogenisation during such sintering. In the latter case, i.e., liquid phase sintering, the sintering time had to be much shorter because densification occurred much faster with the liquid phase. This required precise control of sintering time, which has to be as short as possible in order to avoid homogenisation of the structure. In case of solid-state sintered FGM tool, to obtain dense material, post-sintering HIP treatment becomes necessary.

All the consolidation methods described in Chapters 8 and 9 are applicable for powder metallurgy FGMs. Recently, spark plasma sintering has been found to be a convenient method, particularly when dissimilar materials are to be sintered simultaneously.

12.15 MPIF PM MATERIAL CODE

Metal Powder Industries Federation of USA issued the MPIF standard 35 in 1965, which is updated from time to time. The latest edition of the standard was issued in 2007. The standards provide design and materials engineers with the information necessary for specifying powder metallurgy materials for structural parts. Similarly, there are standards for PM self-lubricating bearings, powder-forged or metal injection–moulded products. Here, we are more concerned with the standard for structural parts.

The PM material code designation or identifying code in the case of structural PM parts defines a specific material in terms of chemical composition and minimum strength expressed in 10^3 psi. For example, FC-0208-60 is a PM copper steel material containing normal 2% copper and 0.8% combined carbon possessing a minimum yield strength of 60×10^3 psi (60,000 psi; 410 MPa) in the as-sintered condition.

A coding system offers a convenient means for designating both the chemical composition and minimum strength value of any standard PM material. It is based on the system established by the industry with the addition of a two- or three-digit suffix representing minimum strength in place of a suffix letter indicating density range. The density is given for each standard material as one of the typical values.

In the coding system, the prefix letters denote the general type of material. The details are are given below.

The numeric code following the prefix letter code refers to the composition of the material.

In *non-ferrous* materials, the first two numbers in the numeric code designate the percentage of the major alloying constituent. The last two numbers of the numeric code designate the percentage of the minor alloying constituent.

For improved machinability lead is sometimes the third alloying element in a non-ferrous alloy system. Lead will only be indicated by the letter 'P' in the prefix. The percentage of lead or any other minor alloying element that is excluded from the numeric code is represented in the 'Chemical Composition' that appears with each standard material.

In *ferrous* materials, the major alloying elements (except combined carbon) are included in the prefix letter code. Other elements are excluded from the code but are

A	Aluminium	CT	Bronze
C	Copper	CZ	Brass
F	Iron	FC	Iron–copper or copper–steel
G	Free graphite	FD	Diffusion alloyed steel
M	Manganese	FF	Soft magnetic iron
N	Nickel	FL	Pre-alloyed ferrous material except stainless steel
P	Lead	FN	Iron–nickel or nickel–steel
S	Silicon	FS	Iron–silicon
T	Tin	FX	Copper infiltrated iron or steel
U	Sulphur	FY	Iron–phosphorus
Y	Phosphorus	SS	Pre-alloyed stainless steel
Z	Zinc	CNZ	Nickel–silver

represented in the 'Chemical Composition' that appears with each standard material. The first two digits of the numeric code indicate the percentage of the major alloying constituent present.

Combined carbon content in ferrous material is designated by the last two digits in the numeric code. The individual chemical composition tables show limits of carbon content for each alloy.

The range of carbon that is metallurgically combined is indicated by the coding system. The combined carbon level can be estimated metallographically for sintered PM steels that have a well-defined ferrite/pearlite microstructure. For compositions with very low allowable carbon levels (<0.08%), the total carbon determined analytically (ASTM E 1019) is the recommended method.

In the case of PM stainless steels and PM pre-alloyed low-alloy steels, the numeric code is replaced with a designation derived from modifications of the American Iron and Steel Institute alloy coding system, e.g., SS-316L-15, FL-4605-100HT.

When a pre-alloyed steel powder is modified with elemental additions to create a hybrid low-alloy steel or a sinter-hardened steel, an alpha-numeric designator is used, e.g., FLN-4205-40, FLN2-4405-12HT or FLN4C-4005-60. If the base pre-alloyed composition has been modified (slight change to increase or decrease one or two elements), then a numeric designator will be added to the material designation code immediately after the first two digits for the pre-alloyed grade, e.g., FLC-48108-50HT.

In the case of soft magnetic alloys, the phosphorus containing irons are treated differently, since the amount of phosphorus is usually less than 1%. To indicate more accurately the nominal amount of phosphorus, the code takes the nominal per cent phosphorus, multiplies by 100 and uses this number for the first two digits in the code. The last two digits remain '00' since no carbon is required. For example, the iron-0.45% phosphorus alloy would be designated as: FY-4500.

When the code designation 'HT' appears after the suffix digits, it is understood that the PM material specified has been quench hardened and tempered and that the strength represented is ultimate tensile in 10^3 psi.

In the case of soft magnetic alloys, the suffix does *not* designate yield or tensile strength, but rather the maximum coercive field (10 times the value in oersteds) and an alphabetic designator for the minimum density as seen in the table.

For example, a pure iron material at a minimum density of 6.9 g cm^{-3} and coercive field of 2.3 Oe would be designated as F-0000-23W. The iron-0.45% phosphorus alloy at a 7.1 g cm^{-3} minimum density and coercive field of 2.0 Oe would be designated as FY-4500-20X.

Designator	Minimum density, g cm^{-3}
U	6.5
V	6.7
W	6.9
X	7.1
Y	7.3
Z	7.4

Example 12.6: Write down the class of material, composition and minimum yield strength of PM structural materials of following MPIF codes in a tabular form:

CT-1000-13, CNZ-1818-17, CNZ P-1816-13, CZ-1000-11, CZP-2002-12, F-0000-20,

F-0008-35, FC-0208-60, FN-0205-35, FX-2000-25, FX-2008-60, FY-4500-20W,

SS-316 Ni-25, FL-4605-45, FLN-4205-40.

Solution: See Table Ex. 12.6.

Table Ex. 12.6

MPIF code	Material	Composition by per cent	Min. yield strength, MPa
CT-1000-13	PM bronze	Cu-90, Sn-10	90
CNZ-1818-17	PM nickel silver	Cu-64, Ni-18, Zn-18	117
CNZP-1816-13	PM nickel silver	Cu-64, Ni-18, Zn-16, Pb-2	90
CZ-1000-11	PM brass	Cu-90, Zn-10	76
CZP-2002-12	PM brass	Cu-78, Zn-20, Pb-2	83
F-0000-20	PM iron	Fe-99, C-0.2	138
F-0008-35	PM steel	Fe-98, C-0.8	241
FC-0208-60	PM copper steel	Fe-96, Cu-2, C-0.8	411
FN-0205	PM nickel steel	Fe-96, Ni-2, C-0.5	241
FX-2000-25	PM infiltrated iron	Fe-78, Cu-20	172
FX-2008-60	PM infiltrated steel	Fe-77, Cu-20, C-0.8	414
FY-4500-20W	PM phosphorus iron	Fe, P-0.45	138
SS-316N1-25	PM stainless steel (austenitic)	AISI 316 (modified)	172
FL-4605-45	PM 4600 steel (pre-alloyed)	AISI 4600 (modified), C-0.5	310
FLN-4205-40	PM 4200 steel (hybrid low-alloy)	AISI 4200 (modified), 1.5Ni-0.5C	276

SUMMARY

- Copper and nickel are common alloying elements in PM steels.

- Addition of Mn and Cr in steels offers economic advantage in steels, but sintering parameters need better control.

- PM high-speed steels and stainless steels have better properties than many of the cast and wrought grades.

- Liquid phase sintering is the most common method for sintering Al alloys,

- Ag-CdO is the most common silver-based sintered alloy which is used for electric contact applications.

- Nickel-based PM alloys like Nimonics, Inconels and TD-nickel are extensively used in gas turbines for blades, valves and discs.

- Refractory metal powder compacts are generally activated sintered in solid state with a minimum quantity of later transition metals.

- Tungsten-based heavy alloys are an ideal system for studying liquid phase sintering mechanisms.

- Cemented carbides are a group of metal–matrix composites in which tungsten carbide is bonded with cobalt.

- Nanocrystalline sintered metals have given birth to a number of unusual physical and mechanical properties. The main goal is to retain the initial nanocrystalline state while achieving full density.

- Functionally graded materials have deliberately or naturally created gradients in their composition/structure while achieving full density.

- MPIF material code in the case of structural PM parts defines a specific material as to chemical composition and minimum strength.

Further Reading

Alman D and Newkirk J, (eds), *Powder Metallurgy Alloys and Particulate Materials for Industrial Applications*, The Minerals, Metals and Materials Society, Warrendale, PA, 2000.

Anonymous, *Aluminium and Light Alloys for Automotive Applications*, Metal Powder Industries Federation, Princeton, 2000.

Brookes KJA, *Hard Metals and Other Hard Materials*, 3rd edn, International Carbide Data, East Barnet. UK, 1998.

Chung DDL, *Composite Materials: Functional Materials for Modern Technology*, Springer-Verlag, London, 2003.

Exner HE and Danninger H, In *Metallury of Iron*, Vol. 10, Springer-Verlag, Berlin, 1991.

Frazier WE, Koczak MJ and Lee PW, (eds), *Low Density High Temperature Powder Metallurgy Alloys*, The Minerals, Metals and Materials Society, Warrendale, PA, 1991.

Gessinger GH, *Powder Metallurgy of Superalloys*, Butterworth, London, 1987.

Hackl G and Hribernik B, (eds), *First International High-Speed Steels*, Conference Proceedings, Montan Universitat, Leoben, Austria, 1990.

Hausner HH, *Powder Metallurgy in Nuclear Reactor Construction*, International Atomic Energy Agency, Vienna, 1961.

Hench LL and West JK, *Principles of Electronic Ceramics*, John Wiley and Sons, New York, 1990.

Kingery WD, Bowen HK and Uhlmann DR, *Introduction to Ceramics*, 2nd edn, Wiley-Interscience, New York, 1976.

Klar E and Samal P, *Powder Metallurgy Stainless Steels*, ASM International, Materials Park, Ohio, USA, 2007.

Koch CC, (ed.), *Nanostructured Materials: Processing, Properties and Applications*, 2nd edn., William Andrew Publishing, Norwich, NY, 2007.

Konstanty J, *Powder Metallurgy Diamond Tools*, Elsevier, Amsterdam, 2005.

Lassner E and. Schubert WD, *Tungsten: Properties, Chemistry, Technology of the Element, Alloys and Chemical Compounds*, Kluwer Academic, New York, 1999.

Mitchel IV, (ed.), *Nd-Fe Permanent Magnets: Their Present and Future Applications*, Elsevier Applied Science Publishers, London, 1985.

Mostaghaci H, (ed.), *Advanced Ceramic Materials*, Trans. Tech. Publications, Zurich, 1996.

Moulson AJ and Herbert JN, *Electroceramics*, 2nd edn, John Wiley and Sons, Hoboken, NJ, 2003.

MPIF Standard 35, *Materials Standards for PM Structural Parts*, 2007 edition, Metal Powder Industries Federation, Princeton, USA, 2007.

Polmear IJ, *Light Alloys*, 2nd edn, Edward Arnold, London, 1989.

Read JS, *Principles of Ceramic Processing*, 2nd edn, John Wiley and Sons, New York, 1995.

Richerson DW, *Modern Ceramic Engineering*, 2nd edn, Marcel Dekker, New York, 1992.

Salak A, *Ferrous Powder Metallurgy*, Cambridge Int. Science Publishing, Cambridge, 1995.

Schatt W and Wieters KP, *Powder Metallurgy Processing and Materials*, European Powder Metallurgy Association, Shrewsbury, UK, 1997.

Schwarzkopf P and Kieffer R, *Cemented Carbides*, Macmillan, New York, 1960.

Somiya S and Bradt RC, (eds), *Fundamental Structural Ceramics*, Terra Scientific Publishing Co., Tokyo, 1987.

Special issue on the Chemistry of Non-sag Tungsten, *Int. J. Refract. Met. and Hard Mater.*, 13:1–3, 1995.

Upadhyaya GS, (ed.), *Sintered Metal-Ceramic Composites*, Elsevier, Amsterdam, 1984.

Upadhyaya GS, *Sintered Metallic and Ceramic Materials*, John Wiley and Sons, Chichester, UK, 2000.

Upadhyaya GS, (ed.), *Sintering of Multiphase Metal and Ceramic Systems*, Sci-Tech. Publication, Vaduz, 1990.

Upadhyaya GS, *Nature and Properties of Refractory Carbides*, Nova Science Publishers, Inc., Commack, NY, 1996.

Upadhyaya GS, *Cemented Tungsten Carbides: Production, Properties and Testing*, Noyes Publication, Fairfield, NJ, 1998.

Upadhyaya GS, *Manganese in Powder Metallurgy Alloys*, Manganese Centre, Paris, 1986.

Vajpei AC and Upadhyaya GS, *Powder Processing of High T_c Oxide Super Conductors*, Trans. Tech. Publications, Zurich, 1992.

Yih SWH and Wang CT, *Tungsten,* Plenum Press, New York, 1979.

EXERCISES

12.1 A sintered eutectoid steel compact is heated to 800°C, quenched to 600°C, held for 20 s, then quenched to room temperature. What phases and constituents are present at the end of this treatment? (Refer TTT diagram)

12.2 A sintered 0.77% C steel compact is heated to 800°C, quenched to 550°C, held for 2.5 s, then quenched to room temperature. What phases and constituents are present? (Refer TTT diagram)

12.3 A sintered part of eutectoid steel is quenched to 600°C, held for 20 s, quenched to room temperature, and reheated to 700°C and held for 1 s. What phase or phases and constituents are present at the end of this treatment? (Refer TTT diagram)

12.4 A Fe-5% Cu premix green compact was subjected to two sintering cycles:

(a) Isothermal hold given just below α- → γ-phase transformation

(b) Heating carried out continuously right up to the copper melting.

Which cycle will offer reduced copper growth? Give reasons.

12.5 What are the steel requirements for a sinter hardening processing? How is the combination helpful in comparison to sintering followed by heat treatment?

12.6 Fe-Cr binary system forms a brittle σ-phase. You are asked to prepare a sintered steel containing chromium (Fe-0.4C-5Cr) using this σ-phase powder in the powder premix. Suggest a method for preparing the σ-phase powder. How will the properties of the sintered part vary with respect to other alloying methods for chromium?

12.7 Phosphorus in sintered steel is an attractive addition. Why is the same addition in cast steels undesirable? Give another example of a steel where such features are noticed.

12.8 A Fe-0.45% P (premix) sintered part was mass produced. The sintering was carried out in hydrogen at 1290°C. Machining of the parts was required as part of the post-sintering treatment. During production, intermittent high tool wear was noticed. Give reasons.

12.9 In a plant producing cast iron automobile cylinder blocks, a lot of machining swarf is generated. In case you intend to use it in PM forging, what convenient compositional change would you suggest? What heat treatment do you propose for such a ferrous composition?

12.10 The major damage in sintered alloy steels which are used as the cam lobe in diesel engines is due to scuffing and pitting. The table below gives some of the possible compositions for cam lobe and their properties. Suggest the microstructures and resistance to pitting response of these steels.

No.	Alloying additions	Scuffing load limit, N	Hardness, HRC
1	8Cr-1Mo-0Ni	≥3600	40
2	8Cr-2Mo-0Ni	≥3600	51
3	8Cr-2Mo-1Ni	≥3600	57
4	8Cr-2Mo-2Ni	1200	58

Why do compositions 3 and 4 have similar hardness values, but a wide difference in scuffing load? Which alloying addition prevents pitting response?

12.11 Figure P. 12.11 shows the pseudo-binary phase diagram of Fe-C system containing 5% Cr.

    (a) What sintering temperature and atmosphere should be selected for Fe-5Cr- 0.7C PM steel?

    (b) What precautions in sintering parameters should be taken and why?

    (c) In case you intend to decrease the sintering temperature to 1110°C for continuous mesh belt furnace, which additional alloying element would you prefer?

    (d) In case of system (c) above, can you change the sintering technology? If so, what will you prefer?

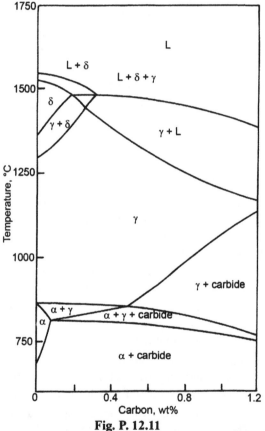

Fig. P. 12.11

12.12 A PM high-speed steel (18W-4Cr-1V-0.36C, bal. Fe) when quenched to form martensite, retains about 5% austenite by volume. Briefly describe the type of tempering treatment required to eliminate the retained austenite. Sketch the resulting microstructure.

12.13 A 300 series stainless steel powder compact was sintered at 1120°C in (i) hydrogen and (ii) 30% H_2 and 70% N_2 gas mixture, respectively, under identical conditions to achieve the sintered density of 6.5–6.7 g cm^{-3}:

    (a) Which atmosphere is preferable from the cost point of view?

    (b) Why is dew point control of the atmosphere vital?

(c) If the dew point is too low the nitrogen absorption was found to be high in atmosphere `B'. Why?

(d) Is the condition (c) preferred? Comment.

(e) If the sintering period is large, will the safe operating dew point window be narrowed?

(f) In case no nitrogen is tolerated in the steel, what other sintering atmosphere apart from hydrogen can be selected?

(g) If the parts are thinner, can a relatively high dew point be tolerated without tarnishing the part?

(h) If the sintered part is to be used for high-temperature series, can a high surface nitrogen content be tolerated?

12.14 Considering Fig. 8.20 (Chapter 8), what minimum degree of vacuum would you recommend for vacuum sintering of 316L stainless steel at three different temperatures: (a) 1200°C, (b) 1300°C and (c) 1400°C? What is the qualitative trend?

12.15 PM Cu-10Sn bronze is sintered as bearing material, but not Cu-15Zn brass. Explain on the basis of binary equilibrium diagrams.

12.16 Describe the production scheme for Cu-2Be precipitation hardening alloy through powder metallurgy. Describe the heat treatment schedule for obtaining maximum hardness in this alloy.

12.17 Cu-ZrO$_2$ (0.1%–1.5%) powder premixes were sintered in hydrogen in the temperature range 850–1050°C at an interval of 50°C. ZrO$_2$ was found to offer dimensional control to copper compacts. Comment on the statement. Can you envisage some correlation between the optimum sintering temperature and the amount of additive particulates?

12.18 Cu-20 mol% Co alloy has been investigated for giant magneto-resistance. What PM technique should be used? Do you suggest any post-sintering heat treatment and if so why?

12.19 Pfizer, Inc., USA, developed a spinodal Cu-15Ni-8Sn wrought PM alloy in strip form for use in electronic connectors. What are the disadvantages if the material is produced by the ingot metallurgy route? Refer to the ternary phase diagram and suggest the composition and crystal structure of the precipitate phase after decomposition.

12.20 Why is Cu-graphite composite mostly produced by powder metallurgy?

12.21 Sintering of Al-powder compact is strongly influenced by oxide layer. Indicate a solution for this.

12.22 An air-atomised aluminium powder was used to prepare the oxide (Al$_2$O$_3$) dispersion-strengthened alloy by mechanical milling. The average particle diameter of as-atomised powder was 15 μm and the oxide surface layer thickness was 30 nm. What is the composition of this dispersion-strengthened alloy, assuming that the final hot-extruded product is of full density? Final analysis of the material showed a greater quantity of oxide than the calculated one from the above data. Explain why.

12.23 An oxide dispersion-strengthened aluminium alloy has unshearable dispersoids of average diameter 200 nm and a shear yield strength of 30 MPa. What will be the shear yield strength if the microstructure consists of a dispersion of the same volume fraction of 50 nm diameter oxide particle?

12.24 For better applications, conventionally sintered Ag-SnO$_2$ (17%) composite used for electric contacts are to be replaced by one processed for obtaining the nanocomposite by reactive milling of Ag$_2$O and Ag$_3$Sn powder mixture. The sintered product revealed the following problems: (a) ex-diffusion of silver from Ag-SnO$_2$ agglomerates and (b) a drastic swelling of sintered pellets. Suggest possible reasons. How will you minimise the problem?

12.25 Why is the potential of PM processing of aluminium alloys based on rapidly solidified powder much greater than for titanium alloys?

12.26 Tungsten powder (average particle size 0.96 µm) compacts were sintered at 1400°C for various periods in vacuum, hydrogen and argon atmospheres. Arrange the atmospheres in ascending order for sintered densities. Give reasons.

12.27 You are required to develop refractory carbide-strengthened tungsten by the sintering route. Which refractory carbide dispersoid will be ideal? Apart from tungsten and the desired strengthening carbide, what other phases may emerge? (Refer Fig. 12.15). Give reasons. Give the flowsheet along with details of sintering temperature, atmosphere and dew point.

12.28 (a) Why is the upscaling of TZC alloy in larger diameters problematic? Refer Table 12.14 for the composition.

(b) Arrange the alloys in Table 12.14 in decreasing order of creep rate (temperature 1200°C, stress 33 MPa). Give reasons.

12.29 W-Re alloy in wire form is used as high-temperature thermocouples. Describe the total PM processing schedule to produce this wire.

12.30 Microstructural control of W-Fe-Ni heavy alloys during sintering has the most prevailing effect on mechanical properties. Elaborate on this statement giving reasons.

12.31 Explain why during sintering hold for a W-Fe-Ni heavy alloy moist hydrogen is preferred as the atmosphere?

12.32 Molybdenum powder compacts containing minor additions (upto 500 ppm) of Ni and Pd were solid-state sintered. What will be the effect of each of these additions on the average grain size and the yield strength of the base metal? What is the relative role of these additions?

12.33 $MoSi_2$ intermetallic compound compact is produced by simultaneous synthesis and densification with powder mixes of elemental molybdenum (<63 µm) and silicon (<47 µm) in stagnant argon (1.2 atm) or stagnant vacuum (5–10 µm).

(a) How will you determine the adiabatic combustion temperature? Is it above or below the melting point of silicon?

(b) In which atmosphere is the reaction zone thickness greater?

(c) In which atmosphere is the maximum combustion temperature lower? Why?

(d) What is its possible application as a structural material?

12.34 Figure 12.18 shows the ternary phase diagram related to porcelain. Suggest what will be the effect of deviating from the designated composition range on the processing and properties of the product.

12.35 The table below gives the composition of three whiteware ceramics A, B and C. Typical sintering temperature for A is 1250°C. What are the approximate values for B and C?

Constituents	A	B	C
China clay	25	48	35
Quartz	40	30	40
Feldspar	35	22	25

12.36 Why is a refractory made of zirconia not made out of pure ZrO_2?

12.37 When YBCO-123 superconductor samples were prepared from the starting materials (Y_2O_3, $BaCo_3$, CuO in molar ratio 1:2:3 for Y:Ba:Cu) by repeated calcining and milling followed by oxygen sintering (950°C for 24 h), the product quality was not uniform. Comment on the statement. What is a better alternative?

12.38 Alnico permanent magnets (27Ni-12Al-4Cu-3Co-Fe) by PM route were prepared after sintering at 1300°C for 3 h. In order to lower the cost can you select water-atomised Fe-Ni pre-alloyed master alloy? How does aluminium help in sintering and how should it be added?

12.39 In MgZn ferrite, the MgO was partially substituted by CuO. The idea was to have lower sintering temperature. What is the effect of such substitution on core loss and microstructure?

12.40 Both soft and hard ferrites are oxide based. Is there any difference in sintering of these two materials?

12.41 Various types of processing methods used for Si_3N_4 ceramics are given below with abbreviations: Identify the processing methods for the following attributes:

Sintered Si_3N_4	SSN
Gas-pressure sintered Si_3N_4	GPSN
Hot pressed Si_3N_4	HPSN
Pre-sintered hot isostatically pressed Si_3N_4	Sinter-HIP-SN
Encapsulated hot isostatically pressed Si_3N_4	HIP-SN
Reaction based Si_3N_4	RBSN

(a) Lowest sintering cost

(b) Highest sintering cost

(c) Highest productivity

(d) Most time consuming

(e) Highest fracture toughness

(f) Products with no or very low additive

(g) Refractories

(h) Cutting tools/ball bearing production.

12.42 The greatest impetus to the development of silicon nitride for engineering applications was its selection by the Department of Defence, USA, in 1971 as the material for the ceramic gas turbine, specified to run at 1370°C. Not only would this be more efficient than the nickel-based superalloy engine running at about 1050°C, it could use poorer fuels and would cause less environmental pollution. Compare the properties of both silicon nitride and Ni-based superalloy in a tabular form. Give reasons why the full development in nitride engine has not yet been achieved.

12.43 Traditional porcelain-based ceramic insulator applications range from high-voltage suspension insulation for power transmission lines to low-voltage shapes for lamp and switch bases. Name some of the exotic insulating materials based on advanced ceramics. Which is the best insulator ceramic material so far?

12.44 In literature there is mention of satisfactory sintering of SiC powder with 0.5% C and 0.5% B additives at 2000°C in inert gas. At that temperature, can you sinter the powder premix compacts in nitrogen or carbon monoxide atmosphere? What about in vacuum? Comment.

12.45 Three Al_2O_3-Cr (25, 50, and 75 vol%) cermets were prepared by attritor milling followed by hot pressing in argon at 1400°C at 30 MPa pressure for 1 h. The per cent sintered densities of the cermets were as follows:

vol%	Sintered density, %	Sintered density, g cm^{-3}
X	97.5	D_1
Y	97.2	D_2
Z	98.8	D_3

What are the compositions of cermets (X, Y, Z) and what are their sintered density values? Given density of $Al_2O_3 = 3.97$ g cm^{-3} and Cr = 7.19 g cm^{-3}. Which alloy will have the highest resistance to wear?

12.46 Figure 12.23 shows the heating time/temperature schedule during sintering cycle of a WC-20Co cemented carbide. In case you are given a WC-6Co milled powder charge, what changes in the cycle would you propose?

12.47 What are the most important factors influencing the properties of hard metals (cemented carbides)?

12.48 Japanese material scientists have researched sintered cemented carbides (WC-Co) by direct milling of tungsten, graphite and cobalt powders followed by sintering. They claim to have produced highly oriented plate-like WC grains. Analyse how the sequence of consolidation and microstructure evolution is different from the WC-Co milling/sintering route. (Refer Konoshita S, Kobayashi M, Taniguchi Y and Hayashi K, *J. Jpn. Soc. of Pow. and Pow. Metall.*, 50(5):377–384, 2003.).

12.49 The following are candidate materials for oxide and non-oxide ceramic cutting tools: tetragonal zirconia polycrystalline (TZP), PSZZ (partially stabilised zirconia), zirconia-toughened alumina (ZTA), SiC and Si_3N_4. From a basic knowledge of bonding, rank them from the viewpoint of application temperature. Also arrange them for stress bearing capability during machining operations.

12.50 Show graphically how cemented tungsten carbide and ceramic cutting tools differ in the following properties:

(a) Hot hardness

(b) Thermal shock resistance

(c) Chemical stability

(d) Abrasion resistance

(e) Toughness.

12.51 You are asked to prepare a mullite-ZrO_2 particulate composite part by reaction sintering of alumina and zircon powders. Draw the sintering time/temperature and sintered density/time profiles. Can you guess from the sintered density value whether the composite has been fully synthesised? Which method should be followed for accurate confirmation? Estimate the theoretical density of the mullite phase, $3Al_2O_3 \cdot 2SiO_2$.

12.52 Why is the densification during hot pressing a whisker-reinforced ceramic–ceramic composite-lower than that for particle-reinforced composite?

12.53 The inhomogeneties in sintered steels produced from powder mixtures are unavoidable, but it should not be regarded as a disadvantage. Write a brief review justifying the above statement.

12.54 You are asked to set up a sintering cycle for a sintered hardenable alloy steel in your laboratory. What special type of electric furnace would you select? How will you specify a range of temperatures for a given steel in order to define a working and practically relevant cooling rate to relate to the transformation products.

12.55 Low-carbon bainitic steels are attractive as controlled cooling is insufficient to provide bainite. A typical steel composition is 0.10-0.15C, 0.5Mo, 0.7-1.0Mn and 0.002B. In order to make such a steel even cheaper, the use of up to 4.5% manganese is advised. Suggest what problems the manganese will introduce during PM processing, including secondary treatment, if any.

12.56 Since copper is much cheaper than silver, attempts are being made to develop Cu–based contact materials. From the table below, it is obvious that Cu-25Cr alloy has better electrical conductivity in comparison to Cu-50Cr alloy. The optimisation of properties required for contact materials, like electrical conductivity, hardness and breakdown voltage, can be achieved by additives X and Y in Cu-25Cr alloy. Suggest some additives. What precautions in processing are required? Is conventional non-PM processing possible?

Materials	Average grain size of Cr phase, μm	Relative density, %	Conductivity, MS m^{-1}	Oxygen content, 10^{-6}	Nitrogen content, 10^{-6}	Brinell hardness, HB	Breakdown field, $10^8 \times$ Vm^{-1}
CuCr25	20–100	99.01	25.0	490	60	80	2.62
CuCr25-X	10–25	99.22	23.1	550	35	86	3.78
CuCr25-Y	5–15	99.27	22.4	460	25	91	4.05
CuCr50*	70–150	99.12	18.7	430	60	97	2.52

* Convenional CuCr50 prepared by sintering and infiltrating method.

12.57 The table below gives various mechanical and thermal properties of two grades of cemented carbides (one metal cutting and the other non-metal cutting grade) and a fully dense PM tool steel. Identify the alloys and give sufficient justification for them.

Alloy	Compressive strength, GPa	Hardness at 20°C, kg mm^{-2}	Hardness at 800°C, kg mm^{-2}	Fracture toughness, MPa m$^{-1/2}$	Young's modulus, GPa	Thermal conductivity, W mK	Thermal expansion, $\times 10^{-6}$ K	Density, g cm^{-3}	Relative abrasion resistance
A	4–6	1,450–1,850	550–750	8–13	550–620	50–120	5.4–6.8	12.5–15.0	70–100
B	2.0–2.8	750–1,000	100	~25	195–205	30–35	12	7.8	5–10
C	2–5	1,000–1,500	300–550	13–17	470–600	70–110	6.0–8.0	13.0–14.5	10–25

12.58 A specified grade of high-speed steel is required for producing a conventional press-sinter and a metal injection moulded part. Will the same grade of pre-alloyed powder suffice for both types of parts? If not why?

12.59 Two sets of PM composites based on 2014 and 6061 aluminium alloy premix powders are prepared, containing SiC particles and continuous fibres (2 vol%), respectively. Liquid phase sintering of the preforms was carried out in vacuum (5×10^{-2} torr) for 32 min at 635°C. Tensile mechanical properties gave the following work-hardening coefficient (n) data. The 2014 alloy was found to exhibit more swelling than the 6061 alloy during sintering.

Material	n
2014 Al-alloy	0.07
2014-2 vol% SiC$_p$	0.08
2014-2 vol% SiC$_f$	0.10
6061 Al-alloy	0.06
6061-2 vol% SiC$_p$	0.07
6061-2 vol% SiC$_f$	0.09

(a) What are the complete specifications of 2014 and 6061 alloys?

(b) Why does the 2014 alloy show more swelling than 6061. [Hint: Refer binary Al-X phase diagrams of these alloy systems].

(c) Which reinforcement particle or continuous fibre would exhibit better densification? Why?

(d) Why does continuous fibre (SiC) reinforcement exhibit more work hardening than the corresponding particulates?

(e) Why do sintered 6061 Al-alloy test pieces show lower n value than 2014 Al-alloy?

(f) Do you expect a monotonous increase in mechanical properties with increase in the amount of reinforcement?

(g) Why is a stress–strain plot generation vital for studying the composites' mechanical stability?

12.60 Why does recrystallised non-sagging K-, Al- and Si-doped tungsten wire show much better ductilities than pure tungsten does?

12.61 Figure P. 12.61 shows schematically the microstructures of rare earth permanent magnets based on melt-spun ribbon flakes taken transverse to the pressing directions. Designate the two constituents in case of processing (a). What will be the magnetic behaviour of these differently processed materials? (Hint: In certain processes you may add a binder)

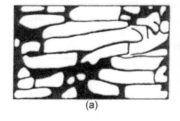
(a)

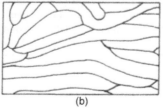
(b)

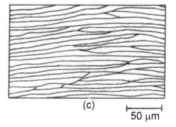
(c)

$\overset{\longmapsto}{50\ \mu m}$

Fig. P. 12. 61

12.62 A composite based on a barium ferrite magnetic ceramic powder, for which the binder is usually polymer, is required. In case you intend to develop a economical super-plastic metallic powder for binder, what will be your choice? Such a composite will be a good conductor as well. What hot pressing temperature do you think will be suitable?

12.63 Self-bonded reaction sintered silicon nitride parts can be prepared from the following green-shaped silicon powder parts containing a plasticiser in nitrogen atmosphere:

(a) Unidirectionally pressed

(b) Isostatic pressed

(c) Slip cast

(d) Extruded.

Describe the differences in the end properties of the parts, if any.

The electro-discharge machining of such parts is very difficult. Suggest the role of some additives to the parts amenable to such machining. Name some cheap additives to enhance sinterability.

12.64 A complex TiC-based cermet had the composition 34.9TiC, 15.0TiN, 16WC, 7Mo$_2$C, 10(Ta,Nb) C, 4.8Ni, 8.5Co and 0.8Ti$_2$AlC. After vacuum sintering at 1450°C, the binder phase of the cermet was extracted chemically and the metallic elements were analysed. The contents were 33.5Ni, 52.8Co, 0.9Ti, 3.5W, 8.1Mo, <0.02Ta and 1.2Al. If the carbon content in the cermet is varied from 9.3% to 9.9%, show the qualitative trend of the variation of the carbide forming elements in the binder phase. Give reasons.

12.65 VC and Cr$_3$C$_2$ are common grain growth inhibitors for WC-Co cemented carbides. Describe how the effect of inhibitors on graphite and η-phase limits is not yet well understood.

12.66 If you prepare WC-W$_2$C composites of varying compositions by reactive hot pressing of WC-W powder mixtures, what will be the effect of the increase in the amount of W$_2$C phase in the composite on sinterability, hardness, Young's modulus, fracture toughness and Poisson's ratio? Plot a graph between C content (mol%) and W/WC molar ratio.

12.67 Often heterogeneous pores are observed on lapped surface of liquid phase sintered WC-Co cemented carbides. Indicate how the amount and size of pores vary with change in sintering temperature, cobalt content and WC particle size. What is the favourable location of large size pores?

12.68 Diamond tools have better wear resistance than cemented carbides, but the cost for the former is much more than for the latter. In order to optimise the cost, if you propose to prepare a sintered WC-Co-diamond composite material, what special problems will be encountered and how will you solve them?

12.69 Al$_2$O$_3$-TiC ceramic matrix composites prepared by hot pressing at 1700°C are used as cutting tools. Due to economic reasons, if the same is prepared by pressureless sintering, suggest what additive can be introduced in the system. What possible sintering atmosphere will be selected?

12.70 According to Mellor "ceramics is the study of arrested reactions." Amplify this statement for the sintering of some of the ceramic systems.

12.71 A PM steel containing 4% Ni and 2% Cu meant for good fatigue resistance has some limitations. How will you circumvent the problem?

12.72 If the microstructure of a sintered steel changes from bainitic to martensitic, how will it improve fatigue properties?

12.73 A set of driven-sprockets (Fig. P. 12.73) have the following specifications:

Size: OD Drive   90–100 mm

                 OD Driven   100–106 mm

Weight: Drive   450–590 g

                 Driven   505–570 g

Tensile strength: 860 MPa

Yield strength: 830 MPa

Elongation: 1.5%

Apparent hardness: 60 HRC

Density: 7.8 g cm⁻³ (surface density); 7.0 g cm⁻³ (core density)

Fig. P. 12.73

Density: 7.8 g cm^{-3} (surface density); 7.0 g cm^{-3} (core density)

Heat treatment: Vacuum carburising

(a) Give the possible grades of PM Steels.

(b) What are the secondary operations which you envisage?

12.74 A powder pressing shop manager in a PM plant commented that PM tool steel dies are somewhat better than the conventional cast and forged ones from the galling resistance view point. Justify the statement.

12.75 In a PM plant producing stainless steels, the sintering furnace did not have the requisite facility for faster cooling. What steps should be taken to mitigate the adverse effect on the product?

12.76 What is the effect of stainless steel chemistry on lowering the critical cooling rates in the sintering furnace?

12.77 What special care is to be taken while sintering stainless steels in hydrogen, nitrogen and vacuum?

12.78 Why is the testing for contaminations in stainless steels done generally on powders and green parts prior to sintering and not after sintering?

12.79 Draw schematically the time–temperature sensitisation curves for austenitic and ferritic stainless steels of equivalent Cr-content. Give reasons for the distinctive differences in these curves. What is the effect of carbon on these curves? Describe the significance of the diagram on the selection of sintering parameters for such steels.

12.80 Why are ferritic stainless steels not sintered in nitrogen-containing atmospheres?

12.81 A 50 kCV grade tantalum powder was used to produce capacitors. Vacuum sintering in a temperature range 1575–1650°C was not found suitable. Give reasons. What change in the sintering parameter should be adopted to achieve the desired goal?

12.82 WC-Co powder premix is proposed to be produced from WO_3, Co_3O_4 and carbon black powders by carbothermic reduction and carburisation. Using thermodynamic data, establish the operating temperature for synthesis in a vacuum of 10 Pa. Which resultant gas will be stable at high temperature, CO or CO_2? Is the particle size of the resultant WC-Co powder smaller than that for the conventional production method? Give reasons.

12.83 TiC and TiN have reasonably good electrical conductivity. What is the engineering implication of this feature?

12.84 CNG (compressed natural gas) engines require valve seat materials to have high wear resistance, due to the environment being drier than in gasoline and diesel engines.What suggestion would you give for the selection of the PM part composition?

12.85 Pre-alloyed and premixed Cu-12Sn bronze were sintered at 880°C. One of the compacts got extensively distorted, whereas the other did not slump at all. Identify the compact that underwent distortion with justification.

12.86 Schematically show the microstructural evolution in Cu-10Sn alloy prepared using (a) premixed and (b) pre-alloyed powders, and sintered at 250°C, 550°C and 850°C.

12.87 90W-7Ni-3Fe alloy can be consolidated to full density through solid-state sintering (1400°C) as well as liquid phase sintering (1500°C). Compare and contrast the microstructures as well as the mechanical properties of the alloys sintered under these two conditions. Justfy.

12.88 Liquid phase sintered W-Cu powder compacts containing up to 80 vol% Cu do not distort during liquid phase sintering to temperatures as high as 1400°C. In contrast, liquid phase sintered W-Ni compacts undergo extensive distortion with as little as 20% Ni. Why?

12.89 Why is the microstructure of a sintered copper–steel more homogeneous than nickel–steel?

12.90 Describe the microstructure of a typical heat-treated hybrid low-alloy sintered steel.

12.91 It was observed that a sintered compact made of a Mo-hybrid alloyed steel had a finer pore structure than a molybdenum pre-alloyed steel powder. Give reasons.

12.92 Why is the processing parameter selection for sinter-hardened steel done in such a manner that no further machining operation is required?

12.93 In an extrusion plant, traditional tool steel dies used for single-shot impact extrusion of discrete compacts in alloy steel and Nimonic 750 produced around 800 components before the die wear became excessive, the extrusion press had to be stripped, the die stack dismantled and refitted, and the tooling reheated before production could recommence. Suggest the substitute PM material for the die to produce say 20,000–40,000 components before having to be replaced.

12.94 Cemented carbides used for sinter-HIP process is invariably vacuum sintered. In case, you have atmosphere sintered lots, what additional operation would you perform?

12.95 List some of the key factors that influence the toughness and strength of a ceramic fibre/ceramic matrix composite?

12.96 5 wt% Al_2O_3 and Al_4C_3 are added to SiC to achieve liquid phase sintering. To achieve the maximum amount of liquid at about 1835°C, how much Al_2O_3 is required for a 1 kg batch? (Hint: Refer Al_2O_3-Al_4C_3 phase diagram)

12.97 A compact made of silicon particles has a bulk density of 1.6 g cm^{-3}. The compact is nitrated to achieve complete conversion of silicon to Si_3N_4. What is the bulk density of the resulting Si_3N_4 part?

12.98 Zirconia(max. 0.5 vol%)-strengthened platinum is a useful material for the production of stirrers for optical glass making. Describe with the help of a flowsheet the production of such products through the PM route. Why is there a limiting amount of ZrO_2?

12.99 Enumerate various methods of preparing oxide dispersion-strengthened alloys through powder metallurgy route. Which method is most promising and why?

12.100 For drawing copper wire up to 2 mm in diameter, alumina guides, which typically draw 1000 tons of wire with no sign of wear are sufficient. However, in case of coated or enamelled copper wire, this guide material was not suitable. Suggest what alternative ceramic will be useful and why?

12.101 For producing a dense Al_2O_3-Ni cermet through powder metallurgy, it was suggested that the Al_2O_3 particles be first coated by chemical vapour deposition. Can you suggest the coating compound and how it helps in better densification?

12.102 A W-20Cu nanocomposite powder was prepared by spray drying through the salt solution route followed by calcining–continuous reduction. After sintering it showed that the presence of some CuO offers better densification. Explain the probable reason.

13

Applications of PM Products

LEARNING OBJECTIVES

Various applications of PM products
- Structural
- Machine tool
- Power generation
- Friction
- Electrical
- Magnetic
- Other
  - Thermal management
  - Hardfacing
  - Bio-implants

There is no area in technology where PM products are not applied. Powder metallurgy parts find application in automotive, aerospace, electrical, electronic, nuclear or bio-implant devices. PM parts can be classified in various ways, e.g., porous products or dense products, metallic or non-metallic products, etc. In this chapter an attempt is made to describe them on the basis of their function. In some cases the function may be more than one. For example, a porous part may act as a structural member with the additional function of filtering a fluid; an electric conductor may be required to have excellent creep resistance behaviour.

13.1 STRUCTURAL APPLICATIONS

For any structural application, strength is an important criterion. A knowledge of the relation between powder compaction and sintering is of utmost importance for high-strength components.

At an early stage of component development, the designer requires engineering data to evaluate alternative materials. These data for PM materials are now available, either through standards (i.e., ISO-material standard), from suppliers of components, from powder suppliers or from newly launched 'Global PM Property Database'—a cooperative project between European Powder Metallurgy Association, the Metal Powder Industries Federation, USA, and the Japan Powder Metallurgy Association.

The ultimate tensile strength levels of recently developed ferrous powders, made by the combination of modern compaction techniques and sintering methods, are more than 1000 MPa. The highest mechanical performance is achieved for materials with high density. A density of more than 7.5 g cm^{-3} has been reached after single pressing and single sintering of a pre-alloyed steel powder with 1.5% Cr and 0.2% Mo. The materials data are equivalent to those for wrought steel. In other words, the PM component can be successfully used for any structural application, provided the production cost is lower than for machined components.

PM structural products are extensively used in automotive, aerospace and defence applications.

13.1.1 Automotive applications

In Chapter 1 a brief account of PM automotive applications is given. Figure 13.1 illustrates current PM applications for automobiles. More than 100 PM components are used, weighing between 7 and 20 kg.

Because of the high price of petrol, the demand for automobiles with internal combustion engines is on the decline, and the demand for hybrid electric vehicles is on the rise. Ultimately, electric vehicles using fuel cells will be used. The future design of parts will have to satisfy highly functional duties. Small-sized efficient solid oxide fuel cells (SOFC) will be in greater demand. In addition, highly effective permanent magnetic materials for motor driven vehicles, for example, applications in rotors for stepper motors, will be extensively used. In brief, the artificial barrier between powder processing of either metallic or ceramic systems, will become more blurred. The significance of complex materials, including composites with highly functional properties, at different critical locations in automobile parts will be more obvious.

Bimetallic bearings are most economic and efficient. They are used for automotive engine manufacture, overhaul and rebuilding. Such bearings also find increasing use in other kinds of equipment, such as hydraulic pumps and motors, machinery, mechanical drives, and various other applications. PM engine bearings are lined mainly with Cu-Pb and Al-based alloys, which are prepared by roll compacting of powders followed by sintering, roll cladding to steel and final bearing fabrication (refer Chapter 9).

The major applications of other PM parts in automotive industries can be classified as:

- Engine timing system, e.g., pulleys, sprockets, hubs
- Engine ignition system, e.g., distribution gears
- Engine valve train, e.g., valve seat inserts, valve guides, rocker arms, rocker arm fulcrums and ball seat bearings, valve lifter, guides
- Oil and water pumps, e.g., gears, gerotors, impellers

- Gearbox/transmission, e.g., parts for gear-selection mechanism, clutch hubs and other clutch parts, synchroniser hubs, cones and sleeves
- Steering system, e.g., rack and pinion mechanism parts, power-steering pump parts
- Suspension, e.g., shock absorber and strut parts
- Braking system, e.g., disc pad supports, pistons, adjusters, ABS sensors
- Various locks, catches, pivots, etc.

Spur, helical, bevel, face, spur–helical and helical–helical gears are produced by PM methods. The PM process is capable of producing spur gears in many tooth forms and modifications. In helical PM gears, the helix angle is limited to approximately 35°. Bevel and face gears differ somewhat from spur and helical gears in flexibility of mechanical properties. Due to the nature of the compaction process, sections such as face gear teeth cannot be made to high densities without resorting to copper infiltration or repressing. To design and specify PM gears, one must understand what tolerances can be achieved, what mechanical properties can be obtained, and what to establish as inspection criteria. The choice of the PM alloy is governed by the magnitude and nature of the transmitted load, the speed, the life requirements, the environment, the type of lubrication and the gear and assembly precision. An analysis of mechanical and PM gears shows that a PM gear can be produced at 68.27% lower cost than machined gear. It is worth noting that 80% of automotive and truck oil pump gears produced annually in the USA are by the PM technique. PM gears can be combined with other components such as cams, ratchets and other gears to form a single component. Powder-forged gears have shown a longer fatigue life than gears machined from bar stock.

In addition, hot-worked fully dense PM parts find application in automotive transmission components, connecting rods and differential gears.

In recent years, materials with improved wear/tribological characteristics in automotive valve train applications have been developed. Some examples are: valve seat inserts, cam lobes, rocker arm tips, tappet shims and valve guides. The most recent developments in this direction have been complex components, e.g., composite camshafts, where the properties of individual elements (e.g., lobes, bearings, tubes) are tailored. In addition, the concept of selective densification has been used in gear and sprocket teeth.

- For fuel saving, the use of PM aluminium alloy parts has become popular. For example, a toothed pulley of sintered aluminium alloy weighs 345 g compared with 920 g for the equivalent sintered ferrous part.
- PM main bearing caps and PM forged connecting rods are other examples. Sintered aluminium is used to make the automotive rear view mirror mount button, and metastable supersaturated aluminium PM alloys are used for IC engine piston applications.
- Sintered stainless steel exhaust flanges have been used due to the combination of high temperature oxidation resistance, corrosion resistance, resistance to thermal cycling, and some degree of stress rupture and creep resistance.
- PM cylinder liners for large diesel engines are prepared by dry cold isostatic pressing, sintering/sizing. Fe-0.45 P-0.9 C is a typical steel composition, which results in a fully pearlitic structure with good wear resistance.
- Brake discs using particulate-reinforced aluminium matrix PM composites containing silicon carbide are in vogue.

Fig. 13.1 Current PM applications in automobiles (Courtesy: Saeron Automotive Co. Ltd., Korea)

(a) Engine parts
  - Camshaft sprocket (timing pulley)
  - Water pump pulley
  - Oil pump Ge-rotor

(b) Steering system parts
  - Rack stopper
  - Support yoke
  - Rotor and cam ring
  - Side plane and pressure plate

(c) Transmission parts
  - Synchronised hub

(d) Extra Parts
  - Fuel injection pump gear and holder
  - Control sleeve
  - ABS sensor ring

(e) Shock absorber
  - Piston
  - Body (base valve)
  - Rod guide
  - Retainer

(f) Automobile electric equipment parts
  - Pump rotor (vacuum pump)
  - Planetary gear and sleeve bearing

The schematic diagram (Fig. 13.2) shows a general pictorial view of a turbo-charged diesel engine. In each portion of the assembly, some sintered ceramic parts are applicable for energy efficiency.

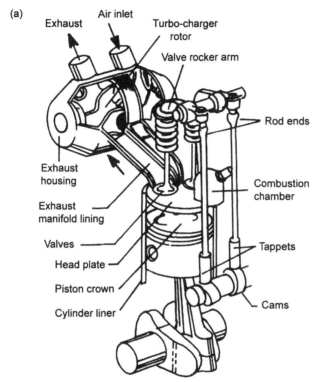

Nomenclature: PSZ– Partially stabilised zirconia; LAS– Lithium aluminium silicate; TTA– Transformation-toughened alumina

(b)

Adiabatic components	Desired characteristics							
	Low friction	Light weight	Insulation	Wear resistance	Heat resistance	Corrosion resistance	Expansion coefficient	High technology ceramics
Piston		•	•		•	•	•	Si_3N_4, PSZ, TTA
Piston ring				•	•			Si_3N_4, PSZ, Coating
Cylinder liner	•			•	•	•	•	Si_3N_4, PSZ, Coating
Pre-chamber			•		•	•		Si_3N_4, PSZ
Valve		•	•	•	•	•		Si_3N_4, PSZ, Composite
Valve seat Insert			•	•	•	•		Si_3N_4, PSZ
Valve guides	•		•		•	•		Si_3N_4, PSZ, SiC
Exhaust/Intake ports			•		•	•		ZrO_2, Si_3N_4, TiO_2-Al_2O_3
Manifolds			•		•	•		ZrO_2, Si_3N_4, TiO_2-Al_2O_3
Tappets		•		•			•	Si_3N_4, PSZ, SiC
Mechanical seals	•			•		•		Si_3N_4, PSZ, SiC
Turbocharger turbine rotor		•	•		•	•	•	Si_3N_4, SiC
Turbine housing			•		•	•	•	LAS
Heat-shield			•		•	•	•	ZrO_2, LAS
Ceramic bearings	•	•		•	•	•	•	Si_3N_4

Fig. 13.2 (a) Potential ceramic components in turbo-charged diesel engine applications; (b) High-temperature ceramics for adiabatic engines

Example 13.1: Various metallic materials are used for the following parts of a diesel engine. List their substitute sintered materials, when the engine is an adiabatic turbo-compound: (a) piston head, (b) head liner, (c) valve, (d) cylinder liner, (e) valve guide, (f) exhaust port, (g) inner surface of exhaust manifold and (h) turbine blade. Describe the development of ceramic materials for a rotary heat exchanger.

Solution: Table Ex.13.1 illustrates the use of ceramic materials.

Table Ex. 13.1

Parts	Present materials	Potential materials for adiabatic turbocharges	Operating temperature
Piston head	Al-alloy	Si_3N_4	800°C
Head liner	Cast iron	Si_3N_4	700°C
Valve	Anti-erosive, heat-resistant special steels	Si_3N_4	800°C
Cylinder liner	Cr-plated steel	PSZ	400°C
Valve guide	Cu-alloy	Si_3N_4	800°C
Exhaust port	Cast iron	Iron tube/Al_2O_2	800°C
Inner surface of exhaust manifold	Heat-resistant cast iron	Fibre/cast iron	1100–200°C
Turbine blade	-	Si_3N_4	1000°C

A rotary regenerator consists of a thin-walled honeycomb-type configuration that continually rotates through the exhaust and inlet gases. The exhaust gases heat the regenerator material to over 980°C. This then rotates into the inlet region, where the heat exchanger gives up its heat to preheat the air prior to combustion. Preheating the combustion air provides a substantial improvement in efficiency and reduction in fuel consumption. The major requirement of the rotary regenerator material is thermal shock resistance. The initial ceramic material selected was lithium aluminium silicate (LAS), because of its near-zero coefficient of thermal expansion and the resulting good thermal shock resistance. However, this material was prone to corrosion problems. Later development was for magnesium aluminium silicate. However, this material has a higher thermal expansion. Presently, the best ceramic structural material for this application is Si_3N_4. Much work in this field has been made by Japanese auto-industries.

Example 13.2: There is great potential for ceramic liners in diesel engines. Describe the advantages. What major material development has been made?

Solution: Although perfect adiabatic (no heat gained or lost) engine operation is not achievable in practice, the insulated or ceramic diesel concept is often generally referred to as the adiabatic diesel. A low conductivity TTZ ceramic has been favoured for the cylinder wall and piston material to minimise heat loss. An analysis shows that the major factor in reducing heat loss from the cylinder is not cylinder material conductivity, but elimination of the coolant. With no ceramic insulation, elimination of the coolant would, of course (Continued on next page) raise the temperature of the cast iron cylinder wall and adversely affect engine life. This is shown in Fig. Ex. 13.2 for a high power condition, where the time-averaged temperature of an all-cast iron cylinder would be increased from approximately 190°C to 630°C, and the heat loss would be reduced by over 80%. Changing the material from cast iron to TTZ would further reduce heat loss.

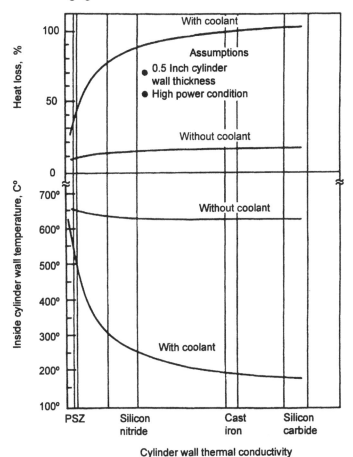

Fig. Ex. 13.2

There are two main approaches to reduce heat loss through elimination of the coolant. The first uses a low conductivity TTZ ceramic liner inside a cast iron block. Here the main function of the ceramic liner is not to reduce heat loss, but to protect the cast iron interior surface from oxidation and high temperature. This arrangement is attractive because of the close match of thermal expansion between the iron and TTZ, which permits a simple and effective ceramic-to-metal attachment.

The other more revolutionary approach is to use free-standing ceramic cylinders. In this case, since there is no metal containment, the ceramic cylinder becomes a structural member. The ceramic material selected would be based, not on its thermal conductivity, but on its contribution to system reliability. Material strength, Weibull modulus, elastic modulus and thermal expansion would be key factors.

13.1.2 Aerospace applications

In aerospace applications the most critical part is the engine, which requires strong, lightweight material. In order to increase thermal efficiency and reduce fuel consumption, aero engines need to operate with high gas pressures and temperatures. These requirements have made the PM metallic and ceramic materials very expensive. The principal areas of applications of superalloys in engine components are shown in Fig. 13.3. The details of PM Ni-based superalloys have been described in Chapter 12, and full density processing in Chapter 9. It may be mentioned that in addition to alloy development, the advancements in PM processing have also played a significant role. Intermetallics, particularly titanium aluminide, are an attractive replacement for conventional disk and spacer assembly, bringing about 75% weight saving.

Ceramic gas turbine stators, made of Si_3N_4 and SiC, are still at the development stage. These ceramics have a combination of high strength, moderate-to-low thermal expansion, relatively high thermal conductivity, and good oxidation–corrosion resistance. The stator on a gas-turbine engine is a non-moving airfoil-shaped component that diverts the hot gas flow from the combustor at an optimum angle to the rotor to achieve peak aerodynamic performance. Figure 13.4 shows some Si_3N_4 stator vanes currently under development for a gas-turbine engine.

Aircraft fasteners are generally made of PM Ti-6Al-4V alloys. The powder is pressed in a doughnut preform, and mechanically hot coined to the twelve-point nut shape. The threads are tapped in the nut after PM forming. When this component is produced from wrought titanium, the cost savings are substantial.

In *satellite structural applications*, beryllium is widely used, as it has high thermal conductivity and heat capacity and a low coefficient of thermal expansion. The specific parts

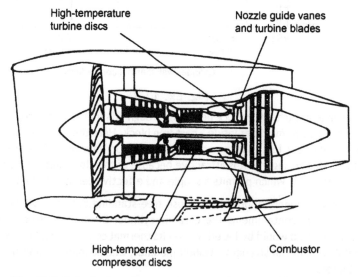

High-temperature
turbine discs

Nozzle guide vanes
and turbine blades

High-temperature
compressor discs

Combustor

Fig. 13.3 Principal engine components using superalloys

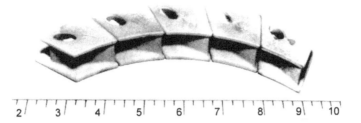

Fig. 13.4 Si_3N_4 stator valves under development for an advanced gas turbine engine

are antenna supports, solar panel booms and superstructures. As a result of lack of fine grain size and good structural integrity of cast beryllium, PM beryllium is the only solution.

Porous tungsten is a suitable application in *ion jet rockets*, which use an electrical potential difference to accelerate ions ejected through a nozzle, thus providing the required thrust. A rocket nozzle throat made of sintered tungsten is another application, where silver infiltrated (20 vol% porosity) tungsten is used. The infiltrant provides a transpirational cooling action as it vaporises during firing. This composite has an excellent thermal shock resistance.

A number of ceramics have been tried for *high-load, high-speed bearings*, where the temperature is beyond the limits of metal bearings. Rolling contact fatigue tests showed that the Si_3N_4 ceramic (NC-132 grade of Norton make in hot-pressed condition) had an L_{10} life of 12,000,000 cycles at 7.83 GPa Hertzian stress loading. This was eight times the life of M-50 CVM steel. Figure 13.5 illustrates some bearings produced from Si_3N_4. These ceramic balls failed slowly with surface spallation very similar to fatigue failure mode of metals.

Another application of sintered metals is in *radome*, which is essentially a protective covering and window for electronic guidance and detection equipment on missiles, aircraft and spacecrafts. Apart from electro-optical properties, this must be resistant to high velocity impact by rain and other possible atmospheric particles. For some applications, it must also be resistant to high temperature and thermal shock.

Fig. 13.5 Si_3N_4 bearings

13.1.3 Ordnance applications

The PM parts used in defence transport vehicles (surface, water or air) are similar to what has been described in earlier sections. The most common PM applications in ordnance are:

- Heavy-alloy penetrators
- Machine-gun parts, e.g., gears, guide fins for projectiles, accelerators
- Missile applications, e.g., nose cones, dome housing for side winder missiles, wings, lens housing for missiles, etc.
- Spin-stabilised medium calibre projectiles
- Fragmentation devices
- Incendiary devices, e.g., zirconium sponge pellets
- Training ammunition, where a partially densified iron powder compact is encased in a low-strength, thermally degradable plastic container.

The earliest application of sintered parts in ordnance was sintered iron rotating bands, which replaced costly copper bands. PM technology has also brought about a change in configuration of booster rotors, which are assembled on the back end of a fuse. Hollow PM titanium bolts with hexagonal heads have replaced high-strength plastic fasteners.

PM metal matrix composites (30% SiC) Al 6061-T6 have been found suitable for torpedo applications in navy.

Armour ceramics, either in monolithic or composite forms are also used in defence. Monoliths are Al_2O_3, B_4C, SiC, TiB_2 and AlN, while the ceramic composites are Al_2O_3/SiC whisker, Ni/TiC, TiB_2/B_4C and TiB_2/SiC particulate composites.

Heavy-alloy, high-velocity kinetic energy penetrations based on W-Fe-Ni alloys are classical examples of PM materials application in defence, which have unqualified prominence. In the past the defence industry used Armour Piercing Discarding Sabot (APDS), where the penetrator cone was surrounded by a ballistic sheath for aerodynamic purposes (Fig. 13.6).

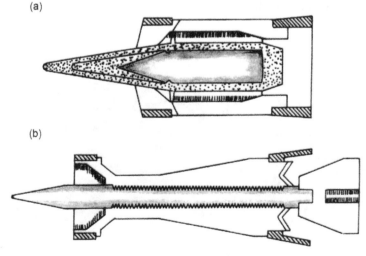

(a)

(b)

Fig. 13.6 Schematics of (a) APDS and (b) APFSDS

However, this is no longer tenable for thicker armour carried by new MBTS on battlefields. The new design is based on the long rod principle and is known as Armour Piercing Fin Stabilised Discarding Sabot (APFSDS) type of ammunition (Fig. 13.6). This sabot functions as before, except that the load is transferred in tension via a system of grooves. The key factor in penetration performance is not whether the penetrator will yield and deform, but the manner in which the deformation and inversion process occurs.

13.2 MACHINE TOOLS

13.2.1 Cutting applications

Cutting tool materials must have the following characteristics:

- Room temperature hardness
- Resistance to thermal softening
- Toughness
- Stiffness
- Wear resistance
- Resistance to chemical reaction
- High thermal conductivity
- Low thermal expansion
- Dimensional stability
- Machinability and grindability
- Cost—both material and processing.

There is no single material which will satisfy all these considerations. The relative importance of each item will shift with the nature of the product machined, the volume of production, the type of machining operation (intermittent or continuous cut, roughing or finishing, high or low speed, etc.), the tool design details (cutting and clearance angles, method of holding, rigidity, etc.), the general condition of the machine tool and the physical characteristics of work material.

The range of tool materials available include steels, non-ferrous alloys containing chromium, cobalt and tungsten, cemented carbides, cermets, ceramics and super-hard materials. In some cases the tool material might have the required properties intrinsically, e.g., diamond (very hard material), or conferred on it by virtue of heat treatment or some other processing, e.g., coatings.

Figure 13.7 lists the materials used for making cutting tools indicating the ranges of hardness, strength and temperature stabilities.

In Chapter 12, the processing details of sintered high-speed steels and cemented carbides have been given. Figure 13.8 shows some typical tool parts of high-speed steels processed through liquid phase sintering.

Tool wear depends on tool and workpiece materials, tool geometry and type of cutting fluids. There are two types of tool wear: crater wear and flank wear. In the crater region (common

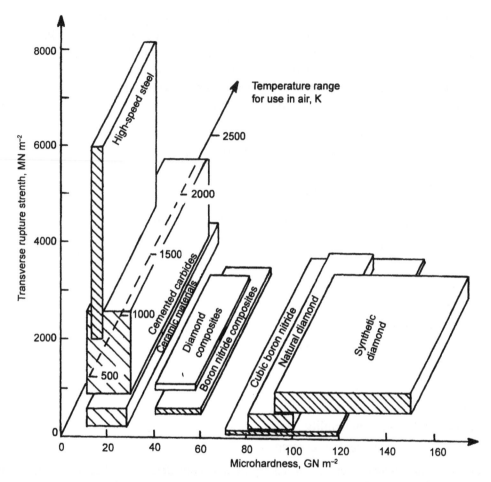

Fig. 13.7 Schematic diagram of the property range of cutting tool materials. The hatched areas show room temperature properties, the temperature axis gives the maximum range of application in air, disregarding temperature dependence of properties.

in steel machining), the tool is subjected to substantial compressive and shear stresses, to temperatures which may reach 1000°C or more, and to very steep temperature gradients in the order of 1°C μm^{-1}.

Crator wear is also known as *diffusion wear*, in which the tool shape is changed by the diffusion of the tool surface by the chip material flowing over it. Clearly this depends on the solubility of the tool material in the work metal and the interface temperature obtained during cutting. The interface temperature is proportional to the cutting speed.

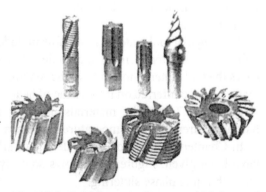

Fig. 13.8 Cold isostatically pressed high-speed steel parts

Table 13.1 Allowable average wear land in steels for cutting tools for various operations

Operation	Allowable wear land, mm	
	High-speed steels	Carbides
Turning	1.5	0.4
Face milling	1.5	0.4
End milling	0.3	0.3
Drilling	0.4	0.4
Reaming	0.15	0.15

Flank wear is a synonym for *attrition wear*, in which the particles of microscopic size are detached from the tool surface and carried away in the stream of work metal. It is due to the cold welding effect which can occur readily in metal cutting, especially at low surface speed. Fresh, clean metal surfaces are formed continuously by the cutting action and the high pressure produced; intermittent welding between tip and chip is inevitable. The motion of the tool piece and chip continually breaks the junctions so formed and discrete particles of tool material are carried away. The wear rate in the category is associated with the grain size of tool materials rather than hardness. A finer grain cemented carbide with 16% Co and a hardness of 1265 HV can show better resistance to attrition wear than a coarser grain hard metal with 4.5% Co and a hardness of 1600 HV.

For various cutting operations the values of average wear land data for high-speed steel and cemented carbide cutting tools are given in Table 13.1.

Diamond tools impregnated with metal matrix is another type of tool which is produced by hot pressing in the temperature range 200–900°C. Common matrices are cobalt or cobalt-containing WC or tin bronze additives. The very wide range of cobalt powder grade available in commercial quantities, offers great flexibility in planning the matrix microstructure and properties. There are a wide variety of natural and synthetic grits of varying properties. The major difference between natural and synthetic diamond is the impurity content. Natural grits made by crushing mined diamonds are free from metallic inclusions and hence show markedly higher thermal stability. They can retain their strength even up to 1400°C, whereas synthetic grits begin to loose strength beyond 800°C. From the matrix retention viewpoint, crushed crystals possess excellent bonding characteristics, which result from the numerous re-entrant surfaces available. This makes natural diamond advantageous for frame sawing to marble limestone and other less hard stones. For more rigorous applications, rounded diamonds are used. However, in this case, higher cutting forces are generated, and therefore, rigid and powerful machines are required.

13.2.2 Non-cutting applications

Metal forming tools fall under non-cutting applications. There are two types: hot forming and cold forming. The first includes hot pressing or extrusion processes, die casting of molten metals, etc. In the latter type, resistance to abrasion wear is the major requirement. Whenever a conventional high-alloy, cold-worked tool steel is used, and more than one tool is required to finish the production run, a PM high-speed steel is economically advantageous, even if it is much more expensive than the conventional steel.

Cemented carbides are extensively used for improved wire drawing dies, and those with minimal binder content, fine grains and maximum hardness are prepared. Cemented carbide

---

Example 13.3: Arrange data on physical properties of cemented carbides and cermets in a tabular form. What are the major differences and how do they affect technological properties and machining details?

Solution: The data are given in Table Ex. 13.3.

Table Ex. 13.3

Cutting materials	Cemented carbide		Cermets	
	P10	P20	Roughing grade	Finishing grade
Density, g cm^{-3}	10.8	12.4	6.1	8.1
Hardness, Vickers	1600	1500	1600	1600
Modulus of elasticity, kN mm^{-2}	540	550	500	500
Bending strength, N mm^{-2}	1700	2100	1500	1700
Compressive strength, N mm^{-2}	4400	4400	4200	4400
Coefficient of thermal expansion, 10^{-6} K^{-1}	7.9	6.9	7.6	7.9
Thermal conductivity, W m^{-1} K^{-1}	27	38	12	10

The major difference is in thermal conductivity which proportionately affects the thermal shock resistance. The lower thermal shock resistance of cermets limits the application of coolants to finish turning and threading.

Cermets are capable of operating over a wide range of cutting speeds. The higher plastic deformation resistance and better chemical inertness of titanium carbonitrides in comparison to tungsten carbides permits higher cutting speeds and reduces edge build up at lower cutting speed.

Cermets achieve excellent surface finishes and close size control in turning, grooving, threading and milling applications. More expensive grinding operations can be replaced by machining with cermet cutting tools. Coolants can be used, except in rough machining operations, and generally results in longer tool life.

Since the density of the cermet material is lower than WC-Co, a relatively large number of inserts per unit mass can be produced.

Example 13.4: Grey cast iron is the most abundantly used material for high-volume manufacture, such as automotive and earth-moving equipment components. Straight cemented carbide cutting tools have been used for machining. With a view to better productivity, would you suggest tool replacement by either oxide or non-oxide ceramic tools? Keep in view the economic factor as well. Using the fracture toughness and hardness data from the *Handbook*, plot schematically a graph between abrasive wear resistance parameter (inverse of volume removed) and $K_{Ic}^{3/4}H^{1/2}$, where, K_{Ic} is fracture toughness and H is the hardness. Justify your answer.

Solution: Alumina tools can be operated at a speed of 300–500 m min^{-1}. To enhance its speed further, Al_2O_3-ZrO_2 and Al_2O_3-TiC can be used in the order given. However, the best solution is to use Si_3N_4-based cutting tool (speed 3000 m min^{-1}), but the price is very high. Hence it is necessary to compromise with Al_2O_3-TiC tool. An alternative is to braze silicon nitride tip on the substrate of a more complex and less expensive cemented carbide.

The schematic plot is shown in Figure Ex.13.4.

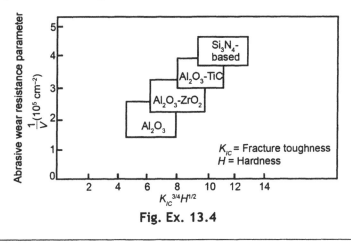

Fig. Ex. 13.4

dies are also used for drawing rods, bars, tubes and special sections. Carbide rolls are of great significance in the rolling of foils, rods, wires and tapes. A high elastic modulus keeps the foil thickness constant and pore-free. HIPed cemented carbides are universally recommended.

Wear parts are vital in the engineering industry. Major examples include nozzles, guides, plungers, balls and many other shapes. Here cemented carbides and technical ceramics are extremely useful.

13.3 POWER GENERATION

13.3.1 Applications based on fossil fuels

In this type of power generation, due to the very large size of the equipment, the wide use of PM parts is restricted, except those requiring a unique combination of properties. In oil-fired boilers the vanadium and sulphur levels in the oil play an important part in fireside corrosion, while in

Example 13.5: Sintered high duty mechanical seals are designed to deal with high and variable pressures, high shaft speeds and often with pump fluids which present unique problems. Describe the design and material aspects of such seals. Also, mention the major applications of such seals.

Solution: Major applications of such seals are:

- Pipeline pump seals for handling fluids as crude oil, finished product and fresh or sea water
- Water injection pump seals
- Boiler feed pump seals.

The running faces constitute the most important part of a mechanical seal. High-pressure fluids create high loads at the running faces and thus high thermal loadings. This can induce distortion in the running faces of the seal. Any deformation is to be radial and no rotation of the cross section occurs. In light section components with low modulus materials, such as carbon, axial buckling can occur at high pressures. The Robust components should be used for any safe residual symmetrical (conical) distortion. Another important feature is that the flexibly mounted rotating components must lie in good dynamic balance so that no swash is induced on the rotating face due to whirling, etc. However, some swash on the rotating face is inevitable, due to manufacturing tolerances, small misalignments and shaft vibration.

The maximum allowed rotary seal swash, X, is given by the expression:

$$X = \frac{2Fr^2}{lw^2}$$

where, F is the net seal closing force, r is the outside radius of sealing, l is the moment of inertia of the stationary seal ring assembly about a tangent to the sealing face and W is the rotational speed.

From the design standpoint, one should aim for low inertia components, a low friction and rotation arrangement and flexibility in the sliding packing.

The major high-duty seal materials are ceramic based. Table Ex.13.5 lists the important properties of silicon carbide, tungsten carbide and hot-pressed alumina as seal face material. The most attractive properties of silicon carbide are its high hardness, high thermal conductivity and low density. The high thermal conductivity makes the fluid film stable; thus higher seal pressures and shaft speeds can be tolerated.

Figure Ex.13.5 shows the theoretical limits of vapourisation for silicon carbide and tungsten carbide mechanical seals in water. This confirms that silicon carbide is a better choice than tungsten carbide. (Continued on next page)

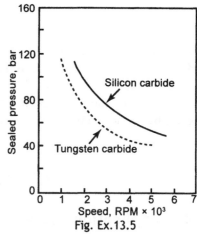

Fig. Ex.13.5

Table Ex. 13.5

Property	Silicon carbide	Tungsten carbide	Hot-pressed alumina 99.5%
Density, Kg m⁻³	3100	14670	3900
Young's modulus, GN m⁻²	400	618	365
Bending strength, MN m⁻²	500	2100	480
Compressive strength, MN m⁻²	3500	5275	2750
Hardness, HV	2500/3500	1500	2500
Poissons ratio	0.24	0.26	0.27
Weibull modulus	10	7.5	10
Thermal conductivity, W m⁻¹ K⁻¹			
at 20°C	210	85	32
at 500°C	84	66	8.4
Expansion coefficient, μm m⁻¹K⁻¹	4.4	5.5	9
Specific heat, J kg⁻¹ K⁻¹	900	218	1040
Heat shock factor, W m⁻¹	20800	21100	1200

coal-fired boilers the chlorine content in the coal promotes corrosion. Hot-extruded special PM tubes are the requirements at extreme environment locations. HIPed PM rings with enhanced toughness, fatigue strength and creep rupture strength are very useful. The diameter ranges from 300–1100 mm and the material is 12% Cr steel. High-performance PM superalloys will be in great demand in coal conversion processes. Ceramic parts are extensively used in heat exchangers meant to recover waste heat. The material most commonly used is SiC in tubular form.

13.3.2 Applications based on nuclear fuels

In Chapter 12, the sintering process for oxide nuclear fuels is described. Another group of ceramic fuels is $(U, Pu)C$. Figure 13.9 illustrates the schematic densification/swelling behaviour

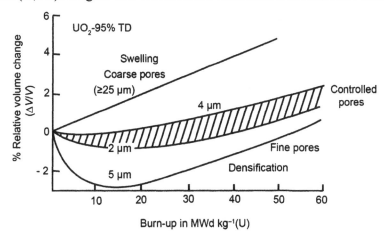

Fig. 13.9 Schematic densification/swelling behaviour of UO_2 fuel of different pore sizes

of UO_2 fuels of different pore characteristics. The study of this behaviour is important in determining the performance in the reactor. The level of open porosity must be low (<1%) to ensure low adsorbed moisture and low fission gas release. The pore amount should be sufficient to accommodate swelling. For less than 2 μm pore size, the porosity level should be minimum in order to avoid in-reactor densification. For minimising the amount of radiation damage in the core of nuclear fuel elements, dispersion-type fuel elements in a metal-matrix prepared by PM process is common. The matrices are aluminium, stainless steel or even refractory metals such as tungsten, molybdenum and niobium. The latter type of dispersed fuel is applicable to very-high-temperature gas-cooled reactors, thermionic converter reactors with fuel cladding at thermionic emitter temperatures (~1800°C) and nuclear rocket auxiliary power and propulsion. Some reactors use UO_2-Gd_2O_3 (6 wt%) fuel, where Gd_2O_3 serves as burnable poison for neutron absorption. The widespread selection of oxide as fuel for nuclear power reactors is because of its excellent properties with regard to compatibility with coolant on account of its chemical inertness, irradiation stability due to its open CaF_2-type crystal lattice and high fission product retention capability. The cross section of an irradiated UO_2 fuel pin shows concentric zones of altered microstructure which are indicative of temperature of the fuel during irradiation. With the increase in temperature from 1100°C to more than 1600°C, the microstructure changes from intergranular porosity and equiaxed grain growth to columnar grains and finally to central void. Such a change in microstructure contributes in redistribution of plutonium and fission products in the fuel pellet. The important factors governing the release of fission gas in oxide fuel are temperature, fuel burn-up, grain size and fission rate.

Oxide dispersion-strengthened ferritic alloy (a typical composition Fe-13Cr-1.5Mo-3.5Ti-2TiO$_2$) tubes prepared by full density PM consolidation method have been successfully tried. Y_2O_3-containing ODS alloy appears to be somewhat superior to TiO_2 dispersion. These alloys fulfill the creep requirements around 700°C and also limit swelling.

13.3.3 Fuel cells

In fuel cells, the chemical reaction provides the power and does not involve either of the electrodes in the primary cells. Two common fuel cell reactions are

$$2H_2 + O_2 \rightarrow 2H_2O$$
$$C + O_2 \rightarrow CO_2$$

The device comprises an anode (exposed to fuel), electrolyte, and a cathode (exposed to oxidant). The electrolyte separates anode and cathode and facilitates the ionic transport required for the oxidation of fuel. Electron flow in the external circuit produces usable electrical power. Table 13.2 gives information on various types of fuel cells and their applications. Among all the fuel cell technologies, solid oxide fuel cell (SOFC) technology offers the advantage of all solid-state construction, multifuel operational capability, high electrical conversion efficiency, and a simpler balance of plant. The electrolyte in SOFC is typically ZrO_2 with 8–10 mol% Y_2O_3. In order to minimise the voltage losses in the electrolyte, the electrolyte layer should be as thin as possible. In case of a flat electrolyte supporting the electrodes, a minimum thickness of 100–150 μm is required for ZrO_2-8 mol% Y_2O_3 ceramic. In case of reducing fuel

Table 13.2 Charge carriers and fuel cell operation temperatur

Type of fuel cell	Electrolyte	Operating temperature	Charge carrier	Application and fuel comments
AFC	Potassium hydroxide	$\sim 80°C$	OH^-	Space, Pure H_2, CO, CO_2 intolerant
MCFC	Molten carbonates	$\sim 650°C$	$CO_3^=$	Stationary power Fuel flexibility
PAFC	Phosphoric acid	$\sim 200°C$	H^+	Stationary power, Transportation Relatively pure H_2
PEMFC	Ion exchange membrane	$\sim 50°C$	H^+	Transportation Pure H_2, CO intolerant
SOFC	Solid metal oxide	$600-1000°C$	$O^=$	Stationary power, APU Fuel flexibility

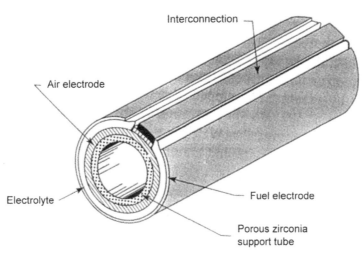

Fig. 13.10 Sealless tubular design for SOFC assembly developed by the Westinghouse Power Corporation

environment, nickel and nickel-based cermets are suitable as the anode. The common cathode material is electronically conducting perovskite ceramic lanthanum magnetite $LaMnO_3$, in which La is partially replaced by Sr.

Other parts in SOFC assembly are interconnection and sealing. In order to connect multiple cells, an electronic conductor is required, which is exposed to both highly reducing and oxidizing conditions at the electrodes. Another requirement is the dimensional stability and matching of the thermal expansion coefficient with other components. At temperatures above 850°C, Cr-alloy or $LaCrO_3$ ceramic interconnects are used. For sealing, glasses are used, but it must be ensured that they do not undergo recrystallisation at elevated temperature.

Generally, SOFCs have two designs: tubular and flat plate. Figure 13.10 shows the sealless tubular design developed by the Westinghouse Power Corporation, USA.

13.4 POROUS PM PRODUCTS

Some porous products like bearings and filters are common in any engineering application. In porous bearings or oil impregnated self-lubricating bearings, the oil is forced out with the running of the shaft due to the temperature rise. When the rotation stops and the bearing cools, the oil is re-absorbed by capillary action. For many self-lubricating bearings, the lubricant contained in the pores of the bearing lasts for the life time of the bearing. The common shapes of such bearings are hollow cylinders which are force-fitted into a housing, flanged bearings, thrust washers and self-aligning bearings which have a partly spherical external shape to allow an aligning movement after assembly in a housing. The K-factor of the bearing is a measure of the theoretical radial crushing force that the bearing can withstand. Since the material in porous bearings has a lower compressive strength than solid materials, the bearings can be inserted into the bearing housing by force-fittings.

Controlled porosity is employed in the manufacture of metal filters and diaphragms. Close control over the pore size and permeability is achieved by using powders with a narrow range of particle sizes. The filter profile is formed by a loose packing of the powder in the mould and the inherently poor compressibility of spheres is not a disadvantage. Metal filters are available in a wide range of materials including bronze, copper, nickel, stainless steel, titanium and monel, and are widely used for the filtration of fuel oils, chemical solutions and emulsions.

These elements must be characterised by precise particle removal, predictable pressure drop across the element, strength and impact resistance of metals as well as ease of fabrication.

The limit or wall thickness that can be produced is relevant to the size of the powder spheres involved. In general, it varies from a minimum of 1.5 mm for a 300 mesh powder to a maximum of 3 mm for 50 mesh particles. Tolerances vary considerably.

Filters can also be prepared from irregular powders, and the most commonly used are stainless steel and nickel. However, these lead to products of inferior quality. They may be processed using moulds, by pressing and sintering, by a slurry technique, or by roll compacting. The powders can also be mixed with a material which will form pores during manufacture (e.g. crystalline ammonium carbonate).

Table 13.3 compares various filters manufactured by different processes and materials. Sintered cupro-nickel is widely used in the bottled gas industry as a filter in the flow direction and a flame arrestor in the other direction. It is also used as a filter in aircraft cabin air circuits. Stainless steel filters are employed in the production of nuclear fuels and for flame arrestors in electronic and gas-detection equipment. They are also used in de-icing devices on aircraft and to protect windscreens and engine intakes. Porous nickel sheets are widely used in the production of batteries and fuel cells, and porous sintered bodies can act as flow restrictors which dampen surge flow, so that irregular flow speeds (e.g. pressure drops or peaks) are equalised and attenuated.

Figure 13.11 shows stainless steel filter cylinders produced by PM processing.

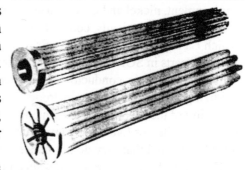

Fig. 13.11 Stainless steel filter cylinders

Table 13.3 Advantages of various types of filter media

Filtration characteristics and requirements	Sintered filters			
	Metal screen	Bronze filter	Stainless steel	Ceramics
Weight	x			
Size	x			
Flow rate	xx	x	x	
Pressure drop	xx	x	x	
Re-cleanability (economy)	xx	x	x	
Ability to operate at temperatures over 177°C	x	x	xx	x
Ability to capture and retain small particles below element rating		x	x	x
Ability to remove irregular-shaped particles		xx	x	x
Ability to permit passage of additives	x	x	x	x
Production volume (readily available)		x	x	x
Adaptability to manufacture in a variety of shapes		x		x
Strength	x	x	x	
Price		x	x	x

xx Superior, x Good.

Since the porosity and pore sizes of nanocrystal materials can be closely controlled, it is possible to manufacture a metal–ceramic microfiltration membrane for high-temperature and high-pressure applications. A filter membrane material has been produced using a stainless steel substrate coated with a very fine layer of porous nanocrystal TiO_2 material. The resulting membrane has a pore size of 10^{-1} μm. Industrial applications of the filter include chemical and industrial processes such as the separation of water from oil in the petroleum industry and for similar purification in food and beverage processing. Applications for filters with pore sizes smaller than one-tenth of a micron are also being developed for filtering blood and for use in gas filtration systems.

13.5 FRICTION ELEMENTS

Friction elements fitted with sintered metals are produced in the form of discs, sectional linings and blocks of different shapes. Table 13.4 shows some of the classifications of applications of friction materials. Disc-shaped elements are mainly used in friction assemblies working with

Table 13.4 Classification of applications of sintered metal friction materials

Dry applications				Wet applications		
Clutch		Brake		Clutch		Brake
Mild (farm tractors, lorries)	Moderate (industrial material movers)	Moderate (auto-motive)	Severe (aircraft)	Mild (light automatic)	Severe (power shift in tractors and military vehicles)	Moderate/severe (off-road and military vehicles)

Table 13.5 Some compositions of iron base and copper base PM friction materials

	Fe	Cu	C	Ni	Other additions, wt%	Country
Iron-base	60–75	-	5–15	-	$MoS_2 < 10$, MoSi 5–20, SiC < 5	USA
	67–80	12–25	3–4	-	Pb 1–3, Zn 1–2.5, SiO_2 0.3–0.9	Germany
	62–72	-	4–8	15	SiO_2 5–13, Co 2, Cr 2	Japan
Copper-base	20	50–80	5–15	-	SiO_2 5, MoS_2 20, Ti 2–10	USA
	5–15	50–65	< 25	-	Al_2O_3 5, Sb 4–6	Germany
	4–6	62–67	5–9	-	SiO_2 4.5–8, Sn 6–10, Pb 6–12	Japan

lubrication. Efficient operation of wet-running friction materials requires high friction and a rapid build-up of friction force during clutch operation. A grooving system is needed to ensure the quick break-up of any oil film which may form on the friction area, while at the same time ensuring constant oil flow through the plates.

The composition of the friction material is critical, as some metals are speed sensitive, others are energy sensitive and some are selected for ease of fabrication. The sintered material should have a high thermal conductivity and may be used over a wide range of temperatures, e.g., Cu-base materials up to 800°C, and Fe-base materials up to 1000°C. Typical compositions of iron-based and Cu-based sintered friction materials are given in Table 13.5. It may be borne in mind that numerous compositions are in use, which are often kept a trade secret.

13.6 ELECTRICAL MATERIALS

13.6.1 Electrical contact materials

Contact materials are important for electrical devices. Electrical contacts may operate under no load, for example, electrical connections or plugs, or under load, for example, switching contacts used in contactors and switchgears as well as sliding contacts in electrical machines. The switching medium is important with respect to arc quenching: generally the operations are performed in air, sulphur hexafluoride, in liquids such as oil or oil-containing water and in vacuum.

Table 13.6 Electrical, mechanical and environmental conditions for electrical contact materials

Electrical	Mechanical	Ambient
(1) Current:	(1) Frequency of operation:	(1) Gases or fumes
• AC or DC	• Low: up to 1 mm^{-1}	(2) Corrosion (films)
• Frequency	• Intermediate: 1 min^{-1}	(3) Contact contamination
• Magnitude:	up to 1 s^{-1}	• Manufacture
Light duty:	• High: 1/2 up to 5 s^{-1}	• Assembly
Up to 1.5–2 A	• Very high: above 5 s^{-1}	(4) Special atmosphere
Medium duty:	(2) Speed of operation	(5) Humidity
2–5 A	• Closing	(6) Temperature
Heavy duty:	• Opening	
5–20 A	(3) Contact gap	
Extra heavy duty:		
> 20 A		
(2) Voltage	(4) Wiping, sliding or rolling motion	
• AC or DC	(5) Open contacts (shelf life)	
• Magnitude:	(6) Method of operation	
110, 220, 440 AC		
6, 12, 28 DC		
• Above or below critical		
arcing voltage		
(3) Type of load	(7) Contact bounce or chatter	
• Resistive	(8) Contact force	
• Inductive		
• Capacitive		
• Motor load		
• Lamp load		
(4) Overload requirements		
(5) Auxiliary protection		

The basic categories of contact devices are: make–break, remountable, sliding and fixed. Each of these types has common characteristics, with a wide range of different requirements, depending on the electrical and mechanical conditions of use.

Table 13.6 lists some of the major factors involved in the selection of an electrical contact material.

The ideal contact material should have good electrical and thermal conductivity, and resistance to arcing and abrasion. They can be grouped in various categories, e.g., metal–metal systems (Ag-Ni, W-Ag and W-Cu), metal–metal compound (silver with oxides of Cd, Zn, Sn and Cu) and finally metal–non-metal (Al-10% graphite).

W-20-40% Cu system is characterised by excellent burning resistance (Fig. 13.12). Experimental results show that the burn-off rate of tungsten–copper is lower in SF_6 than in air and is further reduced by the formation of a tungsten skeleton. Infiltrated tungsten-based composites have a high hot skeleton strength of the refractory metal which will retain its

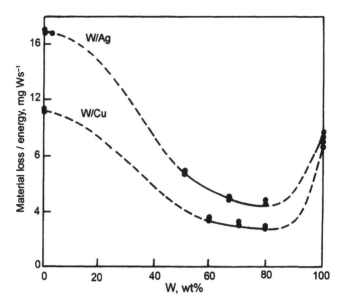

Fig. 13.12 Burn-off of infiltrated W-Cu and W-Ag contact materials

Table 13.7 Typical applications of brush materials

Automotive	Auxiliary motors
	Alternators
	Starters
Fractional horsepower	Food mixers
	Power tools
	Vacuum clearners
Power machines	Generators
	Mill motors
	DC motors
Aircraft machines	Starter generators
	Invertors
	Fuel pump motors
Diesel locomotives	Generators
	Traction motors
	Auxiliary motors

form even when the copper or silver is molten. W-Cu sintered alloys are widely used as tool electrodes for electro-spark machining.

Among sliding electrical contacts, 'brush' materials occupy a significant position. Copper-containing sliding electrical contacts are used widely in the automotives sector (Table 13.7). Table 13.8 lists many of the typical applications for copper-containing brushes, along with typical copper contents, from which it is clear that those applications requiring low power dissipation or involving large currents are largely concentrated in automotive and industrial

Table 13.8 Typical applications for copper-containing brush materials

Major area	Application	Nominal copper content* %
Automotive	Windshield wipers	30–50
	Starters	80–95
	Alternators	30–50
	Seat movers	70–90
	Winch motors	70–90
	Air conditioners	30–50
	Heaters	30–50
	Defogger motors	30–40
Industrial	Collector rolls	85–95
	Plating generators	70–90

* Typical values only; can vary depending on specific application requirements

collector and plating generator applications. While graphite bronzes carry, at most, $100 \, kA \, m^{-2}$, those produced from metal–graphite material are rated at 195–$230 \, kA \, m^{-2}$.

13.6.2 Electric lamps

The tungsten filaments in incandescent electric lamps are a classic example of the use of PM materials. The major requirements of such filaments are:

- Maintenance of a pre-calculated light output
- Smallest deviation of lifetime from the predetermined value
- Non-sagging at incandescent temperatures
- No breakage or deformation due to shock and vibration.

In Chapter 12, an account of tungsten, including doped-tungsten, is given.

Another application of a sintered product in lamp applications is in sodium vapour lamp, where a fully dense polycrystalline alumina tube is used to contain the corrosive plasma. Figure 13.13 shows the construction of a typical high-pressure sodium lamp. This lamp is a two-envelope device, where the inner envelope is partially filled with a halogen gas, within which sodium vapour of 3700°C plasma temperature is struck. The alumina tube itself operates at a wall temperature of about 1200°C and is enclosed by an evacuated outer envelope of glass. This highly dense alumina is known by the trade name Lucolox™ and was discovered by Professor RL Coble at MIT, USA. As a matter of fact this is the MgO-doped alumina, the sintering fundamentals of which are described in Chapter 7. This is an outstanding example to demonstrate how sintering theory has transformed the whole illumination industry. The doped alumina envelopes continue to be produced worldwide by a number of manufacturers at an estimated rate of 16 million per year.

Example 13.6: Diesel engines have an advantage of better fuel economy over gasoline engines. On the other hand, small diesel engines have difficulty in engine starting. The solution for this was to use ceramic glow plug. Describe the assembly and material aspects of such plugs.

Solution: The compressed air temperature is not raised easily in winter. Glow plugs are installed and preheated to 800°C to promote fuel ignition. Diesel car drivers have to wait for 4 to 20 s—depending upon temperature controlling systems—prior to engine starting. Ceramic glow plugs have been developed to minimise such waiting time.

Conventional metallic glow plugs have nickel heating elements electrically insulated from steel sheath. The insulating ceramic powder also limits thermal flow and 4 to 20 are necessary before the plug surface is heated to 800°C. With higher power, the time necessary can be shortened but the large temperature difference within a plug may destroy the plug.

Figures Ex.13.6a and Ex.13.6b show ceramic glow plugs which have tungsten heating elements embedded in dense silicon nitride by hot pressing. Although the thermal conductivity of silicon nitride is lower than that of steel, these plugs have no porous insulating layer, and as a whole, they have less thermal flow resistance than conventional ones. The plug surface is heated to 900°C within 2.5 s, because of dense ceramic structure and better refractoriness of tungsten heating elements. Since there is a time lag of approximately 0.6–1.5 s when a driver turns the engine switch from 'on' position to 'start', this preheating time of 2.5 s virtually eliminates the waiting time and gives diesel car drivers the same starting action as gasoline car drivers.

With the introduction of turbo charging, the durability of metallic plugs is becoming a serious problem because of metal sheath corrosion at higher combustion temperatures. Ceramic glow plugs show better durability. This characteristic is also effective in the use of after-glow operation after the engine has started, which has advantages of lower idling noise and less white smoke emission immediately after starting.

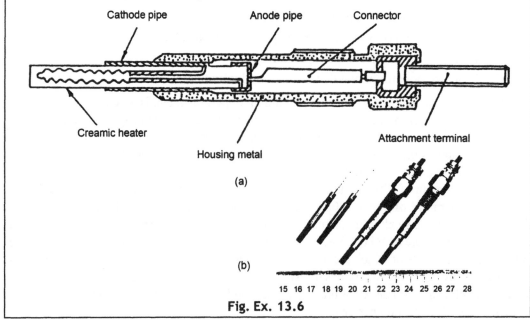

Fig. Ex. 13.6

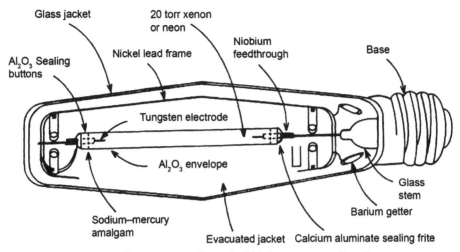

Fig. 13.13 Construction of a typical high-pressure sodium lamp

Example 13.7 Tungsten wires are extensively used in electrical applications. The fracture of this brittle material is crucial. Describe in a tabular form the various parameters that affect the fracture during wire drawing under the following heads:

(a) Machine type
(b) Tools
(c) Process
(d) Materials properties

Solution: The details are given in Table Ex. 13.7. (For further details, refer: Schade P, Wire drawing failures and tungsten fracture phenomena, *International Journal of Refractory Metals & Hard Materials*. 24 (4): 332)

Table Ex. 13.7

Machine type	Tools	Process	Material properties
Single	Type of dies cemented carbides	Wire temperature	Yield strength
Duplex/triple	Single crystalline diamond Polycrystalline diamond	Drawing speed	Ultimate tensile strength
Multiple	Geometry of dies Entrance angle Bearing length Transition angles Surface finish Drawing drum Lubrication	Reduction in area per pass True strain of drawing Drawing stress Back pull stress Friction stress Intermediate anneals Die temperature Die series	Fracture strength Fracture toughness Elongation/ductility Ductile–brittle transition temperature (DBTT) Polygonisation temperature Texture Residual stress Stress gradients Microstructure

13.6.3 Dielectrics

Piezoelectric ceramics are extensively used in various applications like phonograph pickups, band pass filters, and control of oscillator frequencies in communication industry, sonic delay lines, sonar, high-voltage step-up transformers, ultrasonic cleaning, medical ultrasound uses, industrial non-destructive inspection and accelerometers. Some of the applications are based on the conversion of a mechanical force to an electrical signal and vice versa. Barium titanate is a classic example of hard ferroelectric. At the Curie temperature, the crystal transforms from a disordered pure electric structure to an ordered ferroelectric structure. In the past, practically all ferroelectric devices were prepared from slices of crystals (primarily Rochelle salt) cut in a specific direction to optimise the ferroelectric characteristics. Polycrystalline barium titanate has given us sintered devices. The production steps involve:

- Preparation of dense polycrystalline barium titanate by powder pressing/sintering route;
- Application of conductive electrodes to the two surfaces perpendicular to the desired polarisation direction;
- Heating the part to above 120°C and application to large enough electric field between the electrodes to force many of the domains to align parallel to the direction of the applied field;
- Cooling the part below the Curie temperature and removing the electric field.

Figure 13.14 illustrates a variety of ceramic piezoelectric elements used in different applications.

Fig. 13.14 Piezoelectric ceramics and assemblies for a variety of applications

13.6.4 Capacitors

The manufacture of a solid tantalum capacitor involves making a porous tantalum pellet by pressing and sintering tantalum powder. The pellet, with an attached tantalum lead wire, is electrochemically oxidised to grow a thin layer of insulating anodic tantalum oxide on the surface of the tantalum. Next, the anodised pellet is impregnated with manganese nitrate, which is then thermally decomposed to leave a deposit of semiconducting manganese dioxide on the tantalum oxide. These processes create the conductor(Ta)/insulator (Ta_2O_5)/semiconductor(MnO_2) configuration needed for a capacitor. Finally, the unit is encapsulated (usually) in the chip configuration. Many of the solid tantalum capacitor manufacturing steps are sensitive to the physical properties of the tantalum powder and wire.

Single-layer ceramic capacitors remain the least expensive capacitors and are widely used in consumer applications. However, their usage is declining as multilayer ceramics are substituted in newer designs. Multilayer ceramics have higher volumetric efficiency, are rapidly reducing in cost, and are ideal for the new labour and space saving surface-mount technology. These capacitors are displacing not only single-layer ceramics in new applications, but also small tantalum capacitors and film capacitors. The growth in multilayer ceramic capacitors is also being accelerated by electronic applications moving towards higher frequencies and digital techniques which reduce the electronic size of capacitance required in the application.

Table 13.9 shows some of the qualitative design factors used by equipment designers for choosing a dielectric and style to fill a capacitor function requirement.

Table 13.9 Capacitor design factors

Factor	Single-layer ceramic	Multilayer ceramic	Tantalum	Aluminium	Film
Reliability	+	+	+	+	+
Volumetric efficiency	–	+	+	–	–
Electronic characteristics		+	+		
Available as 'Chip'	–	+	+	–	–
High temperature	–	+	+	–	–
Cost	+		–	+	

Source: *Proceedings of International Symposium Ta and Nb*, Nov. 1988, Tantalum-Niobium Study Centre, Brussels, Belgium

Example 13.8: The tantalum capacitor is regarded as a system. It is often noticed that the tantalum lead wire, another PM product in the capacitor system, fails, due to embrittlement. Discuss the reason for such a failure.

Solution: There are two types of embrittlement in tantalum leads. One is primarily related to the high oxygen content in tantalum powder. The oxygen that diffused from tantalum powder into tantalum wire during sintering, as well as the oxygen picked up in later stages of the capacitor manufacture process is responsible for this type of tantalum wire embrittlement. The other type of embrittlement is associated with the damage to the wire caused by welding. Hydrogen absorption in the damaged area plays a large role in the second type of embrittlement. In addition, oxygen can be easily picked up in the damaged area at elevated temperature. Both types of wire embrittlement involve diffusion of gaseous elements in tantalum. They are of a cumulative nature.

13.7 MAGNETIC MATERIALS

13.7.1 Soft magnetic materials

Among metallic systems, sintered Fe-P alloys are very promising. Fine iron powder encapsulated by polymer has been used in DC electromagnetic systems. Because the polymer serves as an insulator between the iron particles, the mould part can be used for AC and pulsed-DC higher frequency electromagnetic applications.

Oxide ferrites have found an increasing number of applications, because, in many cases they yield higher efficiency, smaller volume, lower costs and greater uniformity. In the microwave field, there is no competition as metallic soft magnetic materials have large eddy current losses at these frequencies, which cannot be permissible. Among the linear B/H, low flux density application of ferrites are filter inductors, IF transformers, antenna cores, adjustable inductors, tuners, miniature inductors and loading coils.

13.7.2 Hard magnetic materials

Hard or permanent magnets are used in instruments such as galvanometers, ammeters, voltmeters, flux meters, speed meters, watt meters, compasses and recorders. They are also used in synchronous and hysteresis motors, fan and toy motors, aircraft motors and generators, alternators and magnetos, dynamos and magnetising yokes. For electronic devices, permanent magnets are used in telephones and tape recorders, loud speakers and hearing aids, TV tubes and electron beam focussing and positioning and microwave devices such as travelling wave tubes, isolators, circulators and phase shifters. In mechanical systems these are used in chucks and latches, tool holders, conveyors, mineral separators, vending machines, magnetic bearings, eddy current brakes and frictionless drives using magnetic levitation.

Hard ferrites are extensively used. This is attributed to the very favourable ratio of flux density to cost. Bonded ferrites are widely used, due to the improved development of high-energy ferrite powders and technologies for injection moulded products. Figure 13.15 illustrates some examples of ferrite permanent magnets.

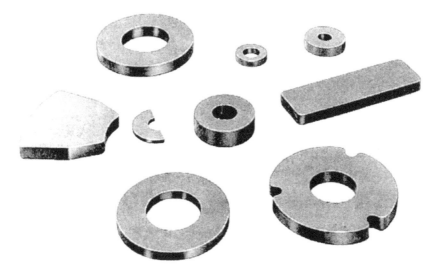

Fig. 13.15 Some ceramic permanent magnets (hard ferrites)

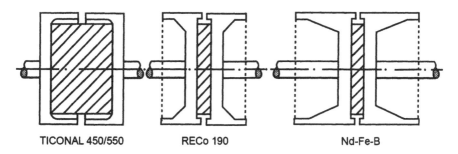

TICONAL 450/550   RECo 190   Nd-Fe-B

Fig. 13.16 Rotors for stepper motors with different permanent magnets

Rare earth–based intermetallic permanent magnetic material, i.e., Nd-Fe-B, has been described in the previous chapter. In the case of stepper motors, its high performance can be gauged by the comparison with other classes of materials (Fig. 13.16). The left-hand side construction has an Alnico 550 magnet. The rare earth-Co 190 magnet is shown in the middle. The magnet made of Nd-Fe-B shown on the right, has the same size as RECo 190, but with 50% more flux compatibility.

13.8 OXYGEN SENSOR

Increasing emphasis on fuel efficiency and pollution control in automobiles, has led to the development of combustion control systems based on oxygen pressure sensors. The oxygen sensors are typically conductive ceramics, e.g., non-stoichiometric TiO_2, where the defects in the lattice cause electron conductivity. Figure 13.17 shows a titania exhaust sensor for automobiles, where the conductivity of a small piece of polycrystalline TiO_2 exposed to the exhaust gas is continuously monitored. In order to shorten the response time of conductivity

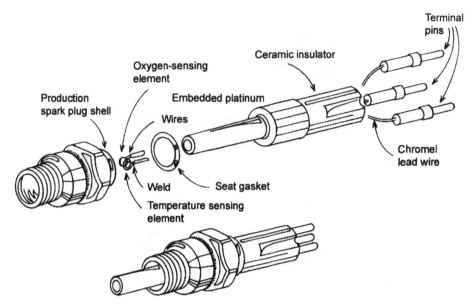

Fig. 13.17 Temperature-compensated titania exhaust sensor

with change in oxygen activity, a porous TiO_2 element is required. This increases the surface area for gas exchange, and decreases the effective cross section across which the non-stoichiometry must change. Since the electrical conductivity is temperature dependent, a temperature-compensated device is used. Ford Motors uses a piece of dense TiO_2 which operates at the same temperature as the porous sensor, but does not equilibrate quickly with the gas stream to provide a reference point for the conductivity. A comparison of the resistivities of the two ceramic elements allows the oxygen pressure dependence of conductivity in the porous TiO_2 to be isolated.

13.9 THERMAL MANAGEMENT MATERIALS

Miniaturisation in electronic components has given rise to an urgent requirement for thermal management materials with the ability to remove excess heat in traditional materials such as Kovar. The need for increased thermal conductivity and controlled coefficient of thermal expansion has encouraged the development of a number of composite materials, most of which are based on refractory metals, i.e., tungsten and molybdenum. Figure 13.18 shows the schematics of a pin grid array (PGA) package in which molybdenum has been used as a heat sink.

Ceramic substrates used to mount the semiconductors are often made from alumina, beryllia or aluminium nitride. Some relevant data are given in Table 13.10. The heat dissipation materials must be thermally compatible with the semiconductors and substrate materials. The target thermal expansion coefficient is in the range 7 ppm $(°C)^{-1}$. Depending on the design and assembly procedure, there are the additional criteria of strength, microwave absorption and electroplating compatibility.

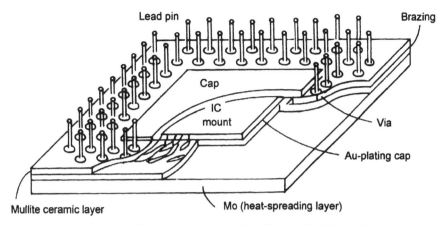

Fig. 13.18 Mullite–molybdenum pin grid array (PGA) package

Table 13.10 Data important for electrical package substrate materials

Material	Dielectric constant, 1 MHz	Dielectric loss, tan δ	Thermal expansion, × 10⁻⁶ (°C)⁻¹	Thermal conductivity, W cm⁻¹ K⁻¹	Bending strength, MPa	Withstand voltage, kV cm⁻¹
Mullite	7	0.0003	4.4	0.06	300	-
Alumina	9.5	0.0003	8	0.3	500	100
AlN	8.9	0.001	4.5	2.6	400	150
Steatite	6.0	0.0002	7.8	0.03	130	-
BeO	6.5	0.0005	7.8	2.4	200	140
Zircon	7.8	0.001	4.8	0.05	320	-
Cordierite	5.0	0.003	1.8	0.01	140	-
SiC	45	-	4.0	1.5	600	-
Si$_3$N$_4$	8.1	0.0007	3.8	0.3	1000	100
Diamond	5.7	-	2.3	20	-	-
Polyimide	3.5	-	66	0.002	-	-
Glass ceramic	5.8	0.005	11.5	0.02	100	400

Light–weight metal matrix composites (for example SiC-Al) have attracted much attention. Since density is a matter of concern for aerospace applications, such MMCs are of interest. The objective is to combine a metal with a high thermal conductivity with a reinforcement which has a high elastic modulus and a low density to achieve a light-weight, high-strength, high thermal conductivity composite.

13.10 HARDFACING RODS

Bare rods for hardfacing applications are very hard and non-deformable. For these reasons, they cannot be produced by methods such as rolling, swagging and drawing. They are generally cast into rod shape followed by grinding to remove surface defects. Airco of USA, as far back

Table 13.11 Nickel base hardfacing rods by PM

Alloy	Nominal chemical composition, %					Hardness, R_c	Melting range, °C
	Ni	Cr	B	Si	Fe		
MH # 1	0.6	14	3.0	4.5	4.5	60	977–1050
MH # 5	0.5	12	2.5	4.2	4.0	50	970–1066
MH # 3	0.3	9	1.5	3.0	2.8	40	1025–1105
MH # 6	-	-	2.7	4.3	-	50	980–1025
MH # 7	-	-	1.7	3.6	-	30	980–1066

as 1969, developed PM rods by direct particle liquid phase sintering process. Powder particles were poured into a suitably designed mould, which was vibrated to increase powder packing. The sintering temperature is between the solidus and liquidus temperatures of the alloy being sintered, and thus, it can be termed as super-solidus sintering.

The advantages of PM process over casting are as follows:

- Higher production yields
- Higher length-to-area ratios possible
- Alloys with less fluid and larger solidification range possible
- Rods with up to 60 v/o of WC; CrB particles possible
- No secondary finishing and cleaning operation
- Clean and free of pollution, dust and sand
- Flexible, potentially economical, high quality, wide range of alloys feasible.

Table 13.11 gives some typical Ni base hardfacing rod compositions prepared by PM route.

13.11 METAL FOAMS

Solid metallic foams, especially the ones based on light metals are known to have interesting combinations of different properties such as stiffness in conjunction with very low specific weight on high compression strengths combined with good energy absorption characteristics. Of the two methods of production—direct foaming of melts by gas injection and indirect foamable precursor method—the latter is achieved by PM technology. In this method, aluminium and titanium powders are mixed and compacted, followed by hot pressing, extrusion or powder rolling to form a dense precursor. The product is heated to the semisolid state. The decomposition of titanium hydride promotes foaming process. In a way, the process cannot be singularly assigned to PM, since a remelting stage is also involved. This remelting stage could be taken as a liquid phase bonding process. The use of metal foams in conjunction with dense material has given way to sandwich panels consisting of a foamed metal core and two metal face sheets. This is achieved by roll cladding conventional aluminium sheets to a sheet of foamable precursor material manufactured from powders. The ability to make 3D-shaped panels and the high stiffness-to-weight ratio are a clear advantage over competing technologies such as honeycomb structures. Thus, aluminium foam sandwich panels could replace conventional stamped steel parts in an automobile and lead to significant weight reductions. Other applications of metal foams are in foam-filled tubes/sections, and as reinforcement for polymer structures. Aluminium foams have also been used as cores for castings. There is need for further developments, due to the following limitations:

- High cost of foams
- Lack of ability to make foams of a uniform quality
- Lack of understanding of the basic mechanism of metal foaming
- Insufficient understanding of morphology/structure relations with properties.

13.12 BIO-IMPLANTS

Sintered metallic and ceramic implants are useful as bio-implants and they must fulfill the following criteria:

- Sufficient static and fatigue strength, and wear resistance for the application
- Non-toxic, non-allergenic and non-carcinogenic
- Non-corrosive in the physiological environment
- No effect on the host tissue, bone or body fluids.

Depending on the type of application the materials may be classified as:

- Bio-inert
- Bio-active/surface reactive
- Biodegradable/resorbable.

Among metallic materials, stainless steel, titanium alloys, niobium and tantalum alloys and Co-Cr-Mo alloys are very common. Among stainless steels, 316L and 17-4 precipitation-hardened alloys are common. PM stainless steel has been used in making an artificial prosthetic device that replaced the head and neck of a femur (Fig. 13.19). Another application of stainless

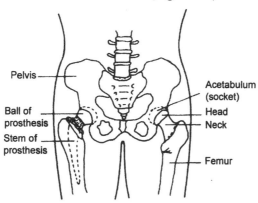

Fig. 13.19 A human hip joint showing the relative position of an artificial prosthesis replacing the head and neck of a femur

steel is in the area of dental brackets. The application of ODS niobium for intramedullory nails and cement-free total endoprostheses has also been reported.

Sintered porous titanium has been used for dental implants—a stomatological system meant for fixing dentures in which a partial or complete defect exists in the dental row. The implant consists of a porous part with a threaded outlet, a screw plug and a head for fixation. The implants were produced by dynamic compaction in an impulse electric field, without using an inert atmosphere.

Among bio-ceramics, alumina and hydroxyapatite are common materials. The mechanical properties of hydroxyapatite (UTS 25-120 MPa, K_{Ic} 1.0 MPa m$^{1/2}$, and E 30-120 GPa) are of the order of those for bone. The theoretical value for the calcium-to-phosphorus ratio of pure hydroxyapatite is 1.67. Compounds with a ratio less than 1:1 are not suitable for biological implant. The temperature for hydroxyapatite is between 1100°C and 1300°C. However, care must be taken to avoid the risk of decomposition during sintering.

SUMMARY

- Recently developed ferrous powders used in combination with modern compaction techniques and sintering methods reach ultimate tensile strength level of more than 1000 MPa. PM structural products are extensively used in automotive, aerospace and defence applications.

- Machine tool applications of PM parts include cutting and wear applications.

- Porous products like bearings and filters are common in any engineering application. The K-factor of the bearing is a measure of the theoretical radial crushing force that the bearing can withstand.

- Sintered materials for friction applications must have high thermal conductivity.

- An ideal electric contact material should have good electrical and thermal conductivity, and resistance to arcing and abrasion.

- Oxide ferrites have increasing number of magnetic applications because they yield higher efficiency, smaller volume, lower cost and greater uniformity.

- A number of composites based on refractory metals are widely used for thermal management applications, because of increased thermal conductivity and controlled coefficient of thermal expansion.

- Among metallic materials, stainless steel, titanium alloys, niobium and tantalum alloys and Co-Cr-Mo alloys are common in bio-implant applications. In bioceramics, alumina and hydroxyapatite are common materials.

Further Reading

Anonymous, *Power Transmission Proceedings*, Metal Powder Industries Federation, Princeton, 2001.

Brownell WE, *Structural Clay Products*, Springer-Verlag, New York, 1976.

Clark DE, Folz DG and McGee TD, *An Introduction to Ceramic Engineering Design*, American Ceramic Society, Westerville, Ohio, 2002.

Creyke NEC, Sainsbury J and Morrell R, *Design with Non-ductile Materials*, Applied Science Publishers, London, 1982.

Design Solutions '99, Metal Powder Industries Federation, Princeton, USA.

Dieter GE, *Engineering Design*, 3rd edn, McGraw Hill Book Co., New York, 2000.

Jenkins I and Wood JV, (eds), *Powder Metallurgy: An Overview*, The Institute of Metals, London, 1991.

Powder Metallurgy Design Manual, Metal Powder Industries Federation, Princeton, NJ, 1995.

Upadhyaya GS and Upadhyaya A, *Materials Science and Engineering*, Viva Books, New Delhi, 2006.

EXERCISES

13.1 Why are oil-impregnated sintered bearings not appropriate at higher sliding speeds and higher loads?

13.2 The future trend in automotive vehicles is towards hybrid vehicles, where diesel, fuel cells, electricity and gasoline will be used in one form or the other. Currently, about 6 kg of powder metallurgy parts are used per car. What do you think will be the potentiali for PM parts in hybrid vehicles? List the names of various new parts and their possible processing through PM route for such high power density vehicles.

13.3 Name some sintered engineering components from the automotive and aerospace industries where the following fatigue conditions are applicable:

(a) Classical fatigue

(b) Wear, rolling contact and classical fatigue

(c) Impact and fatigue

(d) Creep and fatigue.

13.4 Distinguish clearly between performance, reproducibility, reliability and quality of a sintered product.

13.5 Why are full density PM consolidation methods used in the case of nickel-based superalloys instead of normal PM pressing/sintering technology?

13.6 What are the advantages of using Si_3N_4 ball bearings as compared to other ceramics in aerospace applications?

13.7 Why does an APDS made of tungsten carbide fail in severe conditions, while the same made of W-Fe-Ni heavy alloy penetrates successfully?

13.8 What are the effects of high porosity on the service response of refractory bricks and whiteware in unglazed condition?

13.9 Circular wheels containing abrasive grains (e.g., alumina, emery or diamond) are frequently used to cut metals. The abrasive grains are bonded together with bronze or by a polymeric or vitreous material. Describe how such cut-off wheels could be made.

13.10 Discuss how the controlled porosity sintered UO_2 fuel pellets are crucial.

13.11 In the third stage of the Indian nuclear power program, Th-U^{233} fuel cycle will be adopted. Discuss whether the significance of PM processing route should prevail or not.

13.12 What are the major problems encountered at extended nuclear fuel burn-up. When and how does the fuel grain size have a significant effect?

13.13 Currently, Zircaloy-2 (Zr-2Nb) tubes for nuclear reactors are produced by vacuum arc melting, hot extrusion, cold finishing and heat treatment. It is suggested that the tubes be produced through the PM consolidation route. What will be the advantages and disadvantages of the proposed route? Give a flowsheet of the process.

13.14 For molten carbonate fuel cells, describe the process for making electrolyte tiles. What are the crucial parameters?

13.15 In case of porous sintered metals, the bending strength is higher than the tensile strength. Give reasons.

13.16 Comment on why SOFC are not applicable in space/military purposes, where hydrogen is used as the fuel.

13.17 Silver–graphite (2%–5%) contact material is used in miniature circuit breakers.

(a) What will be the effect of composition on weld resistance during arcing?

(b) What consolidation method should be adopted?

(c) Graphite particles are highly anisotropic in shape (flaky). Why are these preferred to the contact surface?

(d) What is the disadvantage of this material during service?

(e) What is the major advantage?

13.18 It is noticed that 54% $MoSi_2$-β-SiC composites are metallic in nature with an almost constant temperature coefficient of electrical conductivity. Describe the production of heating elements based on this alloy.

13.19 You are to develop a cast iron PM filter based on hypo-eutectic composition. Give the full flowsheet starting from the powder production stage.

13.20 What is the best optimisation of stresses for the use of a sintered filter in an engineering assembly?

13.21 W-Cu sintered electrodes are used in electro-spark machining. How far can this alloy be substituted by a less costly molybdenum?

13.22 W-Cu (75/25) PM electrodes are extensively used for electro-discharge machining. It was found that a TiC-based PM electrode containing W and Cu gave rise to electrodes with good surface finishing performance. Describe the difference in the processing for these two materials. It was suggested that a minor addition of a certain metal enhances the sintered density of the latter type of material. What is that metal and how does it aid densification? Is there any adverse effect of this metal on other physical properties?

13.23 Which of the following applications requires hard 'PZT' ceramics: (a) sensor, (b) sonar, (c) igniters, (d) actuators, (e) supersonic generators and (f) ultrasonic cleaners?

13.24 A producer of wristwatches wishes to place an order for bracelet parts manufactured from atomised 18-carat gold alloy powder. What processing method(s) should the PM plant evolve? Remember the cost is not a hindrance in evolving the processing route.

13.25 Question 8.24 is related to sintering of TiB_2-Fe cermet. Can this cermet replace WC-Co cemented carbide, since both titanium and iron are much cheaper than tungsten and cobalt? What special problem will be faced in TiB_2-Fe cermets, which is not met in WC-Co system?

13.26 A ceramic nozzle is to be prepared by PM technology. The nozzle's entrance section suffers severe abrasive impact and generates the largest tensile stress. The stresses along the axial direction of the nozzle decrease from entry to centre, and increases from centre to exit. Explain how you can tailor the composition of the ceramic matrix composite nozzle to fulfill the application.

13.27 Why is the use of refractory nitrides and carbides (TiN, AlN, Si_3N_4, SiC) in bulk not as common in electronic and optical application, as in coatings? What is the future of bulk ceramic materials in electronic applications?

13.28 A silver-based PM contact material containing 12 mass% WC and 8 mass% of other cubic carbides was developed. Explain the role of following the performance (wear resistance and welding resistance) of the contact materials:

(a) WC particle size

(b) Amount of other cubic carbides

(c) Sintering temperature

(d) Sintering additive.

13.29 Name some applications of nanometal/ceramic powders in engineering industries.

13.30 Name the powder material(s) needed for the production of the following railroad parts:

(a) Brake linings

(b) Circuit breaker contacts

(c) Compound for sealing coach joints

(d) Friction strips on pantograph

(e) Signal flares.

13.31 Name the powder material(s) needed for the production of the following ordnance parts:

(a) Ammunition   (f) Incendiary bombs

(b) Armour piercing cores   (g) Missile filters

(c) Anti-personnel bombs   (h) Projectile rotating bands

(d) Frangible bullets   (i) Proximity fuse cup

(e) First-fire mixes and fuses (j) Solid missile fuel.

13.32 Two ferrous components of heavy duty trucks (22,700 kg), namely, internal ring gear (1355 Nm torque) and synchroniser ring are ordered in a PM plant. Which PM processing method is more economic for the manufacture, and also fulfills the stress requirements? Will the processing method be the same for car components of similar nature?

13.33 Suggest some applications of low sintering temperature of nanocrystalline powders in evolving a useful product.

13.34 What are the advantages and limitations of insulated iron powder–based electric motor starter?

13.35 A cast and machined handle of 17-4 PH stainless steel is to be replaced by one manufactured by PM route.

(a) What type of steel will be selected?

(b) Which PM process will be adopted?

(c) What can be done in part configuration in order to reduce the weight and hence the price?

13.36 What information will be given to PM part suppliers by the users of gears, splines, sprockets or similar products before purchase?

13.37 Why are PM exhaust stainless steel flanges better than those prepared after conventional stamping and welding operations?

13.38 Among the many applications tungsten-based heavy alloys are used for (a) die casting moulds and (b) radiation shielding plates. Mention the properties affecting the functioning of these parts.

13.39 A PM plant wants to manufacture helical gear and cam assembly for dual action electric hedge trimmers. The assembly transfers the rotary motion of the motor into the reciprocating action of the trimmer blades. Suggest an economical alternative to produce this assembly.

13.40 A hub is manufactured by PM route for a C-steel copper infiltrated to meet strength requirement. In order to improve the corrosion resistance of this part, what additional treatment(s) are required?

13.41 The internal ring gear of a planetary gear transmission is used on heavy duty cordless electric drills. This is produced from FLC-46-08-50 HT (MPIF designation) ferrous materials. The sintered and heat-treated part was found to give excessive noise in drill performance. Why is it so? What alteration in the production schedule would you suggest?

13.42 Because of its light weight aluminium is used widely in automobiles. When designing an aluminium connecting rod, excess clearance is included so that when the engine cools to a low temperature the rod and bearing do not seize to the crank due to shrinkage of aluminium. To solve the problem, what materials' design would you suggest? Describe the preparation of the product from PM route. (Hint: Tailor the alloy so that the expansion is similar to steel.)

13.43 Name the sintered materials used in the following components of a modern automobile:

(a) Exhaust gas oxygen sensor

(b) A turbo-charger rotor

(c) High-performance fuel injector

(d) Honeycomb structure for catalytic conversion.

14

Techno-economics of PM Processing

LEARNING OBJECTIVES

- Techno-economic aspects of metal/ceramic powder production and sintered parts
- How energy efficient PM processing is
- Techno-economic aspects of full density consolidation of powders
- How economic powder injection moulding is
- Recycling of PM parts
- Strategies for PM part business

P owder metallurgy parts are widely used in various engineering industries and this has been discussed in the previous chapter. The production volume of a particular PM part also influences its application. Powder injection moulding process is being accepted on a part-by-part basis and is expected to penetrate almost all market segments in competition with investment castings. Various techno-economic aspects are described in the following sections.

14.1 COST OF METAL AND CERAMIC POWDERS

The cost of metal and ceramic powders varies widely, and the choice depends on the purity of the powder and the production route adopted to produce them. Figure 14.1 gives a rough comparison of the relative costs for several commercial metal powders. No absolute values are given since prices fluctuate. It is evident that iron powder is still the cheapest and the share of ferrous-based PM products is the biggest. In addition, it must also be borne in mind that the price of powder is dependent on purity level and particle size. This is one of the reasons why PM products, which use very fine size powder, are often more expensive than conventional products.

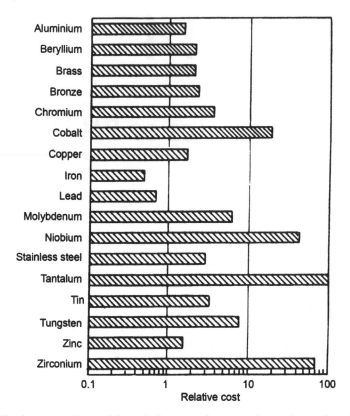

Fig. 14.1 Logarithmic comparison of the relative cost for several metal powders (Reproduced with permission from German RM, *Powder Metallurgy Science*, 2nd edn, Metal Powder Industries Federation, Princeton, USA, 1994)

In order to cut the cost of PM steel, the selection of cheap alloying additions in steel is desirable, e.g., chromium and silicon. Ancorsteel 4300 is a patented powder of Hoeganaes Corporation with nominal composition of 1% Cr, 0.85% Mo, 0.6% Si, and is commercially available as binder-treated product with nickel and graphite levels that can be tailored for specific applications. The advantage of these grades of steel is that it can be sintered in a continous mesh belt furnace in the temperature range 1120–1150°C. Table 14.1 gives a comparison of this new steel grade with wrought steel grades. At 7.0 g cm^{-3} the properties are equivalent to AISI 8620 steel (0.1–0.2C, 0.1–0.35Si, 0.6–1.0Ni, 0.4–0.8Cr, Fe bal.) in quenched and tempered condition. With this grade of PM steel higher densities are achievable, and the performance is likely to exceed the wrought steel grades, certainly to AISI 8620 grade.

In case of clay-based ceramics the cost is fairly low when compared to non-clay–based ceramics. Unlike metal powders, the cost of special electrical ceramics may vary considerably, depending on the tight specifications in demand. The costs of advanced ceramics vary greatly depending on the grade and quantity. On the whole, alumina is cheaper than zirconia and silicon carbides, the latter being cheaper than silicon nitrides and sialons.

Table 14.1 Comparison of as-sintered and rapid cooled 4300 with quenched and tempered wrought steel grades

Steel	Processing	YS, MPa	UTS, MPa	Hardness, HV_{10}
4300 + 0.5C	7.25 g cm^{-3} 2.2°C s^{-1}	1075	1365	440
AISI 4340	Wrought Q and T	1260	1725	510
AISI 8620	Wrought Q and T	965	1325	450

Q – quenched, T – tempered

14.2 TECHNO-ECONOMICS OF METAL POWDER PRODUCTION

Various metal powder production routes were discussed in Chapter 2. The atomisation method is the most flexible one. Water atomisation/nitrogen gas atomisation methods are preferred from the economic viewpoint. In case such atomising media are not suited for high-value materials, e.g., tool steels, inert gas atomisation is the choice.

14.2.1 Pyro-metallurgical methods

These processes are endothermic in nature and heat is supplied in the reactor—for example, production of sponge iron powder and calcination in ceramic systems. The total investment cost of a rotary kiln includes handling into and out of the kiln, gas cooling, gas cleaning and gas blowing. As a continuous process operating under steady-state conditions, the reduction process has a low requirement for process labour, but maintenance can be an appreciable operating cost.

14.2.2 Hydro-metallurgical methods

The rates of leaching reactions are generally limited by diffusional processes and therefore tend to be slow. The volumetric capacity of equipment required for leaching is thus high. However, investment capital required is not so great since relatively simple facilities can be used, e.g., stirred tanks or vats. Leaching requires reagents and these can constitute indirect operating costs. Leaching rates can be considerably enhanced by increasing temperature. Particularly in an aqueous leach system with a high ratio of liquid to solids, the energy required to raise the solution temperature can be substantial. The selected particle size for the leach and the residue disposal are included in the consequential costs of leaching.

The energy demand for solution purification is not great but the cost of reagents employed may be significant. pH adjustments, from say 1.5 upto 7, to precipitate impurities from large volumes of solution can be a costly operation. The recovery in a precipitation process is a function of the efficiency with which the precipitate of impurities can be separated and washed. The efficiency of solution purification reflects in the purity of the final product. Often, in order to meet the market specification, the investment and operating costs of the purification process increase considerably.

14.2.3 Electrolytic methods

This process is a capital intensive operation, because the production rates are relatively slow; there is a physical limit to the area of electrodes that can be fitted in a given volume. There is also a limit to how much power one can get into a cell. Other factors contributing to a highly capital intensive process are:

- Cost of equipments required to service anodes and cathodes
- Power supplies, often from AC sources, have to be rectified and distributed through massive pure copper or aluminium bus bar connectors
- Ventilation of the area from noxious fumes during electrode reactions.

However, there is an advantage in the electrolytic process—there is least loss of values. All the values entering the cell can be recovered, though not necessarily in a single step.

In general, the electrolytic process for metal powder production is expensive to run since it requires care in both operation and maintenance to achieve high quality performance.

14.2.4 Melt atomisation

The economics of metal powder production by atomisation is governed by several factors, in particular, metal and material costs, energy, labour and capital costs, and the costs associated with melting, atomisation, powder collection, classification, handling and packaging. Dunkley examined the operating cost structure in large-scale commercial water atomisation and continuous or large batch size gas atomisation. Table 14.2 illustrates the conversion costs on a percentage basis in gas atomisation of silver and Ni-alloy melts. It is worth noting that the energy costs are negligible. In case of water atomisation of iron melt (Table 14.3), it is evident that the capital costs are high and energy and materials costs are major factors. Since the operation of the plant is essentially continuous, labour costs are largely indirect. To incorporate the scale of operations (output per unit time), Dunkley multiplies each cost component by S^{-N}. The values of N used in the analysis are: material cost (0.3–0.5), energy costs (0.08–0.15), labour cost (0.5–0.75) and capital cost (0.3–0.5). It is assumed that metal costs are relatively unaffected by the scale of the operation.

Table 14.2 Conversion costs on a percentage basis in gas atomisation

Cost component	Ag alloys, %	Ni alloys, %
Stock	11*	13**
System capital	11	19
Direct labour	62	51
Metal loss	8	9
Material and energy	8	8
	100	100

* 20-day level; ** 40-day level (Adapted from Lawley A, *Atomization*, Metal Powder Industries Federation, Princeton, 1992)

Table 14.3 Conversion costs on a percentage basis for water atomisation of iron powder

Cost component	% of total
Capital	27
Labour	21
Overhead	9
Energy	24
Metal and materials	19

(Adapted from Lawley A, *Atomization*, Metal Powder Industries Federation, Princeton, 1992)

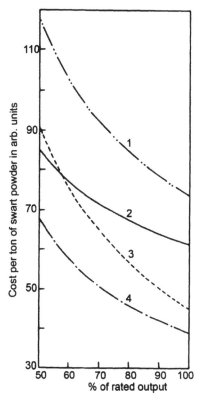

Fig. 14.2 Effect of plant utilisation on the cost of swarf powder: 1 – room temperature processing, non-reclaimable cutting fluid; 2 – cryogenic processing, non-reclaimable cutting fluid; 3 – room temperature processing, reclaimable oil; 4 – cryogenic processing, reclaimable oil

14.2.5 Mechanical milling

In Chapter 2, the coverage of metal powders produced by mechanical methods has been discussed. An example of powders from steel swarf has also been described. Figure 14.2 shows how the cost of the milled powder from swarf is affected by equipment utilisation, particularly in room temperature processing, owing to the larger initial cost. The cost of swarf powder is most sensitive to changes in the price of the untreated swarf and to variations in production costs; the influence of the initial cost of equipment cost and that of the liquid nitrogen price is relatively small.

14.3 TECHNO-ECONOMIC ASPECTS OF SINTERED PARTS

Unless the estimates for sintered parts show a reduction in cost, there is no point in employing PM processing for production purposes. The parts need not be sintered just because of the 'elegance' of the process. It must invariably result in cost reduction, or at least in technical improvement of the material.

There are many factors which influence the cost of sintered parts. The cost of the powder is always higher than that of conventional cast or wrought raw materials. The capital and operating costs for production of metal or ceramic powder and components decrease when the market grows and economy of scale is achieved. Many process advances in the production of uniform quality powder, particularly in advanced ceramics, have occurred on a small scale and are likely to be quite expensive until produced on a large scale. Unfortunately, for large scale production of powder to be profitable, two more steps are required: (i) Somebody has to make a component and (ii) It has to be used in a system to assemble it. Therefore, in the early years, the plant could be too large and have low profitability until the ultimate production is achieved.

The cost of the press tool can also affect the price of the product either directly or indirectly. The most important factor of all, as far as price structure is concerned, is the number of parts to be produced. Different degrees of part complexity will offer different extents of savings. Figure 14.3 illustrates how the cost per piece of a PM clutch ring decreases substantially in comparison to a machined part with increase in the production

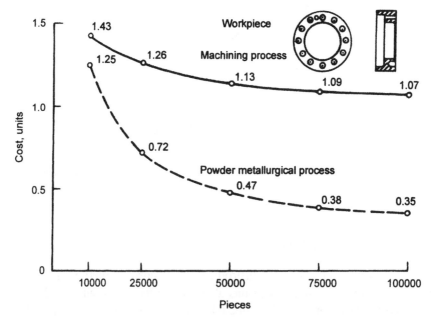

Fig. 14.3 Production costs of a clutch ring in relation to number of pieces

volume. With a minimum production of 100,000 identical parts, the PM process results in an economy factor of 1:3.

Most of the sintered parts are not constituted of a single metal or ceramic and alloying becomes an integral part of overall product economy. Some of the factors which must be considered are:

- Economy in the amount of material used
- Restriction of expensive materials to the probable point of failure
- Use of processes which enhance the performance of cheap materials
- Development and selection of materials with the lowest unit cost
- Specification of materials in a way which will minimise the demands of capital and labour
- Rationalisation of specification to minimise stock holding.

The development of alloys has helped in downsizing the dimensions of structural PM parts, because of the enhancement of their properties, e.g., mechanical, electrical, magnetic properties, etc. A major application of such parts has been in small, but robust, automobiles. The process selection for cheap materials is more important when the usage of powder in a PM part is relatively high, for example, exercise weights. These are not precision parts like bodies of wrist watches, and conventional PM forging can be adequately applied. Regarding the alloying addition with lowest unit cost, a good example is phosphorus-added sintered iron. Here, apart from the addition of cheap phosphorus, the selection of lower sintering temperature to impart liquid phase sintering to enhance mechanical properties is an added advantage. Another example is the Ni/Cr case-hardening steel (Ni 1.5–2.0, Cr 0.75–1.25, Mn 0.5–1.0 and C 0.15–0.2), which can be replaced by straight C/Mn steel (Mn 1.1–1.4, C 0.16–0.21). The mechanical properties of the latter are similar to the former, except for some lowering in the Izod value. Of course, care should be taken in proper selection of sintering atmosphere, since manganese has high affinity for oxygen.

The revision of specifications should be encouraged in order to increase the suitability of materials for their intended purpose, without any loss of service performance. Revision of this kind will significantly increase the ratio of output to input of the production units concerned, and thus make a contribution towards productivity. The essential feature of such changes is that superfluous and irrelevant aspects of specifications are discarded. This will not only affect the economics of the engineering firms but also materials manufacture, permitting greater continuity of production and smaller stock holding. Another contributing factor from the point of view of stock holding and return on capital is the managerial skill of the organisation. Between two firms producing similar types of parts in similar bulks, one with a better managerial skill is more profitable.

A cheap alloying element, like sulphur in steel, is another example. An induction-hardened sulphur-containing plain carbon steel has maintained or improved fatigue life while reducing costs. A phosphor-bronze strip is nearly twice as expensive as a 60/40 brasss strip.

Economy can be achieved by confining expensive materials to the probable point of failure. For example, a costly PM Nimonic 90 disc can be friction welded to an alloy steel shaft.

14.4 ENERGY SAVING IN POWDER METALLURGY

There are two ways in which powder metallurgy contributes to energy saving:

- Through overall efficiency of energy use in powder and component manufacturing
- By substituting materials and parts made by powder metallurgy for conventionally made materials and components.

The main contribution towards the first aspect lies in improvements to the sintering process. The range of reported energy consumption for sintering is large: extending from low-end estimates of 5–8 GJ t^{-1} to over 40 GJ t^{-1} at the upper-end. Typical figures tend to lie in the range 10–18 GJ t^{-1}. A careful combination of factors, like, careful furnace operation and control, improved furnace design, lower thermal mass furnaces and nitrogen sintering to replace the conventional endo-atmosphere derived from natural gas, can yield about 40%–50% energy savings, even allowing for energy to produce and store the nitrogen gas for the atmosphere.

Table 14.4 gives an idea of energy consumption of PM ferrous structural parts and strips from slurry. The latter example is typical for a continuous PM process. The analysis suggests that an overall energy saving of around 5 GJ t^{-1} could be achieved against the standard conventional technology, although, to achieve gains against continuous cast strips would require even further improvements in the slurry powder rolling route. However, it may be noted that the slurry route is not a singular PM route. Other competitive routes are:

- Dry powder rolling used extensively for PM nickel and stainless steel strips
- Continuous atomisation/rolling
- Spray deposition and rolling.

The two latter processes have the advantage of directly integrating atomisation with strip production, and therefore, they offer the opportunity of high energy efficiency, particularly when alloys have to be produced from the molten state.

Table 14.4 Energy consumption (GJ t^{-1}) of PM ferrous structural parts and strips from slurry

PM structural part	
Reduced iron powder	21.7
Blend/compact	0.5
Sinter	5–20.0
Finish and miscellaneous	2.0
Total	29.0–44.0
PM ferrous Strip	
Reduced iron powder	23
Form slurry, dry and compact	3–4.5
Sinter (including nitrogen atmosphere)	6–8
Miscellaneous	1–1.5
Total	33–37

Table 14.5 Energy balance of finished ferrous structural PM parts

Process	Energy, k Wh kg^{-1}
Powder blending and compaction	0.10
Sintering furnace	2.10
Sintering atmosphere gas generator	0.20
Quality control and machining, if any	0.10
Miscellaneous, including heat treatment	0.38
Total	2.88

In case of structural ferrous PM parts, the energy consumption in processing in terms of kWh kg^{-1} of finished parts is shown in Table 14.5. It is evident that the sintering furnace consumes maximum share of energy. In addition, the protective reducing gas also consumes a significant share of power, which is not included in the table. In sintered steels, the diffusivity of carbon in iron is pretty fast, so much so that 1% carbon can be dissolved completely at 1150°C in 3 min. This means that the sintering time to which the green parts are subjected in the furnace, primarily depends on the extent of interparticle bonding, and in turn, on the

Example 14.1: Give some practical suggestions for saving energy in sintering furnaces.

Solution

Some suggestions for saving energy in sintering furnaces are as follows:
- Shut down heating equipment when not in use for an extended period.
- Reduce temperature of furnace to the minimum where atmosphere is not required.
- Use the most efficient furnace at its maximum capacity, and less efficient furnaces only when necessary.
- Rework the less efficient furnaces to more efficient furnaces through reinsulating, and more efficient heating systems.
- Consider 3- or 4-day round-the-clock operation rather than one or two shifts a day for 6 or 7 days.
- Record fuel consumption, gas or electric, on the basis of parts produced. This will give a record of the most efficient equipment and operations.
- Minimise the operation of equipment required to be maintained in standby condition. Planning preventive maintenance shut downs helps cut down on standby operation.
- If gas is used for heating, correct flue gas analyses for maximum efficiency. This could result in a 10% or better saving in fuel.
- Recuperators on radiant tubes might save as much as 30% in fuel consumption when used to preheat combustion air.
- Preheat work loads with flue gases to reduce heating time and conserve fuel.
- Use percentage timers on electric furnaces to save up to 20% of the input energy.
- Consider performing several operations along with sintering to save reheating of the work.

The sintering atmosphere gases have different energy efficiencies. For example, the total energy contained in different gases heated upto 1120°C are: cracked ammonia 6 kWh N^{-1}m^{-3}, 40% nitrogen endo-gas 5 kWh N^{-1}m^{-3}, 60% nitrogen endo 3 kWh N^{-1}m^{-3} and 80% nitrogen exo-gas 2 kWh N^{-1}m^{-3}.

Example 14.2: What measures do you envisage for minimum gas consumption in a mesh belt continuous sintering furnace?

Solution

To ensure minimum gas consumption in the sintering process the following must be ensured:

- Contact between the gas and the parts should be promoted by means of a limited muffle height.
- Counterflow should be promoted by increasing the furnace length to the detriment of the width.
- Any gas leakage on the cooler side should be avoided.

mechanical properties of the product. In fact, except for induction heating, where the product is heated through its bulk, some time is required to bring the green part to the homogeneous temperature, all the more as the product is less conductive due to the presence of porosity. In case there are several layers of parts placed in the tray, it may take a much longer time to obtain perfect temperature homogenisation in the parts.

The introduction of rapid burn-off (RBO) in the dewaxing or preheating zone has an energy saving effect. Here, direct firing with gas burners located in this zone is carried out. As a consequence this zone is separated from the heating zone of the furnace by a curtain. This device offers high heating speed upto 800°C (3 min for a small part and 6 min for a 6-kg part). In addition, relatively large gas volume allows for the lubricant to be removed without depositing on the furnace wall. Another attractive example of enhanced saving has been noticed in those furnaces where the mesh belt was roller driven. For example, in a roller driven belt furnace of width 0.3 m and length 5.2 m, the furnace load can be doubled from 50 kg h^{-1} to 100 kg h^{-1}, thus lowering the energy consumption from 1.1 kWh kg^{-1} to 0.66 kWh kg^{-1}.

Of all the reducing gases, hydrogen is the most expensive, followed by cracked ammonia (75 H$_2$/25 N$_2$). Since N$_2$/H$_2$ mixed gases are now easily available from tanks and bottle bundles, the provision for cracking ammonia gas in the PM plant is getting unpopular. The risk that the steel parts may be decarburised during the sintering process can be avoided by manually adding carburising gas (natural gas, propane). The use of endo-gas in sintering is much more economical. For example, the cost of endo-gas in a particular sintering furnace is 2.85 euros per hour, whereas, for a 90N$_2$/10H$_2$ gas it is 6.88 euros per hour. In an endo-gas sintering furnace, the atmosphere can be easily controlled with dew point meters (°C) or ceramic oxygen sensors (m volts). Figure 14.4 shows the comparison of the costs of various N$_2$/H$_2$ and endo-gas. The figure also shows the data corresponding to cases where RBO provision was made.

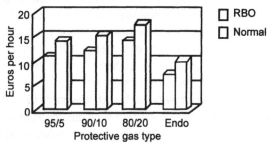

Fig. 14.4 Cost of gases (N$_2$H$_2$ and endo) consumed per hour in a typical sintering furnace with or without rapid burn-off provision (Courtesy: Mahler Gmbh, Germany).

Example 14.3: Give some practical suggestions for saving energy in a sintering atmosphere.

Solution

Some of the energy saving suggestions are as follows:
- Maintain maximum tightness in the furnace through proper maintenance of seals, etc.
- Keep furnace door openings to a minimum height and width or in time on pusher or roller hearth furnaces.
- On belt conveyors or furnaces with constantly open doors, use tight curtains to prevent air infiltration and excessive gas leakage from the furnace.
- Control the atmosphere by water vapour, carbon dioxide or other suitable means to use the minimum quantity of enriching gases.
- Use the lowest dew point gas with the absolute minimum of oxygen pick-up from leaky pipes to reduce the volume of gas needed.
- Blend gases to give the ultimate in reducing properties with the minimum flow.
- Turn off the gases at the highest temperature possible when shutting the furnace down. Even though small at the time it can add up rapidly over the year.

Example 14.4 North American Höganäs and Gasbarre Products Inc., USA, jointly conducted economic studies on sintering ferrous alloys using 24-inch continuous mesh belt (max. 1120°C) and high temperature 12-inch pusher-type (max. 1425°C) furnaces. The initial cost of the former furnace was $225,000, while the latter cost $4,000,000. The capacity of the former was 200 kg h^{-1}, while for the other it was 90 kg h^{-1}.

(a) Which type of furnace is more versatile?

(b) The depreciation cost per hour for the pusher furnace was found to be twice that for the mesh belt furnace. Why?

(c) The cost associated with maintenance, consumables and labour was almost 80% greater for the pusher-type furnace. Why?

(d) How can the process cost for the pusher-type furnace be decreased?

Solution

(a) Mesh belt furnace

(b) Larger initial capital investment and lower utilisation rate

(c) Maintenance cost is increased due to the increased complexity of the high temperature pusher furnace, including the pusher mechanism.

    (i) Consumables, e.g., trays, are expensive, fragile and have limited service life

    (ii) Added complexity of loading, unloading and handling of sintering furniture.
       In addition, the labour costs for high-temperature furnaces are high.

(d) Use cheaper alloying additions like chromium and manganese in PM steels, e.g., Astaloy Cr M.

14.5 TECHNO-ECONOMIC ASPECTS OF FULL DENSITY CONSOLIDATION

Various full density PM consolidation processes have been described in Chapter 9. In this section, we will take up two processes—HIP and hot forging—which are common. The cost of HIP equipment depends on the system component size. The size of the equipment influences installation and building cost as well as operational cost. Workpiece volume, high pressures and temperature increase the cost of a HIP system. Table 14.6 illustrates the cost of HIP as a function of equipment size.

Economic advantage of sinter-HIP over HIP has been highlighted in Chapter 9. The main attention in sinter-HIP should be aimed at surface porosity closure of the parts, which is conveniently achieved through liquid phase sintering. Such sintering would help in added economy due to the requirement of lower sintering temperature.

In case of powder preform forging, it is worthwhile to compare the economic aspects with respect to other competing technologies. Table 14.7 lists all potential competitive processes and compares them with powder preform forging, assuming an average part weight of 0.5 kg. This comparison shows that neither conventional die forgings nor conventional sintered parts are important competitors for powder forgings. In die castings, less emphasis is on tolerance because machining after forging is invariably applied. In sintered parts, due to the presence of porosity, the strength is not to the mark. Powder preform forging, on the other hand, combines the advantages of both the fabrication techniques, i.e., powder metallurgy and conventional press forging. The competition for PM forging appears to be cold forging and precision casting. In the former the main goal is high strength and minimal machining. One should not assume that precision castings are cheaper. In fact, the minimum threshold output for precision casting may not necessarily be required by industry, e.g., aerospace industry.

Table 14.6 Cost of tool HIPers as a function of equipment size for several existing installations at 1200°C

Vessel dimensions	A	B	C	D	E
Diameter, cm	15.2	25.4	50.8	86.6	102.0
Volume, m³	0.0092	0.0386	0.308	0.639	2.08
Cost estimation					
$ per cycle*	800	1700	3500	3500	7500
$ per cm³	0.087	0.044	0.0114	0.0055	0.0036
$ per kg**					
20% packing efficiency	54.30	27.50	7.13	3.43	2.26
50% packing efficiency	21.70	11.00	2.85	1.37	0.90
70% packing efficiency	15.50	7.86	2.04	0.98	0.64

* Based on 710 cycles per month. Cycle costs obtained from tool HIPer

** Based on a density of 8 g cm⁻³

Table 14.7 Comparison of powder preform forging with competing technologies

Process	Powder preform forging	Sintering	Die forging	Cold forging	Precision casting
Part weight, kg	0.1–5	0.01–1	0.05–1000	0.01–35	0.1–10
Height/dia	≤1	≤1	not limited	not limited	not limited
Shape	No large variations in cross section, openings limited	No large variations in cross section, openings limited	Any, openings limited	Mostly of rotational symmetry	Any, any openings possible
Material utilisation, %	100	100	50–70	95–100	70–90
Tolerance	IT 8–10	IT 6–8	IT 13–15	IT 7–9	IT 8–10
Surface roughness, μm	5–30	1–30	30–100	1–10	10–30
No. of parts at which production becomes economical (for 0.5 kg per part)	20000	5000	1000	5000	2000
Main goal	High strength, no machining	Moderate strength, porous materials, no machining	High strength, machining to final shape	High strength, minimal machining	Intermediate strength, minimal machining
Cost of one production unit (sintering = 100%)	250	100	150	150	100
Possibilities for automation	Good	Good	Limited	Very good	Limited

14.6 TECHNO-ECONOMIC ASPECTS OF POWDER INJECTION MOULDING

PIM technology requires fine powder size, which is an economic disadvantage when compared to the material cost in conventional PM. Processing cost is another aspect. For example, stainless steel poses a relatively higher processing cost, because of the critical deoxidation problem of chromium. This is the reason why most users try to substitute this alloy with Fe-Ni. The magnetic soft properties of Fe-Ni are as good as 431 stainless steel. The oxides in PIM stainless steels might have two origins: They could have come from the raw powder or been created during the removal of the binder system. To remove these oxides, a good vacuum furnace is needed, which is capital intensive. In brief, the economics of generating larger size PIM stainless steel is still unfavourable. Statistics show that the typical PIM part is near 25 mm in maximum dimension and

8 g in median mass. In addition to material/processing costs, other factors which discourage large PIM structures are expensive tooling, larger and expensive processing equipment and slower cycle times. The configuration of the product also has an effect on economics. For example, a constant and thin-wall product facilitates rapid debinding, making the overall processing economical. Neither a very simple nor a very complicated multi-dimensional shape is economical. Excellent success in PIM technology has been seen in the production of the wristwatch. For materials which are difficult to machine, e.g. ceramics, stainless steels and titanium, PIM process is still competitive. In general, tolerances are good in PIM products rather than conventional PM parts; some grinding or lapping may be necessary for closer tolerances.

14.7 TECHNO-ECONOMIC ASPECTS OF SECONDARY TREATMENTS

A description of secondary treatments has been given in Chapter 10. The costs of these treatments are the same as for treatments of to-cast and wrought products. Table 14.8 illustrates the cost patterns of various treatments. Machining is a costly operation and should be limited to essential requirements only. As far as the economics of machining is concerned, there are

Table 14.8 Some secondary treatment costs

Operation	Costing basis	Cost, $
Annealing	per kg	1.55
Cadmium plating	per m^2	108
Chromium plating	per m^2	310
Coining	per h	35
Copper plating	per m^2	140
Inspection	per h	40
Machining	per h	25
Nickel plating	per m^2	109
Precipitation hardening	per kg	3
Quenching and tempering	per kg	2.31
Resin impregnation	per kg	2.40
Surface carburisation	per kg	1.50
Surface finishing	per h	24–30
Surface nitriding	per kg	3
Surface steam oxidation	per kg	5
Tumbling	per kg	0.25
Zinc plating	per m^2	78

Data reproduced with premission from German RM, *Powder Metallurgy and Particulate Materials Processing*, MPIF, Princeton, 2005.

two major goals: minimum cost per part and maximum production rate. The total cost per piece (C_p) can be presented as:

$$C_p = C_m + C_s + C_\ell + C_t \qquad (14.1)$$

where, C_m is the machining cost, C_s is the setting up cost, C_ℓ is the cost of loading, unloading and machine handling, and C_t is the tooling cost, which includes tool changing, regrinding and depreciation of the cutter.

Figure 14.5 shows a variation of machining and tooling costs with respect to cutting speed. The material cost, is independent of this variable. It is evident that the total cost for a workpiece is always optimal at a critical cutting speed for a certain material and tool. Accurate data is very important as small changes in cutting speed can have considerable effect on the minimum cost or time per piece.

As far as the economics of various surface engineering methods are concerned, Table 14.9 summarises the comparison of different processes meant for tribological applications. The group titled as 'nitriding' relates to diffusion or implantation techniques which modify surface behaviour by the penetration of nitrogen into the surface. The other group is titled as 'coatings'. Assuming the base cost for plasma nitriding as 100, relative values for other techniques are mentioned. The column 'fixed costs' refers mainly to investment depreciation of a plant. The variable costs include the material, gases, energy, etc., for the process and the labour costs. In the last column of the table, the profitability of the surface-treated workpieces is considered. It is impossible to find the best surface coating process for all applications, because there are advantages and disadvantages in every case.

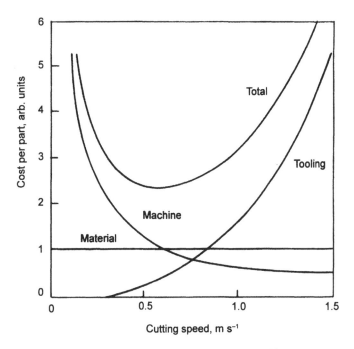

Fig. 14.5 Composite cost of producing a component by machining at different speeds

Table 14.9 Economic comparison of selected surface technologies for wear applications

Process	Fixed costs, %	Variable costs, %	Profitability, %
Nitriding			
Salt bath	30	200	100
Gas	80	120	100
Plasma	100	100	100
Ion implantation	500	100	100
Coating			
Galvano-techniques	30	200	100
CVD	100	150	300
PVD	400	300	500
PACVD	200	150	500

(Source: Adapted from Grum R, *Surface and Coating Technology*, 60: 613–18, 1993)

Shot peening is another important secondary treatment process for PM parts to improve the fatigue strength. The PM parts commonly subjected to this treatment are connecting rods, gears, and other highly loaded transmission components. Part size, quantity and the parameters of the process influence the processes cost. The cost of the shot peening process is also dependent on whether the equipment is located in-house and the process is in-line or the parts are sent out to a service centre for shot peening. Other secondary processes, such as surface densification, can be economically added to shot-peening treatment.

14.8 RECYCLING OF PM MATERIALS

Like any other material, PM processed materials have significant techno-economic contribution during recycling. In considering recycled materials as possible sources of supply, the following factors should be evaluated:

- Physical and chemical state of the material
- Chemical composition of the material, including the presence of unwanted materials or contaminations
- Reliability of the source of supply
- Cost of this source material.

In powder metallurgy, the prompt industrial scrap is of significance. This may be generated because of faulty processing during powder compaction and sintering. In addition, obsolete scrap which is no longer useful—for example, fired heavy-alloy shots—also contributes to PM scrap. The value of the scrap encourages recovery attempts. Tungsten recycling is taken up here as a case study.

For recycling tungsten alloys, the aim should be to achieve a high level of utilisation. The highest value is reached in all cases where direct recovery is possible. For example, WC and cobalt powders are reclaimed from hard metals, as in cold stream or zinc processing, where

Table 14.10 Recycling of tungsten from the most significant scraps

Scrap treatment	Hard metal	Heavy alloy		W-Cu	W-Ag	Pure tungsten
		Hard	Soft			
Zinc process	1	-	-	-	-	-
Cold stream	1	-	2	-	-	1
Bloating	1	-	-	-	-	-
Oxidation/Reduction	-	-	1	-	-	1
Alkali leach	-	1	3	1	1	1
Electrolytic	-	1	3	1	2	1
Chlorination	3	1	3	3	3	1
Melting (ferro-tungsten, stellite)	2	2	2	-	-	2

1: best economic use; 2: good economic use; 3: least economic use. (Adapted from. Schubert WD, *Int. J. of Refract. Metals and Hard Mater*, 11(3):151–57, 1992).

very high material purity is not mandatory. (For details of the zinc process, readers can refer to Upadhyaya GS, *Cemented Tungsten Carbides*, Andrew Williams, Norwich, NJ, USA, 1998)

A number of chemical processes of different degrees of process economy exist for both hard and soft heavy-alloy scrap. Investigations are being carried out on electro-chemical recycling of W-Cu and heavy-alloy scrap (sheets, penetrators, turning). Pollution aspects of waste reclamation, e.g., recycling of thoriated tungsten or oil contaminated soft scrap, have an increased impact on both process technology and economics. Table 14.10 summarises the recycling aspects of tungsten.

14.9 STRATEGY FOR PM PART BUSINESS

Various PM processing variables have already been described in the earlier chapters of this book. For successful exploitation of PM potentiality, a meaningful dialogue between the PM part consumer and vendor is essential. During negotiations between these two, various parameters emerge. They are:

1. *Price*: What is the cost of tooling and test fixtures? What is the piece price? What quality is this price buying?

2. *Weight spread of parts*: Are these parts used in a balancing mechanism where weight spread is of extreme importance and must be held to a very close tolerance, or is the weight spread of no real importance per se?

3. *Density range of parts*: Are the parts of low-, medium-, or high-density levels? The material and tool maintenance costs in their own turn are affected by the density level.

4. *Dimensional tolerances*: Does the vendor know which tolerances are critical and how the users are going to check them?

5. *Mechanical and physical properties*: Do the parts meet the acceptance criteria for various properties? The test(s) should appear on the drawing or be spelled out in the purchase order.

Therefore, it is amply clear that the last four points mentioned dictate the price of the PM part. Often, there is wide divergence in the quotations, as much as 100%. In addition, there may be other reasons too for price differentials. Some fabricators with competent, well-trained tool designers, tool makers, production engineers and quality control personnel are used repeatedly. Some fabricators have a variety of sophisticated quality control equipment, and would therefore quote a high price. Lastly, there are also business tricks which make the price variation elusive. Some fabricators may quote a very high price instead of simply declining to entertain the quotation enquiry, while others may quote too low to be sure that they do get the final order.

It is not economical to accomplish all PM operations, including secondary operation, in one organisation. The reasons may be many:

- Working with less manpower
- Cost of a particular manufacturing operation, say coating, may be high
- Customer feedback warrants standards which may not be met in the existing set up
- Machine or equipment condition may not permit machining within close tolerances.

Sometimes, even the heat treatment operation may get outsourced. The management should not suffer from any ego problem in outsourcing some jobs, in case the overall economy factor makes it necessary.

Example 14.5: Generally, sintered components are difficult to machine, due to internal porosity which induces microvibrations in the cutting tool and lowers the thermal conductivity of the material. To solve this, green machining is proposed. What are the other positive techno-economic features of green machining?

Solution: Machining of sintered hardenable parts is more difficult due to the presence of hard microconstituents (martensite/bainite). Thus, green machining is attractive.

Green machining does not require the use of any coolant/lubricant. Thus, chips formed during green machining are clean and do not require degreasing before reclamation.

Green PM chips exhibit a size distribution that is similar to that of the base powder. Therefore, they can be directly mixed with a base powder and reprocessed into components, without the milling and annealing operations which are required in case of sintered PM chips. This is a major techno-economic advantage.

It is reported that up to 20 w/o of green chips can be added to a compatible powder mix without significantly affecting compaction behaviour and sintered properties.

The dimensional change during sintering decreases as the weight fraction of green chips increases. This may be attributed to the higher surface area and high dislocation density that contribute to enhanced sintering kinetics.

(For further details refer: Robert-Perron E et al., Chip reclamation in green machining for high performance PM components, *Int. J. Powder Metall.*, 43(3):49–55, 2007.

Example 14.6: Both aluminium and titanium belong to the group of light metals. The density of pure aluminium (2.7 g cm^{-3}) is lower than that of titanium (4.5 g cm^{-3}). Can you replace some of the PM titanium applications by less costly PM aluminium metal matrix composites (MMC)? Give reasons.

Solution: Aluminium reinforced by discontinuous hard particles is more cost effective than fibre-reinforced composites. Since the machining of Al-MMC is problematic, diamond tooling is required. This makes the netshape PM-MMC more attractive than cast Al-MMC. PM forging of MMC is a more attractive proposal as compared to PM hot extrusion, because of higher wear of the extrusion tooling.

The other area of development may be directed towards making the aluminium matrix serve better at elevated temperatures. For this, proper alloy design is required. Al-8Fe-2X(Ce, Mo, V) is an appropriate choice. Here, high tensile strength at room and elevated temperatures and improved creep resistance can be attained by a microstructure containing fine intermetallic particles. The correlation of hot working temperature and yield strength, along with the influence of particle size, allows for a certain amount of flexibility in tailoring the alloy to the requirements of the application. Particulate PM composites based on such alloys would further contribute in strengthening, depending on the particle size of the reinforced phase, say SiC. The added advantage would be in significantly enhanced modulus of elasticity. The ultimate tensile strength of hot PM forged Ti-6Al-4V alloy is 920 MPa, whereas for HIPed Al-8Fe-2Mo, it is 500 MPa. The modudlus of elasticity for Ti-6Al-4V (115 GPa) is equivalent to that of 6061 Al-25 vol% SiC. In brief, the Al alloy MMC would approach the mechanical properties of Ti-alloy used in aerospace industry. However, a set of other properties need to be assessed and compared, before coming to a firm conclusion. The major limitations of titanium are high cost and very poor machinability.

Example 14.7: Figure Ex. 14.7 shows the detailed drawing of a gear for a lorry oil pump. The conventional steel (wrought) was low-alloy case-hardening steel of AISI 5120 grade (0.17-0.2C, 0.1-0.35Si, 1.0-1.4Mn, 1.0-1.3Cr) with minimum strength of 980 MPa and Izod impact strength of minimum 38 kg. A PM plant was consulted to produce the same with appropriate composition through pressing/sintering route to offer energy and material saving. Describe the possible PM steel grades and manufacturing cost saving for an annual requirement of 60,000 units.

Solution: Keeping in view the minimum strength achieved by wrought steel grade, the possible steel grades may be as follows (Reference: *Materials Standard for PM Structural Parts*, MPIF Standard 35, 2007):

FN-0405-155HT (3.0-5.5Ni, 0-2Cu, 0.3-0.6C)

FN-0208-155HT (1-3Ni, 0-2.5Cu, 0.6-0.9C)

FL-4405-150HT (0.05-0.3Mn, 0.75-0.95Mo, 0.4-0.7C)

FLN2C-4005-170HT (1.55-1.95Ni, 0.4-0.6Mo, 0.05-0.3Mn, 1.3-1.7Cu, 0.4-0.7C)

FLN4C-4005-170HT (3.6-4.4Ni, 0.4-0.6Mo, 0.05-0.3Mn, 1.3-1.7Cu, 0.4-0.7C)

FLN4C-4005-165HT (1.35-2.5Ni, 0.49-0.85Mo, 0.2-0.4Mn, 0.4-0.7C)

(Continued on next page)

The first two grades of steels are premix grades. The third grade is pre-alloyed steel. The grades number 4, 5 and 6 belong to hybrid low-alloy steels, where pre-alloyed low-alloy steel powders using Ni, Mo and Mn are the major alloying elements to which varying amounts of elemental metal powder is admixed. The costliest alloying element is Mo followed by Ni, Cu and Mn. If the amount of Ni is increased, concurrently, the amount of Mo can be decreased. It is evident that in case we increase the carbon content, the amount of nickel required could be decreased. It may be emphasised that often a trade-off is to be made between the powder cost and processing (compaction and sintering) cost. The dimensional control during sintering is an important aspect and for trouble-free processing some compromise becomes necessary. The energy requirements for machining route and PM route are given in Table Ex. 14.7a. The manufacturing cost is elaborated in Table Ex. 14.7b. These tables are adapted from the *Lecture Series on Powder Metallurgy* 1992, The European Powder Metallurgy Association, Shrewsbury, UK. (Continued on next page)

Table Ex. 14.7a

Machining method			
Work plan	Machine	Energy kWh per piece	Energy as % of total energy expenditure
Turning	Machine	0.635	24.26
Washing	Washing machine	0.015	0.57
Broaching	Rotating automatic machine	0.041	1.57
Washing	Continous washing machine	0.029	1.11
Re-turning	Semi-automatic lathe	0.084	3.21
Milling	Hobbing machine	0.967	36.94
Washing	Continuous washing machine	0.029	1.11
Grinding	Grinder	0.015	0.299
Nitriding	Nitriding unit	0.504	19.25
Deburring	Deburring machine	0.015	0.57
		2.618	~100.00
Powder metallurgy method			
Work plan	Machine	Energy kW h per piece	Energy as % of total energy expenditure
Pressing	Powder press 180 t	0.079	3.09
Sintering	Belt furnace	0.325	12.41
Pressing	Sizing press	0.104	3.97
Sintering	Walking beam furnace	0.180	6.88
Pressing	Sizing press	0.104	3.97
Turning	2-Spindle automatic lathe	0.147	5.61
Grinding	Table grinder	0.093	3.55
Deburring	Deburring machine	0.029	1.11
		1.061	40.52

Table Ex 14.7b: Comparison of the manufacturing costs for an oil pump gear for a lorry (cost basis 100,000 pieces). All cost patterns are based on those prevailing in 1990.

Milled gear	Proportion of manufacturing costs, %	PM gear	Proportion of manufacturing costs of milled costs, %
Work cycles		Work cycles	
Turning, trimming and drilling on a multiple spindle automatic machine	8.49	Pressing (100 t press)	2.37
Broaching of the groove	3.17	Sintering	2.56
Milling teeth	47.50	Heat treatment	1.92
Heat treatment (hardening)	1.92	Flat grinding of ends at right angles to the pitch circle	5.93
Flat grinding of ends at right angles to the pitch circle	5.93	Deburring	0.53
Deburring	0.53	Inspection	0.26
Inspection	0.26	Wear of tools and gauges per piece	8.19
Wear of tools and gauges per piece	17.10		
	100.00		31.73

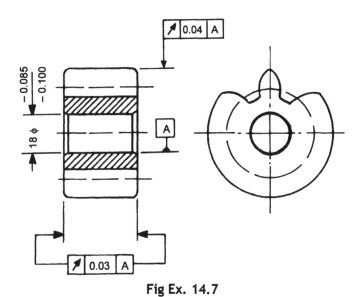

Fig Ex. 14.7

SUMMARY

- In pyro-metallurgical continuous reduction process, there is low requirement for process labour, but maintenance may be a major operating cost.
- The cost of electrolytic metal powder production is high as it requires much care in both operation and maintenance.
- All sintered parts should be associated with cost reduction, or at least in technical improvement of material.
- Alloy developments have helped in downsizing the dimensions of structural PM parts, because of the enhancement of their properties.
- The major fraction of energy consumtion in PM processing is at the sintering stage.
- The competition for PM forging lies in the cold forging and precision casting.
- Configuration of the product influences the economics of powder injection moulding.
- Machining of PM parts is a costly operation and should be limited for essential requirements only.
- For successful exploitation of PM potentiality, a successful dialogue between PM part consumer and vendor is essential.

Further Reading

Lawley A, *Atomization: The Production of Metal Powder*, Metal Powder Industries Federation, Princeton, 1992.

Upadhyaya GS, Energy Consevation in PM Industries, Newsletter, Powder Metallurgy Association of India, 2(3):11, 1976.

Upadhyaya GS and Upadhyaya A, *Materials Science and Engineering*, Viva Books, New Delhi, 2006.

EXERCISES

14.1 In lead–tin solders tin is considerably costly. What cheaper metals can partially replace tin without greatly widening the melting range of the solder?

14.2 List these ceramics in order of increasing cost processing:

(a) MgO   (b) Al_2O_3 (c) ZrO_2 (d) Si_3N_4 (e) BN.

14.3 Taking into consideration the chemistry and processing of refractories, rank the following refractories in the ascending order of their prices: Tar-bonded magnesite, bloating clay, 50-70% Al_2O_3, Tar-bonded dolomite, 80% Al_2O_3, MgO-20 graphite and direct-bonded 60% MgO.

14.4 Steel powder preforms were prepared by two routes: (a) isostatic compaction and (b) die compaction. Compare the technological and economic advantages and disadvantages.

14.5 Solid cemented carbide hubs can be used at very high cutting speeds. However, they are expensive and have low toughness. On the other hand, powder metallurgy high-speed steels have excellent toughness, but have reduced life due to high-temperature operation. What further improvements would you suggest for making the HSS applicable?

14.6 A batch size 10,000 of spark plug insulators is to be manufactured in a plant. The weight of the plug is about 0.05 kg and its minimum section thickness is 1.2 mm. The mean precision has to be high (< ±0.2 mm). What possible methods are available in principle to manufacture such a part? Which one would you finally select? Justify. (Adapted from Ashby MF, *Selection of Materials in Design*, 2nd edn, Butterworth-Heinemann, Oxford, 1999).

14.7 Compare the product cost of a tantalum capacitor to the cost of raw materials using the same weight basis.

14.8 What is the relative dependence of profits on cost of raw materials when the value added is low or high? Explain.

14.9 A PM part vendor intends to order parts of two compositions Fe-4Ni and Fe-4Ni-0.45C, both in heat-treated condition. What would you advise from the techno-economic viewpoint?

14.10 Figure P. 14.10 shows the view of a PM blade adapter for a lawn mower. The vendor suggests that the part be infiltrated for better mechanical properties, but at the same time he insists that the end product must be economical. What infiltration strategy will you apply?

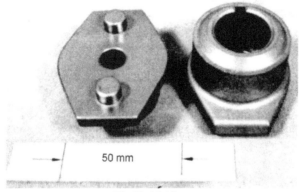

Fig P. 14.10

14.11 Under what conditions can you make the selection of HIP process an economic attraction? Name some structural automobile parts you may select for mass production. In case you intend to altogether eliminate the conventional sintering step during processing, what type of compaction should be selected before final containerless HIP and what special care has to be taken?

14.12 In some countries, for example, Italy and India, ferritic stainless steel coins produced from rolled ingot metallurgy stock are extensively in use. In case you are asked to prepare an alternate PM route, discuss variants. Which one of these can be adopted and why? What are the advantages of PM coins? (It may be noted that the intrinsic value of a coin should be less than the face value).

14.13 An endo-gas with the following composition is produced from natural gas and is used as a carburising medium: Co 19.8%, H_2 40.4%, CO 0.3%, CH_4 0.5%, N_2 balance. Because of fuel insecurity, it is proposed to produce a similar composition using a mixture of N_2 and cracked methanol. What should be the ratio of N_2 and methanol? In case you intend to use the gas mainly as a sintering atmosphere, what ratio of N_2 to methanol would suffice? Which gas mixture will be cheaper?

14.14 In the heat treatment shop of a PM plant, if you have 300 tons of ferrous hardening load and 300 tons of annealing load of gears, suggest your preference from the economic viewpoint: two small furnaces or one big combination furnace of 1000 kg per hour capacity.

14.15 In laboratories, a planetary ball mill is very useful for particle size reduction; but in industries, attritor mill or rotary ball mills are preferred. Why?

14.16 As cost effectiveness forces a reassessment of the PM alloy steel selection, it is more cost effective to invest in additional processing to reduce the content of high-priced alloying elements. Comment on the statement: "Presently, Mo is roughly three times the price of Ni".

14.17 The upper limit of MIM part size is established by processing economics, not technical limitations. Explain.

14.18 Cast alloys are generally less expensive then PM alloys. In the case of refractory metals alloys, it is the reverse. Give reasons.

14.19 A marketing manager said: "I guarantee that I can walk in any factory and find a money saving application for ceramics." Give same examples for justifying this statement.

Index

For Product Safety Concerns and Information please contact our EU
representative GPSR@taylorandfrancis.com Taylor & Francis Verlag GmbH,
Kaufingerstraße 24, 80331 München, Germany

Printed and bound by CPI Group (UK) Ltd, Croydon, CR0 4YY
01/05/2025
01858478-0003